DEFINING AND ASSESSING ADVERSE ENVIRONMENTAL IMPACT FROM POWER PLANT IMPINGEMENT AND ENTRAINMENT OF AQUATIC ORGANISMS

Defining and Assessing Adverse Environmental Impact from Power Plant Impingement and Entrainment of Aquatic Organisms

Editors:

Douglas A. Dixon
Electric Power Research Institute (EPRI), Palo Alto, CA, USA

John A. Veil
Argonne National Laboratory, Washington, DC, USA

Joe Wisniewski
Wisniewski & Associates, Inc., McLean, VA, USA

A.A. BALKEMA PUBLISHERS LISSE / ABINGDON / EXTON (PA) / TOKYO

Cover art by Judy Hall-Griswold, Biolines Environmental Consulting, Stanley, ID; jhgris@ruralnetwork.net

Published by: A.A. Balkema, a member of Swets & Zeitlinger Publishers
www.balkema.nl and www.szp.swets.nl

ISBN 90 5809 517 7

Printed in the Netherlands

Table of Contents

Preface

The Electric Power Research Institute (EPRI), headquartered in Palo Alto, California, USA, is a non-profit energy research consortium for the benefit of the energy industry, its customers, and society. The mission of EPRI's Environment Sector is to be the premier provider of timely, credible scientific and technical knowledge, tools and services to (1) inform critical policy and regulatory deliberations, (2) support cost-effective compliance, stewardship, strategic issue management and business decision-making, and (3) address longer-term sustainability issues.

A current issue of major importance to the U.S. electric power industry is the development of regulations to address Section 316(b) of the Clean Water Act of 1972. Section 316(b) addresses the protection of aquatic life at power plant cooling water intake structures (CWIS). CWIS affect fish and invertebrates via impingement of organisms on intake screens and entrainment of organisms, particularly early life stages (eggs and larvae), into the cooling system where they are exposed to physical, chemical and thermal stress. Historical §316(b) demonstration studies have shown that billions of aquatic organisms are annually exposed to these stresses. In accordance with our mission, EPRI has a program dedicated to providing science and technology-based solutions for aquatic life protection at CWIS.

Section 316(b) states:

> *Any standard established pursuant to section 301 or section 306 of this Act and applicable to a point source shall require that the location, design, construction, and capacity of cooling water intake structures reflect the best technology available for minimizing* ***adverse environmental impact****.*

Over the 30 years since its enactment, there has been considerable discussion and debate among stakeholders regarding the definition of terms and implementation process for this section. Neither the legislation, nor its legislative history, defines "adverse environmental impact (AEI)." In 1976, the U.S. Environmental Protection Agency (USEPA) proposed regulations for implementing §316(b). However, these regulations were challenged on procedural grounds and, subsequently, were formally withdrawn by USEPA. Nevertheless, in the absence of formal regulations, permit applicants, scientists, and regulators continued to rely on USEPA draft guidance publications, and also on administrative decisions in several permit proceedings, to define the §316(b) requirements for permitting CWIS during the 1970s, 1980s, and 1990s.

In the early 1990s, a coalition of U.S. environmental groups sued USEPA for failing to promulgate §316(b) regulations. In 1995, the parties entered into a Consent Decree directing USEPA to issue final regulations. USEPA divided the rulemaking process into three phases. Regulations for new facilities were issued in November 2001; regulations for power plants with intakes exceeding 50 MGD will be finalized in February of 2004; and regulations for CWIS at non-power plants with intake flows exceeding a volume yet to be determined will be issued in June of 2006. The proposed regulations are intended to minimize the potential AEI associated with CWIS. Minimizing AEI may include requirements affecting the design, construction, location, and capacity of CWIS that are determined to reflect the "best technology available" (BTA).

One central issue in the rule-making process is the definition of AEI, including how it is assessed, endpoints for decision-making, and how it can be minimized. EPA has not defined AEI, nor have they proposed an approach for assessing environmental impact. Several alternative definitions and assessment approaches have been offered for public consideration and comment.

To facilitate an exchange of information among all stakeholders in the §316(b) issue, EPRI organized a national symposium to discuss the meaning of AEI and methods for its assessment. The symposium was held in conjunction with the Annual Meeting of the American Fisheries Society, August 23, 2001 in Phoenix, Arizona, USA. Technical experts in federal and state resource agencies, academia, industry, and non-governmental organizations attended and made presentations on AEI issues including:

- Definition of AEI (including consideration of the full range of options such as individual losses, population-level impacts, fishery opportunity foregone, and disruption of aquatic community structure and function).
- AEI assessment endpoints and thresholds.
- Predictive and retrospective methods for assessing AEI (e.g., conditional mortality, equivalent adult losses, production foregone, biocriteria, trend analysis of fishery-independent and dependent data).
- Role of ecological risk assessment in assessing AEI.

The peer-reviewed accepted papers herein were presented at this symposium. EPRI and the editors are making this information available to the scientific community and specifically to the stakeholders in the §316(b) issue, particularly EPA, for consideration during the rule development effort.

Finally, the symposium and papers reflect an enormous effort by many individuals and organizations. For co-sponsorship of the original symposium, we express our appreciation to the American Fisheries Society and its Western Division. Development of the symposium objectives and selection of papers for presentation was supported by John Veil, Argonne National Laboratory; William Richkus, Versar Inc.; and James Wright, Tennessee Valley Authority. John Veil also served as symposium co-moderator. Completion of this book involved sustained and extensive effort by all of the authors, who were aided by the thoughtful and constructive reviews and comments of many others. We are grateful to all these individuals for the diligence and patience they have shown in bringing this project to fruition.

Douglas A. Dixon, Ph.D. and Kent D. Zammit
Managers, Fish Protection Research, EPRI

Maryland Power Plant Cooling-Water Intake Regulations and their Application in Evaluation of Adverse Environmental Impact

Richard McLean[1], William A. Richkus[2,*], Stephen P. Schreiner[2], and David Fluke[3]
[1]*Power Plant Research Program, Maryland Department of Natural Resources, Annapolis, MD 21401;* [2]*Versar, Inc., Columbia, MD 21045;* [3]*Maryland Department of Environment, Baltimore, MD 21224*

Received December 6, 2001; Revised January 28, 2002; Accepted February 19, 2002; Published February, 2003

Maryland's cooling-water intake and discharge regulations, the Code of Maryland Regulations (COMAR) 26.08.03, stem from Sections 316(a) and (b) of the Clean Water Act (CWA). COMAR 26.08.03.05 and litigative and administrative rulings stipulate that the location, design, construction, and capability of cooling-water intake structures must reflect the best technology available (BTA) for minimizing adverse environmental impacts (AEIs), providing that the costs of implementing the BTA are not wholly disproportionate to the expected environmental benefits. Maryland law exempts facilities that withdraw less than 10 million gallons/day (MGD) and less than 20% of stream or net flow by the intake. If not exempt, BTA must be installed if the cost of doing so is less than five times the value of fish impinged annually. Through site-specific studies and the use of a Spawning and Nursery Area of Consequence (SNAC) model applied to Representative Important Species, several power plants were evaluated to determine if they have had an adverse effect on spawning and nursery areas of consequence. Examples of application of the Maryland law to a number of power plants in the state are presented, together with the outcome of their evaluation.

KEY WORDS: entrainment, impingement, environmental impact, cooling water regulation

DOMAINS: freshwater systems, marine systems, ecosystems and communities, environmental monitoring

---

INTRODUCTION

Maryland takes pride in its strong commitment to environmental protection. A cornerstone of this commitment has been the state's efforts to restore and protect

Corresponding author. Email: Richkuswil@versar.com

the Chesapeake Bay and all of its diverse natural resources. One of the initial steps toward protecting the bay was the creation in the early 1970s of the Power Plant Research Program (PPRP). PPRP was created by legislation in response to public controversy that arose when the Baltimore Gas and Electric Company (BG&E) announced plans to construct the large Calvert Cliffs Nuclear Power Plant along the shoreline of the bay. This plant would withdraw large volumes of cooling water from the bay and discharge the heated water back into bay waters. The public was concerned about the potential for the plant to adversely affect the bay and its fisheries resources, and the state could not respond to these concerns because it did not have adequate technical expertise with regard to the potential impact that these power generating facilities might have on the bay.

As a result, the legislature created PPRP to ensure that, in the future, all existing and proposed power generating and transmission facilities in Maryland would operate in a manner that ensured protection of the state's natural resources and at the same time made electric power available to the public at reasonable rates. With regard to proposed new generating and transmission facilities, PPRP is charged with assessing and advising the Maryland Public Service Commission on the environmental and economic considerations associated with the siting, design, and operation of the proposed facilities. For existing facilities, PPRP provides technical assistance in permit review and evaluation to the Maryland Department of Environment (MDE), which is the state's permitting agency with responsibility for writing national pollution discharge elimination system (NPDES) permits and enforcing compliance with permit provisions.

Since its inception, PPRP has provided technical reviews of issues and developed recommendations concerning requirements associated with Maryland's regulations for cooling-water intake structures (CWIS) for all its generating stations. PPRP works cooperatively with MDE in reviewing all data and information required from plant operators by MDE. In many instances, the state has conducted research independent of permittees in order to assess impacts and technologies to reduce those impacts. The information presented in this paper is based on PPRP's experience in addressing CWIS issues and on the results of the program's very diverse yet comprehensive studies of the manner in which cooling-water withdrawals have impacted aquatic biota in Maryland's waters.

MARYLAND REGULATIONS FOR CWIS

Impingement

As generating stations draw water in for the cooling cycle, aquatic organisms near the intake can be caught in the suction and trapped (impinged) on the intake screens. Large power plants often have systems that wash the screens and return impinged organisms to the water thereby reducing injury and mortality. Injury and mortality, however, can still be significant depending upon species, water temperature, and other site-specific factors. The best technology available (BTA) for impingement was deemed by Maryland to be the technology that was the most cost effective for reducing the magnitude of impingement impact, as established by the value of the fish lost to impingement. Thus, as established in the Code of Maryland Regulations (COMAR) 26.08.03.05.D(1) and D(2), the dollar value of the organisms killed by impingement is to be calculated, and the plant operator is required to implement technologies to reduce impingement only to the extent that the cost to the facility does not exceed the total value of lost organisms over a 5-year period (in practice, generally five times the value of fish lost to impingement in a single year). In essence, Maryland's BTA is based on a simplified cost-benefit assessment.

The technical basis for the regulation is not documented in the state's regulatory records. We believe that the dollar values of fish presented in Section 08.02.09.01 of COMAR were taken from the American Fisheries Society's (AFS) listing of fish values at the time the regulation was promulgated. AFS has regularly updated its fish values, and those values are used throughout the country to establish the costs of fish kills due to many causes. Maryland's regulation did not specify changes of those values over time (for example, to account for inflation or devaluation). Thus, the values in Section 08.02.09.01 have not been modified since they were first promulgated. Plants using cooling water in the state have been evaluated under these regulations since the 1970s. MDE and DNR reviewed the issue of static fish values in their assessments throughout the period.

Entrainment

Aquatic organisms that are drawn through the generating facility through cooling systems, intake valves, and turbines may be injured or killed as they are pulled (entrained) through the station. The general concepts underlying a determination of BTA for entrainment by Maryland are as follows:

- The evaluation of impact should be carried out to a specified level of biological significance, i.e., representative important species (RIS) and spawning and nursery areas of consequence for the RIS.
- The consequences of the cooling-water withdrawal effects should be based on the extent to which they impact the viability of the RIS population and the ecosystem necessary to support its life history functions. The effect of the cooling-water intake itself (i.e., the number of fish impinged or entrained) should not be the major focus; it is the consequence of that effect to the biological entity of concern, whether at the species or the ecosystem level, which establishes what actions the state will take.

The state determined that a sequential approach to entrainment impact assessment is a good, generic approach to the issues involved, with the steps in that sequence being (1) to quantify the effects of the cooling-water withdrawal (i.e., estimate the numbers of organisms lost to entrainment), (2) to establish the biological entity at risk (i.e., select RIS), and (3) to assess the significance of the effects for causing adverse harm to the target entity.

The CWIS operator is required to determine if the entrainment loss results in a significant adverse environmental impact (AEI), which is defined as a statistically measurable effect outside the plant's mixing zone. Entrainment evaluation modeling has been applied in Maryland assessments[1,2].

WATER WITHDRAWAL RATE THRESHOLD

Maryland regulations also establish a water withdrawal rate threshold below which impacts are assumed to be sufficiently small as to not require regulation for BTA. The state defines that threshold as 10 million gallons/day (MGD), if that volume of water is less than 20% of the defined flow for the providing water body: design stream flow (7Q10) for nontidal waters (rivers), and the annual average net flow past the point of discharge which is available for dilution for tidal waters. Note that this exemption takes into account site specificity (i.e., the size of the body from which the water is withdrawn), reinforcing the regulation's intent that facilities be evaluated on a site-specific basis. No documentation exists within Maryland's regulatory archives to explain the technical basis for the 10-MGD threshold. However, discussions with individuals involved in the development of the regulation suggested that the threshold value was selected based on knowledge of the various facilities in the state that withdrew cooling water from the state's surface waters; the status of the ecosystems from which that water was being withdrawn; and the professional judgment of the resource managers and permit regulators with management and regulatory authority at that time. Since then, the state has not modified that threshold, and no impacts have occurred that have supported the need for its reassessment.

Maryland CWIS regulations do not vary according to specific water body type except with regard to the way in which the allowable percentage withdrawal threshold is calculated. Two reasons underlie that decision. First, a site-specific assessment approach was adopted, which makes generalizations related to water body type moot. Second, a site-specific approach was established because the potential for adverse impact was not consistent within each water body type. For example, the regulation did not differentiate between estuarine and fresh waters, recognizing that not all locations within an estuary or a freshwater body are equally sensitive or productive.

IDENTIFYING AND ADDRESSING IMPACTS

Defining Adverse Environmental Impacts

Approaches to minimizing adverse impacts must be based on strong technical data and information. Maryland regulations do not specify the types of studies required to provide the data needed to comply with the regulations. However, because the PPRP existed at the time the regulations were put in place, utility study designs and results of studies were evaluated in a fairly consistent manner, and the state's approach to such evaluations was increasingly refined over time. Also, most of the generating stations in the state were owned by two major utilities – BG&E and Potomac Electric Power Company (PEPCo) – and the utility approach to satisfying the state's requirements became fairly standardized; the same utility staff worked with the same state agency and contractor staff for more and more facilities.

Continued or periodic monitoring is required to measure the effectiveness of a given technology's performance. If the state's CWIS determination required that a facility take some action, monitoring of the required action was made a requirement of the permit issued. The performance measures that would be used to measure BTA effectiveness were made facility- and site-specific, and a function of the type of action required. Thus, the state did not establish any type of standardized monitoring requirement related to CWIS determinations.

Quantification of the effects of water withdrawal is necessary but not sufficient to determine whether additional measures may be necessary to reduce these effects. As noted above, the key is whether the effects caused by the water withdrawal have significance to the biological entity of concern. If the effects are not significant, existing structures and operations are sufficient since there is no truly adverse impact to be minimized. Thus, clearly defining what constitutes adverse impact is crucial. Maryland considers all costs to the citizens of the state in making regulatory determinations, and factors include impacts to the state's living resources and economic costs to the utilities (and, beyond, to the consumers) of measures that could be taken to reduce the effects of water withdrawal. Maryland's regulations thus balance these considerations so that any measures required of the utilities are commensurate with the estimated significance of the effects being reduced. We believe that as 316(b) rules are developed for the nation, the U.S. Environmental Protection Agency should define AEI and place AEI into context with the costs of protecting natural resources.

Defining Best Technology Available

Based on extensive research and data, Maryland has determined that the extent of impacts of cooling-water withdrawal is site specific, as are the need for and the nature of various ameliorating intake technologies. Factors that directly affect the decisions on what constitutes BTA at a particular facility include a determination of an impact, the nature of that impact, the design and location of the facility on the water body, and life stages of affected species. Maryland's regulations do not specify a design intake velocity; Maryland facilities generally have a 1 to 2 ft/s screen face velocity. Impingement rates at Maryland plants with similar intake designs within the Chesapeake Bay have varied widely, and they appear to be related more to the plant's location and the location of the intake than to intake velocity or volume of water withdrawn.

Our assessments of generating facilities in Maryland resulted in BTA determinations that ranged from a decision that the existing intake structure is BTA to recommending mitigative technologies such as wedgewire screens, modifications to intake structures, and installation of barrier nets. Therefore, we believe there is no single technology or suite of technologies that can be applied on a state-wide or nation-wide basis. We believe, however, that it is important to have a consistent national process for identifying BTA at the site-specific level.

Cumulative Impacts

Cumulative effects of impingement and entrainment are not specifically addressed in the regulations, but they have been evaluated in Maryland in a limited and somewhat cursory manner. Most Maryland facilities are relatively far apart spatially, and the biological populations exposed to the effects of these widespread plants are often distinct, with only some intermingling. For example, the major tributaries of the Chesapeake Bay support their own spawning populations of striped bass (Morone saxatilis), and impacts to the Potomac River stock would have no significance to the Nanticoke River stock. Maryland has tracked cumulative impingement losses across all power plants for some species, such as Atlantic Menhaden (Brevortia tyrannus), that may occur over a wide range of salinity regimes and are thus exposed to the effects of all of the power plants located on tidal waters of the state. These assessments have suggested that the cumulative magnitude of impingement is a small fraction of the commercial harvest of the species and a small fraction of the amount of the species consumed by predators. On that basis, the state concluded that the levels of impingement by Maryland's power plants do not represent a significant adverse impact to important resource species in the bay. With regard to Maryland's experience, long-term monitoring of the status of important resource species have temporally addressed cumulative impacts. None of these diverse monitoring programs has suggested any adverse cumulative impact from the power plants operating in Maryland[3,4].

Mitigation

While mitigation is not identified or mentioned in Maryland's regulations, out-of-kind mitigation has been incorporated into some state NPDES permits issued after a CWIS evaluation, as is discussed further below. The state believes that mitigation can play a valuable role in the resolution of 316(b) issues on a site-specific basis. The term mitigation as used here refers to actions aside from alternative intake technologies or operating strategies that might be used to minimize ultimate impacts of cooling-water intakes to the state's resources. Mitigation may include alternative measures that can indirectly compensate the public for resource losses due to CWIS effects.

DISCUSSION

Maryland Facilities' Regulation Compliance

Before reviewing the permitting actions at various facilities, some general observations can be made about how facility permitting often proceeded. COMAR 26.08.03.05D addresses impingement and requires a facility owner to estimate the value of fish lost to impingement over a 5-year period as a basis for determining if modification of the CWIS to achieve BTA would be required. As a result, some quantification of magnitude and composition of impinged organisms was performed at all Maryland plants at which the water withdrawal rate exceeded the 10-MGD threshold. For those facilities where impingement was anticipated or known to be low, a relatively limited sampling effort was often sufficient to confirm that judgment. Conversely, at large plants where very substantial numbers of organisms were expected or known to be impinged, impingement studies in a number of cases were conducted over many years (e.g., 21 years at Calvert Cliffs) to ensure that an accurate characterization of impingement was made[4].

COMAR 26.08.03.05E, which addresses entrainment, does not provide guidance and requirements as detailed as those specified for impingement. Also, data and information that would be required for a rigorous empirical quantification of entrainment impact was most often unavailable and frequently was costly to acquire. For these reasons, initial estimation of whether a facility impacted a Spawning or Nursery Area of Consequence (SNAC) was often done through modeling. PPRP developed a SNAC model for that purpose that was used to estimate entrainment losses of vulnerable RIS, the consequences of those losses to the ecosystem, and the economic value of those losses[1]. An overview of that model was presented by Richkus and McLean[3]. PPRP applied the SNAC model to many of the generating stations in Maryland, and decisions on permitting and permit conditions were often based on the model outcomes. In many cases, results of the SNAC model suggested that impacts were not significant and that existing CWIS could be considered to be BTA. In cases where the SNAC model results suggested that significant impact might be occurring, but where the modeling was conducted using limited data or information from the literature, permits were issued that required the facility owner to conduct studies sufficient to reliably estimate entrainment impacts. Results of such studies were then used as a basis for subsequent permitting decisions.

PPRP assessments of the type just described established that many of the power plants in Maryland were causing minimal impacts due to entrainment and impingement. For example, at the R.P. Smith plant, which is located on the mainstem of the nontidal portion of the Potomac River, annual impingement losses were valued at $90 using COMAR-specified values, and the overall projected ecological impact from entrainment was estimated at less than 0.1% of system net primary production. Small impacts were also estimated for the Dickerson plant, which is also located on the nontidal Potomac River. Similarly minor impacts were found for some of the smaller facilities located on estuarine waters (Baltimore City), such as the Baltimore Refuse Energy Systems Company (BRESCO) waste-to-energy incinerator and the Gould Street Plant, an older facility seldom run at full capacity. For these types of projects, the existing plant CWIS was determined to be BTA and no CWIS modifications or other 316(b) action by the facility owner were required in the permit.

At some facilities, initial estimates of entrainment impacts, derived from SNAC modeling, suggested that significant impacts may be occurring, but no data were available to validate those estimates and confirm the impacts. The H.A. Wagner facility on Baltimore Harbor presents an example of such a situation. At Wagner, SNAC modeling suggested that up to 49% of the local population of bay anchovy and 17% of the silverside population could be lost to entrainment. Because of the uncertainty regarding the validity of modeling results, the facility owner was required to conduct extensive ichthyoplankton studies according to a study design reviewed and approved by the state. These studies would provide the data needed to make a more rigorous impact assessment. Analysis of data from the studies suggested that impacts were not as great as the modeling suggested, and not sufficient to warrant major CWIS modification. Thus, no modification to CWIS was required in the permit for that facility.

At large facilities utilizing once-through cooling systems, such as the Chalk Point Generating Station on the tidal Patuxent River and the Calvert Cliffs Nuclear Power Plant on the mainstem Chesapeake Bay, the large volumes of cooling water withdrawn (e.g., 3,456 MGD at Calvert Cliffs) suggested a high potential for significant impacts. Extensive and comprehensive impact assessment studies were conducted at Chalk Point by PPRP and PEPCo, the owner of the facility at that time. Similarly, at Calvert Cliffs, BG&E, the facility owner at that time, was also required to conduct comprehensive studies to comply with technical specifications in their Nuclear Regulatory Commission (NRC) license for this nuclear facility. PPRP conducted many complementary studies, which were well-coordinated with the BG&E studies.

Ichthyoplankton studies at Chalk Point indicated the potential for significant losses of forage species (bay anchovy, naked goby, silversides) in the Patuxent River estuary. Such losses could adversely affect the successful completion of the life cycles of other important species that use the Patuxent as a spawning and nursery area[5]. Based on field studies, PEPCo concluded that the reduction in anchovy recruitment for the Patuxent was 4% and that entrainment mortality could cause a reduction in forage fish biomass of about 3,000 to 15,000 lb (dry weight)[6]. These estimates were based on field measurements of population size in the Patuxent and entrainment by Chalk Point. An independent analysis of the same data by PPRP indicated that loss of bay anchovy in the estuary due to entrainment might range from 14 to 51% of the population (most probably 20 to 30%) annually[7]. PEPCo calculated the value of the entrainment losses at $150,000/year (1989 dollars) based on its loss estimates. PEPCo also calculated the cost of BTA alternatives (cooling towers and wedgewire screens) as ranging from $10,000,000 to $288,000,000 (1989 dollars). According to PEPCo, the alternatives that were evaluated varied in effectiveness in reducing entrainment from almost none to 100%. As is evident, there was substantial disagreement between the state and the utility regarding the magnitude of entrainment losses and costs of various BTA alternatives.

The substantial magnitude of the scientific and economic disagreements between the parties led to the initiation of negotiations that resulted in a mitigation alternative that was agreeable to both the state and the utility. A major factor leading to the conclusion that the mitigation option was appropriate was the substantial difference between the cost of effective BTA (such as cooling towers) and the projected environmental benefits. In 1991, MDE issued to PEPCo a national pollution discharge elimination system (NPDES) permit that required PEPCo to spend $200,000/year through 1997 on aquaculture of striped bass or other species as requested by the Maryland Department of Natural Resources (DNR), and $50,000/year for aquaculture of yellow perch or other species as specified by DNR. This permit condition called for the production of 200,000 striped bass and 50,000 yellow perch per year, with those fish being used to enhance and restore stocks in the Patuxent River. The permit also required PEPCo to provide $100,000/year to the state for environmental education or for projects to remove obstructions to anadromous fish in the Patuxent River watershed. The state of Maryland believes that a sound decision was made based on the success of the mitigation program. In this case, this program included creating a fish hatchery for potentially impacted fisheries and provision of funds for removal of obstructions to migratory fishes on tributaries by removing dams or providing fish passage facilities. The hatchery and stocking program resulted in the production and release of 3.5 million juvenile striped bass to date, the total estimated weight of which exceeded the estimated weight of forage fish lost from entrainment at Chalk Point. At the end of 1997, 750,000 American shad had also been produced. This species is currently the focus of fishery restoration efforts in Maryland. Each of these benefits is directly related to the enhancement of the state's fisheries. While continuation of the aquaculture program is not mandated in the current Chalk Point permit, the facility owner has continued production and release of fish in cooperation with the state.

At the Calvert Cliffs Nuclear Power Plant, which is located on the mainstem of the Chesapeake Bay, nearly 2 decades of studies were conducted during the construction and initial operation of the two units that comprise the facility. Entrainment at the plant was determined not to be a major concern because the cooling water intake was not located in a spawning area of significance. SNAC model estimates of economic loss due to entrainment were $200 annually, with overall ecological loss being 0.1% of net primary productivity. Naked goby eggs and larvae made up a large proportion of the icthyoplankton entrained, primarily because this species colonized the rip-rap used to line the intake embayment, and their eggs and larvae were being released directly into the cooling-water withdrawal flow.

Impingement at Calvert Cliffs was initially substantial with the numbers of menhaden impinged in several 1975 episodic events sufficiently high to cause intake screen collapse and plant shut-down[4]. Those initial large impingement episodes were associated with low dissolved oxygen in the intake embayment, a problem resolved in part by removal of several skimmer wall panels. Monetary value of fish lost to impingement averaged less than $25,000/year as a result of the relatively high survival of many species impinged and as a result of the relatively low value of the dominant species[4], and no CWIS modifications were required in the Calvert Cliffs permit. However, over a 14-year period, BG&E optimized their intake, screening structures, and operations such that impingement losses in the early 1990s were 10 to 50% of the losses recorded in the 1970s.

CONCLUSIONS

The overview of impact assessment results and the detailed discussions of permitting actions at different categories of generating facilities in Maryland reinforce the basis for Maryland's perspectives on AEI presented earlier in the paper:

- Quantification of the effects of water withdrawal (i.e., numbers of organisms lost due to entrainment and impingement) is necessary but not sufficient to determine whether AEIs are occurring; the key is whether these effects are of consequence to a biological entity of concern (e.g., RIS population).
- Costs to the living resources and economic costs to the utilities and, ultimately, to the consumers must be taken into account when making permit decisions.
- The extent of impact of cooling-water withdrawal should be evaluated on a site-specific basis.
- In some instances, mitigation of some type may be the best way to ensure that the public's interests are addressed when CWIS decisions are made and permits are issued, approved, and enforced.

REFERENCES

1. Polgar, T.T., Summers, K.J., and Haire, M.S. (1979) Evaluation of the Effects of the Morgantown SES Cooling Systems on Spawning and Nursery Areas of Representative Important Species. Prepared for the Maryland Department of Natural Resources Power Plant Research Program. PPSP MP 27.
2. Summers, J.K. and Jacobs, F. (1981) Estimation of the Potential Entrainment Impact on Spawning and Nursery Areas Near the Dickerson Steam Electric Station. Prepared for the Maryland Department of Natural Resources Power Plant Research Program. PPSP D 81 1.
3. Richkus, W.A. and McLean, R. (2000) Historical overview of the efficacy of two decades of power plant fisheries impact assessment activities in Chesapeake Bay. *Environ. Sci. Policy* **3**, S283–S293.
4. Ringger, T.G. (2000) Investigations of impingement of aquatic organisms at the Calvert Cliffs Nuclear Power Plant, 1975–1995. *Environ. Sci. Policy* **3**, S261–S273.
5. MMES (Martin Marietta Environmental Systems, now Versar, Inc.). (1985) Impact Assessment Report: Chalk Point Steam Electric Station Aquatic Monitoring Program. Prepared for the Maryland Department of Natural Resources, Power Plant Research Program. CPC–85–1.
6. Loos, J.J. and Perry, E.S. (1989) Evaluation of Forage Fish Entrainment at Chalk Point Station (Appendix A). Prepared by Potomac Electric Power Company, Washington, D.C.
7. Versar, Inc. (1989) Review and Evaluation of PEPCo's 1989 Fractional Entrainment Loss Estimates for the Chalk Point SES. Prepared for the Maryland Department of Natural Resources, Power Plant Research Program. TR89–20.

BIOSKETCHES

Richard McLean is Manager of Nuclear Programs, Power Plant Research Program, Maryland Department of Natural Resources. He holds a B.S. in Biology and has 30 years experience in power plant impact assessment and regulation. Mr. McLeans's research interests include anadromous fish restoration; power plant impact assessment; nuclear power plant regulation and monitoring; and fate of radionuclides in the environment.

William A. Richkus is Vice President and Operations Manager, Versar, Inc., in Columbia, Maryland. He holds a Ph.D. in Oceanography from the University of Rhode Island (1974), an M.S. in

Oceanography from the University of California-San Diego Scripps Institute of Oceanography (1968), and a B.S. in Zoology from the University of Rhode Island (1966). Dr. Richkus held the positions of Assistant Professor at Trenton State College in 1972, Assistant Professor at Wilkes College in 1973, Research Scientist and Senior Scientist at Martin Marietta Corporation from 1974 to 1986, and Senior Scientist, Division Director, and Vice President of Versar, Inc. from 1987 to the present. His research interests include anadromous and catadromous fisheries biology; fisheries resource management; ecological impact assessment; and assessment of power plant impacts.

Scientific and Societal Considerations in Selecting Assessment Endpoints for Environmental Decision Making

Elizabeth M. Strange*,[1], Joshua Lipton[1], Douglas Beltman[1], and Blaine D. Snyder[2]

[1]Stratus Consulting Inc., P.O. Box 4059, Boulder, CO 80306-4059; [2]Tetra Tech Inc., 10045 Red Run Blvd., Suite 110, Owings Mills, MD 21117

Received November 15, 2001; Revised February 6, 2002; Accepted February 13, 2002; Published February, 2003

It is sometimes argued that, from an ecological point of view, population-, community-, and ecosystem-level endpoints are more relevant than individual-level endpoints for assessing the risks posed by human activities to the sustainability of natural resources. Yet society values amenities provided by natural resources that are not necessarily evaluated or protected by assessment tools that focus on higher levels of biological organization. For example, human-caused stressors can adversely affect recreational opportunities that are valued by society even in the absence of detectable population-level reductions in biota. If protective measures are not initiated until effects at higher levels of biological organization are apparent, natural resources that are ecologically important or highly valued by the public may not be adequately protected. Thus, environmental decision makers should consider both scientific and societal factors in selecting endpoints for ecological risk assessments. At the same time, it is important to clearly distinguish the role of scientists, which is to evaluate ecological effects, from the role of policy makers, which is to determine how to address the uncertainty in scientific assessment in making environmental decisions and to judge what effects are adverse based on societal values and policy goals.

KEY WORDS: ecological risk assessment, assessment endpoints, measurement endpoints, population assessment, natural resource value, environmental value

DOMAINS: ecosystems and communities, organisms, environmental toxicology, environmental management and policy, ecosystems management, environmental modeling, environmental monitoring

---

INTRODUCTION

Ecological risk assessment is a process for evaluating the likelihood of adverse ecological effects[1,2]. It is designed to provide environmental decision makers

* Corresponding author. Emails: estrange@stratusconsulting.com; jlipton@stratusconsulting.com; dbeltman@stratusconsulting.com; blaine.snyder@tetratech.com

with a scientific evaluation of the risks posed to ecological resources by alternative management actions, ranging from the regulation of hazardous waste sites to the management of entire watersheds affected by multiple stressors.

A critical component of the risk assessment process is the selection of assessment and measurement endpoints. Assessment endpoints are the environmental entities that are targets of the risk assessment, and measurement endpoints are the attributes that are actually measured[1,2]. For example, the reproductive success of Coho salmon is an assessment endpoint, while egg survival is a measurement endpoint.

Although numerous documents provide guidelines for endpoint selection[1,2], there remains some confusion about the role of science in the process. Some investigators argue that, from a scientific point of view, population- and higher-level endpoints should take precedence based solely on their ecological relevance[3,4,5]. However, as the EPA's ecological risk assessment guidelines make clear, scientific considerations are only part of the overall process of endpoint selection[2]. In many cases, social, economic, and policy considerations argue for the assessment of individual-level endpoints, as is the case for legally protected habitats or organisms, such as endangered species[6].

Even from a scientific perspective, there are compelling reasons for concluding that higher-level endpoints are not always appropriate or sufficient for assessing ecological risks. Whereas the measurement of higher-level endpoints may provide information about ecological condition, it may provide little information about the causes of observed effects. In contrast, individual-level endpoints are often preferred for ease and reliability of measurement and their relatively high statistical power to detect effects[7,8]. Moreover, individual effects are precursors to population and ecosystem effects, and thus individual-level effects help inform risk managers about potential future risks to higher levels of biological organization.

In this paper, we consider how endpoint selection is constrained by the need to balance ecological and management relevance with measurement validity and practicality, including the amount of time and money needed to complete a scientifically valid study. We outline key scientific, social, and policy considerations in the selection of endpoints and discuss some reasons why individual-level endpoints are sometimes preferable. We conclude by proposing that it is important to consider all of these factors to ensure that the risk assessment process will support the overall goal of environmental protection.

SCIENTIFIC CONSIDERATIONS IN SELECTING RISK ASSESSMENT ENDPOINTS

According to the EPA's *Guidelines for Ecological Risk Assessment*, selection of assessment endpoints should consider (1) susceptibility to the stressor, (2) ecological relevance, and (3) policy goals and societal values[2]. In this section, we consider issues related to ecological relevance.

Although important for evaluating overall ecological condition, there can be ambiguity and uncertainty in population-, community-, and ecosystem-level assessments resulting from natural variability, measurement difficulties, lack of data, and limitations of scientific understanding[9].

Detection of higher-level effects is difficult in large part because of the natural variation inherent in biological populations[7,8]. For example, studies show that it can take at least a decade or two to detect a "signal" from the "noise" in fish population data[10]. Natural variation also means that it is often difficult to establish "baseline" or "average" conditions against which the significance of impacts can be evaluated[7,8,11]. Long-term monitoring can help reduce uncertainties, but this is costly and impractical in many contexts[9,12].

Cause-effect relationships are also difficult to establish at higher levels of biological organization[13], although the stressor identification process has advanced in recent years[14]. Populations, communities, and ecosystems reflect effects of multiple stressors interacting in complex ways[15]. Characteristics of these entities integrate all stressor effects, and therefore it can be very difficult to attribute population- or higher-level ecological effects to any particular stressor. For example, distinguishing the relative impacts of various environmental stressors on declines of salmon (*Oncorhynchus* spp.) in the Pacific Northwest, lake trout (*Salvelinus namaycush*) in the Great Lakes, and many other fish species has proven to be very difficult despite years of study by numerous researchers[16].

Defining the spatial and temporal boundaries of higher-level ecological entities is also difficult and often arbitrary[17]. For example, a fish population can be defined on the basis of the local stock or in terms of its regional extent. Mortalities of individuals may significantly reduce the local population, while effects on the regional population may remain undetectable.

A prominent example of conflicts over population-level impacts has been the ongoing debate over the impacts on fish populations caused by larval entrainment in the cooling water intakes of power plants[18,19]. Most assessments of power plant entrainment have been based on population models with significant uncertainties, such as the potential role of density-dependent compensation in response to power plant mortality. As a result, there has been little agreement about whether or not adverse impacts are occurring, despite the enormous losses of aquatic organisms at power plant intakes.

There is much less uncertainty in individual-level assessments[20]. In most cases, individuals can be defined with less ambiguity and greater ease. Measurement and sampling errors at the individual level are also less than those associated with estimates of populations[7,8]. As a result of greater data availability and reliability, environmental effects are more likely to be detected at the individual level than at higher levels of biological organization.

For example, Bennett et al. [21] found a high percentage of abnormalities in larval striped bass that were thought to result from herbicide use in rice fields, as indicated by the absence of abnormalities following changes in culture practices that reduced herbicide release into rivers with striped bass. In addition, Bailey et al.[22] found that the decline of striped bass in California was correlated with

Entity	Response Time	Ambiguity	Difficulty of Measurement	Relevance for Ecosystem
individual	short	low	low	less direct
population	↓	↓	↓	↓
community	↓	↓	↓	↓
ecosystem	long	high	high	more direct

FIGURE 1. Tradeoffs in endpoint selection.

increased herbicide use. Nevertheless, Kimmerer et al.[23] could find no evidence of a population-level response.

Environmental decision makers must often balance the need for ecological relevance with the need for measurement ease and reliability in deciding what endpoints to evaluate (Fig. 1). In cases where a stressor directly affects individuals, but population or higher-level effects are unclear though potentially important, individual-level endpoints may need to take precedence. Indeed, effects on individuals can be important predictors of potential effects on populations or communities that cannot be measured directly.

The Role of Social Values and Policy Goals in Endpoint Selection

While scientific considerations are important, they are not the only factors that environmental decision makers must take into account in evaluating the potential for adverse effects. In fact, the EPA's *Ecological Risk Assessment Guidelines* stress that the appropriate level of biological organization for an assessment depends on societal values and policy goals as well as data availability and ecological relevance[2]. Indeed, society clearly places value on ecological attributes that are not necessarily captured by assessing only higher levels of biological organization, and thus individuals may warrant protection even *in lieu* of population-level effects.

For example, a survey following the Nestucca oil spill in the state of Washington found that local residents believed that preventing the death of seabirds from oil spills is important, even if seabird populations appear unaffected[24].

Similarly, in a regional survey conducted as part of a natural resource damage assessment for Green Bay, people expressed high value (hundreds of millions of dollars) for restoring bird and fish injuries from PCBs, even though they were explicitly told that there may not be population-level effects[25].

Regulatory Guidance

The value that society places on individual organisms is reflected in many current regulations and statutes. As described below, the Clean Water Act (CWA), the Migratory Bird Treaty Act, the Comprehensive Environmental Response, Compensation and Liability Act (CERCLA), the Oil Pollution Control Act (OPA), the Federal Insecticide, Fungicide and Rodenticide Act (FIFRA), and relevant case law authorize that effects at the individual organism level be assessed in making regulatory decisions.

In some cases, risk assessments and regulatory programs consider effects on individuals to be important as indicators of effects on populations. In these cases, individual-level effects are a measurement endpoint for the population, which is the assessment endpoint. An example is provided by the National Pollution Discharge Elimination System (NPDES) permit program. Under section 301(b)(1)(c) of the CWA, effluent limits must be placed in NPDES permits as necessary to meet water quality standards. To implement this requirement, the EPA and most states rely on toxicity tests that determine the effects of discharges on individual organisms[26]. By evaluating the effects of pollutants on growth, reproduction, and mortality of individuals, the EPA uses individual impacts as surrogates and precursors of population and ecosystem impacts.

In other cases, risk assessments and regulatory programs are intended to protect individual members of a species, regardless of potential effects on the population of the species. For example, the Migratory Bird Treaty Act, 16 U.S.C. §§ 703-712, prohibits, among other things, the killing of individual migratory birds [16 U.S.C. §703]. The act does not require evidence that bird mortalities affect a bird population; effects on individual organisms are the only test.

Another example is provided by CERCLA [42 U.S.C. Section 9601 *et seq.*] and OPA [33 U.S.C. Section 2701 *et seq.*], which require that the public be compensated for natural resource injuries resulting from an oil spill or hazardous substance release. These regulations stipulate that the value of lost resources can include the value of injured individuals of marine species as well as the value that society places on just knowing that a natural area exists.

A final example of regulations designed to protect individuals is provided by FIFRA, 7 U.S.C., which regulates the manufacture, distribution, and use of pesticides. The act is intended to protect the "water, air, land, and all plants and man and other animals living therein, and the interrelationships which exist among these" [7 U.S.C. §136 (j)] from unreasonable adverse effects [7 U.S.C. §136 (d)]. Under FIFRA, effects on biological populations are not a required element of risk assessment. A 1989 decision by the U.S. Court of Appeals for the Fifth Circuit illuminates how "unreasonable adverse effects" are interpreted under FIFRA. In 1988, the EPA canceled registration for the pesticide diazinon unless registration was amended to prohibit use on golf courses and sod farms, based on the EPA's determination that the use of the pesticide in these cases posed an unreasonable risk to birds [53 Fed. Reg. 11119].

Ciba-Geigy Corporation, diazinon's major producer, petitioned the EPA's determination for review by the courts. Among other issues, Ciba-Geigy presented the argument that a risk is unreasonable only if it endangers bird populations, not just individuals [55 Fed. Reg. 31137]. The court rejected Ciba-Geigy's argument, stating that "FIFRA gives the Administrator sufficient discretion to determine that recurring bird kills, even if they do not significantly reduce bird populations, are themselves an unreasonable environmental effect" [874 F.2d 277].

The court clearly sided with the EPA in its determination that effects at the individual organism level can be interpreted as unreasonable environmental effects.

Risk Assessment in the Overall Context of Environmental Decision Making

Current guidelines by the EPA and other environmental agencies indicate that whether estimated risks are considered "adverse," "undesirable," or "unacceptable" should be based on a range of factors, including management goals, policy considerations, societal values, and legal mandates, as well as underlying scientific understanding[2]. Thus, there is no universal definition of "adverse environmental impact," nor can there be. Ultimately, the decision of what is "adverse" rests with policy makers, not scientists. As Rykiel[27] noted: "... science deals with true and false, whereas society deals with good and bad." While someone must decide what ecological conditions are good or bad, it should not be scientists if we are to maintain scientific impartiality[28,29].

Environmental decision makers face a difficult task in choosing from among what are often competing social values. Even cost-benefit comparisons of management options provide few clear-cut answers. As Lackey[29] pointed out: "The marketplace, the most common adjudicator of societal preferences, is never totally unconstrained, nor do most participants have much understanding of the long-term ecological consequences of their individual market decisions. Thus, economics has an important role in resolving competing societal preferences, but is insufficient in itself." Moreover, many biological resources that are valued by society are not traded in markets, and failure to account for these assets can seriously bias environmental decision making[30].

When individual-level effects are considered, the regulatory scope for minimizing impacts to environmental resources is greater than it is for minimizing higher-level impacts. This is because individual effects are more likely to be detected. A focus on the most readily detected effects allows risk managers to undertake actions to reduce impacts before more serious damage to higher levels of organization can occur.

Many resource agencies recognize that if protective measures are not initiated until effects at higher levels of biological organization are apparent, natural resources that are ecologically important or highly valued by society may not be adequately protected. This has led these agencies to exercise a "precautionary approach" to environmental management[31]. The precautionary approach aims to prevent irreversible damage to the environment by implementing strict conservation measures even in the absence of unambiguous scientific evidence that environmental degradation is being caused by human stressors[32].

The precautionary approach is now being applied in fisheries management. For example, in a recent publication, the National Marine Fisheries Service (NMFS) noted that "all fishing activities have environmental impacts and that it is not appropriate to assume that these impacts are unimportant until proven otherwise[31]." The report concluded that the collapse of fish stocks worldwide has resulted in part because corrective actions were often delayed or not implemented when scientific information on stock status was in doubt. NMFS noted that, in 1995, the Food and Agriculture Organization (FAO) of the United Nations drafted an International Code of Conduct that emphasized that "the absence of adequate scientific information should not be used as a reason for postponing or failing to take conservation management measures."[31]

CONCLUSIONS

While the purpose of an ecological risk assessment is to provide environmental decision makers with a scientific evaluation of the risks posed to ecological resources, science cannot answer the difficult question of how much impact is acceptable[29,33,34,35,36]. The distinction between the role of scientists in evaluating ecological effects and the role of policy makers in judging the adversity of effects is important, but often overlooked. To avoid unnecessary conflicts, it is critical to clearly separate the roles of scientists and policy makers in the risk assessment process. Failure to do so may not only undermine the objectivity necessary for valid risk assessment, but can ultimately interfere with the overriding goal of environmental protection.

ACKNOWLEDGEMENTS

Support for this work was provided, in part, by the U.S. EPA to Stratus Consulting Inc. under Contract No. 68-W6-0055 and to Tetra Tech under Contract No. 68-C-99-249. However, the views expressed in this paper are those of the individual authors, and do not represent the official position of the U.S. EPA. The authors wish to thank John Boreman, James Andreason, and Peter Moyle for their helpful comments and suggestions on an earlier draft of this manuscript.

REFERENCES

1. Suter, G.W. (1993) *Ecological Risk Assessment*. Lewis Publishers, Chelsea, MI.
2. U.S. EPA (1998) Guidelines for Ecological Risk Assessment. EPA/630/R-95/002B. U.S. Environmental Protection Agency, Washington, D.C.

3. Clements, W.H. and Kiffney, P.M. (1994). Assessing contaminant effects at higher levels of biological organisation. *Envron. Toxicol. Chem.* **13**, 357–359.
4. Martin, M. and Richardson, B.J. (1995) A paradigm for integrated marine toxicity research? Further views from the Pacific Rim. *Mar. Pollut. Bull.* **30**, 8–13.
5. Dey, W.P., Jinks, S.M., and Lauer, G.J. (2000) The 316(b) assessment process: evolution towards a risk-based approach. *Environ. Sci. Policy* **3**, S15–S23.
6. Schmitt, R.J., Osenberg, C.W., Douros, W.J., and Chesson, J. (1996) The art and science of administrative environmental impact assessment. In *Detecting Ecological Impacts: Concepts and Applications in Coastal Habitats.* Schmitt, R.J. and Osenberg, C.W., Eds. Academic Press, Inc., San Diego, CA. pp. 281–293.
7. Osenberg, C.W., Schmitt, R.J., Holbrook, S.J., Abu-Saba, K.E., and Flegal, A.R. (1994) Detection of environmental impacts: natural variability, effect size, and power analysis. *Ecol. Appl.* **4**, 16–30.
8. Osenberg, C.W., Schmitt, R.J., Holbrook, S.J., Abu-Saba, K.E., and Flegal, A.R. (1996) Detection of environmental impacts: natural variability, effect size, and power analysis. In *Detecting Ecological Impacts: Concepts and Applications in Coastal Habitats.* Schmitt, R.J. and Osenberg, C.W., Eds. Academic Press, Inc., San Diego, CA. pp. 83–108..
9. Schmitt, R.J. and Osenberg, C.W., Eds. (1996*) Detecting Ecological Impacts: Concepts and Applications in Coastal Habitats.* Academic Press, San Diego, CA.
10. Myers, R.A., Bridson, J., and Barrowman, N.J. (1995) Summary of worldwide stock and recruitment data. *Can. Tech. Rep. Fish. Aquat. Sci.* **2024**, 1–327.
11. Stewart-Oaten, A. (1996) Problems in the analysis of environmental monitoring data. In *Detecting Ecological Impacts: Concepts and Applications in Coastal Habitats.* Schmitt, R.J. and Osenberg, C.W., Eds. Academic Press, Inc., San Diego, CA. pp. 109–132.
12. NRC (National Research Council) (1990) Managing Troubled Waters: The Role of Marine Environmental Monitoring. National Academy Press, Washington, D.C.
13. Attrill, M.J. and Depledge, M.H. (1997) Community and population indicators of ecosystem health: targeting links between levels of biological organization. *Aquat. Toxicol.* **38**, 183–197.
14. U.S. EPA (2000) Stressor Identification Guidance Document. EPA/822/B-00/025, U.S. Environmental Protection Agency, Office of Water and Office of Research and Development, Washington, D.C.
15.U.S. EPA (1992) Biological Populations as Indicators of Environmental Change. EPA-230-R-92-011, U.S. Environmental Protection Agency, Office of Policy Planning and Evaluation, Washington, D.C.
16. Walters, C. (1997) Challenges in adaptive management of riparian and coastal ecosystems. *Conserv. Ecol.* [online] **1**,1–23. URL:http://www.consecol.org/vol1/iss2/art1
17. Levin, S.A. (1992) The problem of pattern and scale in ecology. *Ecology* **73**, 1,943–1,976.
18. Christensen, S.W., Van Winkle, W., Barnthouse, L.W., and Vaughan, D.S. (1981) Science and the law: confluence and conflict on the Hudson River. *Environ. Impact Assess. Rev.* **2/1**, 63–88.
19. Barnthouse, L.W., Klauda, R.J., Vaughan, D.S., and Kendall, R.L., Eds. (1988) Science, law, and Hudson river power plants: a case study in environmental impact assessment. *Am. Fish. Soc. Monogr.* **4**.
20. DeAngelis, D.L., Barnthouse, L.W., Van Winkle, W., and Otto, R.G. (1990) A critical appraisal of population approaches in assessing fish community health. *J. Great Lakes Res.* **16**, 576–590.
21. Bennett, W.A., Ostrach, D.J., and Hinton, D.E. (1995) Larval striped bass condition in a drought-stricken estuary: evaluating pelagic food-web limitation. *Ecol. Appl.* **5,** 680–692.
22. Bailey, H.C., Alexander, C., Digiorgio, C., Miller, M., Doroshov, S.I., and Hinton, D.E. (1994) The effect of agricultural discharge on striped bass (*Morone saxatilis*) in California's Sacramento-San Joaquin drainage. *Ecotoxicology* **3**, 123–142.
23. Kimmerer, W.J., Cowan, Jr., J.H., Miller, L.W., and Rose, K.A. (2000) Analysis of an estuarine striped bass (*Morone saxatilis*) population: influence of density-dependent mortality between metamorphosis and recruitment. *Can. J. Fish. Aquat. Sci.* **57**, 478–486.
24. Rowe, R.D., Schulze, W.D., Shaw, W.D., Schenk, D., and Chestnut, L.G. (1991) Contingent valuation of natural resource damage due to the Nestucca Oil Spill. Final Report prepared for Department of Wildlife, State of Washington, Olympia, WA; British Columbia Ministry of Environment, Victoria, BC; Environment Canada, Vancouver, BC, Canada.

25. Breffle, W.S., and Rowe, R.D. (2002) Comparing choice question formats for evaluating natural resource tradeoffs. *Land Econ.*, in press.
26. U.S. EPA (1991) Technical Support Document for Water Quality-Based Toxics Control. EPA/505/2-90-001. U.S. Environmental Protection Agency, Washington, D.C.
27. Rykiel, E.J. (1998) Relationships of scale to policy and decision making. In *Ecological Scale: Theory and Applications*. Peterson, D.L. and Parker, V.T., Eds. Columbia University Press, New York. pp 485–497.
28. Sagoff, M. (1995) The value of integrity. In *Perspectives on Ecological Integrity*. Westra, L. and Lemons, J., Eds. Kluwer Academic Publishers, Dordrecht, The Netherlands. pp. 162–176.
29. Lackey, R.T. (2001) Values, policy, and ecosystem health. *BioScience* 51, 437–443.
30. Lipton, D.W., Wellman, K., Sheifer, C., and Weiher, R.F. (1995) Economic valuation of natural resources – a handbook for coastal resource policymakers. NOAA Coastal Ocean Program Decision Analysis Series No. 5. NOAA Coastal Ocean Office, Silver Spring, MD, 131 p.
31. National Marine Fisheries Service (1999) The precautionary approach: a new paradigm or business as usual? Our Living Oceans. Report on the Status of U.S. Living Marine Resources. U.S. Department of Commerce, NOAA Tech. Memo. NMFS-F/SPO-41. pp. 61–70.
32. Hilborn, R., Maguire, J.-J., Parma, A.M., and Rosenberg, A.A. (2001) The precautionary approach and risk management: can they increase the probability of success in fishery management? *Can. J. Fish. Aquat. Sci.* 58, 99–107.
33. Salzman, L. (1995) Scientists and advocacy. *Conserv. Biol.* 9, 709–710.
34. Lackey, R.T. (1998) Seven pillars of ecosystem management. *Lands. Urban Plan.* 40, 21–30.
35. Lackey, R.T. (1999) The savvy salmon technocrat: life's little rules. *Environ. Pract.* 1, 156–161.
36. Power, M. and McCarty, L.S. (1997) Fallacies in ecological risk assessment practices. *Environ. Sci. Technol.* 31, 370A–375A.

BIOSKETCH

Elizabeth M. Strange is a Manager at Stratus Consulting Inc., an environmental and energy research firm in Boulder, Colorado. Dr. Strange is an aquatic ecologist with expertise in the assessment of human impacts to marine and freshwater ecosystems. She has developed and assessed ecological endpoints for quantifying benefits of proposed regulations, assessing resource injuries, comparing restoration options, and predicting potential consequences of climate change and other global stressors on aquatic ecosystem services. Her work has included the collection, analysis, and modeling of fisheries and water quality data for regulatory impact assessments and natural resource damage assessments. Dr. Strange has also worked closely with natural resource economists to develop methods for integrating environmental assessments and benefits estimation. She has published results of her research in a number of peer-reviewed journals, including *Environmental Management*, *Ecological Economics*, *Environmental Biology of Fishes*, and *Marine Fisheries Review*. Dr. Strange holds a Ph.D. and an M.S. in ecology from the University of California at Davis and a B.A. in biology from San Francisco State University.

Adverse Environmental Impact: 30-Year Search for a Definition

David A. Mayhew*, Paul H. Muessig, and Loren D. Jensen
EA Engineering Science and Technology, Inc., 11019 McCormick Road, Hunt Valley, MD 21031

Received November 15, 2001; Revised January 7, 2002; Accepted February 13, 2002; Published February, 2003

Since passage of the Clean Water Act in 1972, there has been a long, unresolved struggle to define a key phrase in Section 316(b) of the act: "adverse environmental impact" (AEI). Section 316(b) requires that the best technology available be used in cooling-water intake structures to minimize AEI due to entrainment and impingement of aquatic organisms. Various attempts were made to evaluate and define AEI, including focused national conferences on impact assessment. Unresolved arguments regarding AEI were reinvigorated following the 1995 Consent Decree requiring EPA to propose new rules to implement Section 316(b). This article reviews and compares eight proposed definitions of AEI. Six of the definitions define AEI as impact expressed at the population or higher level of biological organization. The two remaining definitions are unrelated to populations: a 1% cropping of the near-field organisms and "one fish equals AEI". The latter definition is based on the desire of some stakeholders to define AEI as the loss of any public trust resources. Equating loss of public trust resources with AEI hampers consensus on a definition because a societal-based policy concept (public trust resources) is commingled with science-based definitions based on population effects. We recommend that a population-based definition of AEI be incorporated into Section 316(b) guidance and observe that this will not preclude a state from exercising its law and policy to protect public trust resources.

KEY WORDS: adverse environmental impact, Clean Water Act, Section 316(b), best technology available, cooling-water intake structure, entrainment, impingement, public trust resources

DOMAINS: freshwater systems, marine systems, water science and technology, environmental management and policy

---

INTRODUCTION

Soon after passage of the National Environmental Policy Act in 1969, which brought the term environmental "impact" into common usage, the U.S. Congress passed Public Law (PL) 92-500, the Federal Water Pollution Control Act Amend-

* Corresponding author. Email: dam@eaest.com; phm@eaest.com; ldj@eaest.com

ments of 1972 (the "Clean Water Act" or "CWA"). Section 316(a) of the CWA addressed thermal discharges, and Section 316(b) addressed cooling-water intake structures (CWIS). Section 316(b) required that "the location, design, construction, and capacity of cooling water intake structures reflect the best technology available [BTA] for minimizing adverse environmental impact." Such impact can result from entrainment of fish eggs and larvae and other small aquatic organisms into the cooling-water stream (and ultimately through pumps and condensers) or from impingement (trapping) of larger organisms on CWIS screens. Although possibly not the first use of the phrase "adverse environmental impact" (AEI), its incorporation in the federal law solidified it as a litmus test in subsequent CWIS impact assessments. Unfortunately, the phrase was not defined or quantified, and this resulted in much confusion, controversy, and litigation. The confusion has continued. Now, 30 years after passage of the act, and after considering four possible definitions in its draft rulemaking for new CWIS (Federal Register Vol. 65, No. 155, pp. 49060-49121, 10 August 2000), the U.S. Environmental Protection Agency (EPA) declined to define AEI in its final rulemaking of 9 November 2001 (Federal Register Vol. 66, No. 243, pp. 65256-65345, 18 December 2001).

HISTORY

It did not take long after passage of the CWA for scientists, regulators, and resource managers to begin to grapple with the meaning of AEI. In June 1975, a *Conference on the Biological Significance of Environmental Impacts* was sponsored by the U.S. Nuclear Regulatory Commission and held at the University of Michigan[1]. The consensus definition that emerged from this forum was:

> An impact is significant if it results in a change that is measurable in a statistically sound sampling program and if it persists, or is expected to persist, more than several years at the population, community, or ecosystem level[2].

The word "adverse" was not featured in this forum, but we equate it with the word "significance."

Soon after the conference, the EPA published the 1977 *Draft Guidance for Evaluating the Adverse Impact of Cooling Water Intake Structures on the Aquatic Environment*[3]. This guidance contained the following definition of AEI:

> Adverse aquatic environmental impacts occur whenever there will be entrainment or impingement damage as a result of the operation of a specific cooling water intake structure. The critical question is the magnitude of any adverse impact. The exact point at which adverse aquatic impact occurs at any given plant site or water body segment is highly speculative and can only be estimated on a case-by-case basis by considering the species involved, magnitude of the losses, years of intake operation remaining, ability to reduce losses, etc.

Whereas the first sentence of this definition appears to identify any entrainment or impingement as adverse impact, it becomes clear that entrainment and impingement losses are not, in and of themselves, adverse impact, pending evaluation of various other factors.

In its 1980 strategy document for addressing power-plant impacts[4], the U.S. Fish and Wildlife Service defined impact as:

> A change in population structure or dynamics of a species resulting from an activity of man that remains at least as long as the activity continues.

Also in 1980, Voigtlander[5] reviewed prior attempts at defining AEI and proposed the following definition:

> An impact is a significant, long-lasting, man-induced change in the numbers or biomass of a species population.

Voigtlander also highlighted a fundamental problem with the concept of impact that has hampered consensus: "Obviously it [impact] is one of those words that is so familiar to us that we all understand what it means – except that everyone understands it somewhat differently." Similar observations were made by Westman[6] and several participants in EPA-sponsored public meetings on the 316(b) rulemaking (comments available at http://www.epa.gov/ost/316b/).

The four definitions above differ somewhat, but in each, the test of impact (or adverse impact or significant impact) pivots on a level of organization above the individual fish or other organism. Either explicitly[2,4,5] or implicitly[3], that level of organization is at least the population level. That is, impact is not deemed adverse or significant unless it is expressed and measurable at least at the population level.

The longest and most intense effort to identify impacts of CWIS took place on the Hudson River between the mid-1960s and 1980[7,8]. Fishing and conservation interests were concerned that entrainment of striped bass eggs and larvae at several power plants and the proposed Cornwall pumped-storage facility would harm the population. There was also concern regarding the loss of fish due to impingement at CWIS. Detailed field studies, population modeling, and other evaluations were conducted and then debated in a series of adjudicatory hearings. Ultimately, settlement negotiations were held wherein disputes over environmental impacts were suspended and replaced with a series of consensus mitigation programs. The mitigation agreements include ongoing monitoring and preparation of annual year-class reports and special entrainment and impingement studies. In the context of this article, the Hudson River studies were never directed at defining AEI as a regulatory standard or threshold. Rather, the effort was directed at measuring the effectiveness of mitigation measures in reducing mortality rates.

Similar long-term impact assessments were carried out at the Salem Nuclear Station on Delaware Bay between the early 1970s and mid-1990s. These studies and regulatory reviews culminated in the mid-1990s with a negotiated settlement with state and federal regulators based on habitat enhancement to offset CWIS losses and testing of alternative intake technologies to reduce impingement. Although AEI was not defined, the settlement was based on providing opportunities for increased biological production within the estuary to offset losses associated with operation of the CWIS at Salem. The cornerstone of the settlement was the utility's establishment and funding of the Estuarine Enhancement Program (EEP). The terms of the settlement were incorporated into the New Jersey Pollutant Discharge Elimination System permit issued in 1995. The primary component of the EEP consisted of restoration, enhancement, and/or preservation of more than 20,000 acres of degraded coastal wetlands and upland buffer along the Delaware Estuary; these wetlands provide nursery, food, shelter, and habitat for many species of fish affected by the CWIS as well as other wildlife. The EEP also included construction of fish ladders to enhance river herring migration and production, installation of protective intake technologies, and a comprehensive biological monitoring program. The EEP was retained in Salem's permit for 2000.

After the mitigation-based negotiated settlements on the Hudson River and at Salem, discussion of the meaning of AEI was reinvigorated following the 1995 Consent Decree with Hudson Riverkeeper et al., requiring the EPA to propose rules to implement Section 316(b). As evidenced in EPA-sponsored public meetings on the pending rulemaking and ultimately in the proposed rule for new facilities, several definitions of AEI were considered for possible inclusion in regulations or guidance.

THE PRESENT

Two of the definitions in EPA's proposed rulemaking focused on population- or higher-level impacts. One was the same definition previously published in 316(b) guidance[3] and cited above. The second definition would place AEI in a biocriteria context, whereby CWIS affects on an aquatic community would be compared to a reference site without a CWIS. Presumably, measures (metrics) of community abundance, diversity, and other characteristics would be compared between the sites, and if similar, a lack of AEI to the aquatic community at the CWIS site may be concluded. An implementation approach was not provided by the EPA, but comments were invited.

Two additional definitions in the proposed rulemaking diverged from all previous definitions in that they were not related to population-level effects. One of these defined AEI as:

> impingement or entrainment of one (1) percent or more of the aquatic organisms from the area around the cooling water intake structure from which organisms are drawn onto screens or other barriers at the entrance to a [CWIS].

EPA considered this a "reasonable approach" because it was similar to its approach with water quality-based regulatory programs. We consider this a poor approach in that an AEI threshold is arbitrarily *assigned,* and no correlation to environmental damage or AEI was presented.

Another alternative considered by the EPA was to define AEI as "any impingement or entrainment of aquatic organisms." This has been informally referred to as the "one fish equals AEI" definition. In discussing this alternative in the proposed rulemaking for existing facilities, EPA cited public comments by a New York State Department of Environmental Conservation representative regarding its long-term implementation of this definition. In those public comments (available at http://www.epa.gov/ost/316b/), the New York State representative explained that agency's rationale for the approach, including, in part, the statement that "these are our trust resources as states, and we do not feel that [it] is right to allocate any of these resources to industrial mortality." Without debating the concept of trust resources, which has basis in law[9], this definition is unrelated to environmental damage or AEI. Furthermore, under such a definition, no CWIS could be permitted without maximum application of BTA, since none can totally avoid some level of entrainment and impingement – regardless of BTA employed.

Under the trust resources concept, the impingement of one fish during a year would represent AEI. At least outside of the context of threatened and endangered species, no one would construe the loss of a single fish as environmental or ecological damage. The idea of a state's ownership of natural resources – and the intrusion of this concept into the 316(b) process – is not new. For example, during a panel discussion at the *Fourth National Workshop on Entrainment and Impingement* in Chicago in 1977[10], a representative of the state of Michigan made a strong case for the state's ownership of the resources and stated, "even though the losses of fish do not warrant the application of extremely expensive technologies, we feel that we cannot let the utilities off for killing fish that belong to the state." In this discussion, the Michigan representative separated implementation of the federal 316(b) statute from a state's right to "mitigate" for losses of its resources. However, we believe there is a tendency in some areas to substitute any loss of a state's trust resources as a definition for AEI.

The Public Trust Doctrine is a legal concept that has its roots in the Roman Empire and which has evolved into a mechanism to protect natural resources for the public good[11]. The doctrine is considered a legal framework for resource planning and management that has increasingly been used not only to protect natural resources for public use but also to prevent overexploitation of those resources[12]. We do not dispute the public trust concept in general or its potential application in matters of CWIS impacts. However, we do not believe it is appropriate to substitute the protective concept of the doctrine as a definition of AEI. Some people may construe the loss of one fish as a social impact, i.e., a loss of public property. But it is not an environmental impact, and that is the focus of Section 316(b).

In response to the proposed rulemaking for existing facilities, the Utility Water Act Group[13] provided extensive comments, including a proposed definition of AEI:

> Adverse environmental impact is a reduction in one or more representative indicator species that (1) creates an unacceptable risk to the population's ability to sustain itself, to support reasonably anticipated commercial or recreational harvests, or to perform its normal ecological function and (2) is attributable to the operation of the cooling water intake structure.

This is another population-based definition, but it is unique in that its "test" or determination of threshold turns on not just a reduction in population, but whether that reduction represents "unacceptable risk." Further, it appears to address resource allocation issues in that unacceptable risks to fishery harvests may represent AEI outside of the context of population sustainability. The Utility Water Act Group proposed that unacceptable risk be determined in a scientific risk assessment and risk management process wherein a number of biological and social factors would be considered.

On 9 November 2001, a final 316(b) rulemaking for new CWIS was signed. After 30 years of research and debate on the meaning of AEI, the EPA declined to define it, citing the same lack of consensus among stakeholders as described in this article. The EPA assumed that entrainment and impingement were real or potential threats to aquatic populations and formulated the rulemaking as a technology-based approach for minimizing any entrainment or impingement.

DISCUSSION

From the period of the early 1970s to the present, eight definitions of AEI were found in the available record and reviewed (Table 1). Six of these cast AEI in a population- or higher-level context. That is, the impact must be measurable and expressed at the population or higher (e.g., community) level of biological organization. Two of the newer definitions originally considered by EPA – "one fish equals AEI" and a 1% cropping of the nearfield waterbody population – were based on counts of entrained and impinged organisms. Whereas no one should argue the right of any stakeholders to consider these last two definitions, their inclusion in the suite did not make the achievement of consensus any easier. Prior to the 1990s, efforts to define AEI had a common basis – impact at the population (or higher) level of biological organization. Now, there is no common basis among competing definitions of AEI. The various definitions reviewed herein reflect the different values (scientific vs. social) of the various stakeholders involved.

In our view, the failure to define AEI in the final rulemaking for new CWIS will not end the debate. As the rulemaking process moves to consideration of the existing CWIS facilities, there will be renewed calls for inclusion of AEI in the process. Many existing facilities have substantial environmental data sets that can be used to determine the presence or absence of AEI. The EPA's rationale for not defining AEI – essentially that it is indefinable – is not compelling. We acknowledge that even among scientists, differences exist regarding what level of loss of aquatic resources represents damage or impact. This need not preclude establishing a definition based on population-level impacts, in the sure knowledge that the state of science will improve to be able to measure those impacts. Our position is that whereas AEI may not presently be easily measured, it is certainly definable._

In our review of historical and current discussions about AEI, we identified several factors that we believe are important, some of which have seriously hampered consensus on AEI.

TABLE 1
Chronology of 316(b) and AEI Definition Milestones

Date	MileStone	Definitions
1969	Passage of National Environmental Policy Act; term "impact" comes into common use	
1972	CWA Section 316(b); term "adverse environmental impact" codified	
1975	Conference on *Biological Significance of Environmental Impacts*[1]	An impact is significant [adverse] if it results in a change that is measurable in a statistically sound sampling program and if it persists, or is expected to persist, more than several years at the population, community, or ecosystem level[2].
1977	EPA (1977) Draft 316(b) guidance[3]	Adverse aquatic environmental impacts occur whenever there will be entrainment or impingement damage as a result of the operation of a specific cooling-water intake structure. The critical question is the magnitude of any adverse impact. The exact point at which adverse aquatic impact occurs at any given plant site or water body segment is highly speculative and can only be estimated on a case-by-case basis by considering the species involved, magnitude of the losses, years of intake operation remaining, ability to reduce losses, etc.
1980	Hudson River case settlement; culmination of the most studied and contested 316(b) issue	
1980	U.S. Fish and Wildlife Service power-plant impact strategy document	A change in population structure or dynamics of a species resulting from an activity of man that remains at least as long as the activity continues[4].
1980	Fifth National Workshop on Entrainment and Impingement: *Issues Associated with Impact Assessment*[14]	An impact is a significant, long-lasting, man-induced change in the numbers or biomass of a species population[5].
1988	Publication of AFS Monograph 4: Science, Law, and Hudson River Power Plants, a Case Study in Environmental Impact Assessment[8]	
1995	Consent Decree between Hudson Riverkeeper et al. and EPA requiring new Section 316(b) rulemaking	
2000	EPA proposed rule for new CWIS facilities (Federal Register, Vol. 65, No. 155, pp. 49060-49121, 10 Aug. 2000)	<u>Considered by EPA:</u> 1) The definition from the 1977 316(b) guidance (see above); 2) Biocriteria-based definition (see text); 3) Impingement or entrainment of one (1) percent or more of the aquatic organisms from the area around the [CWIS] from which organisms are drawn onto screens or other barriers at the entrance to a [CWIS]; 4) Any impingement or entrainment of aquatic organisms. <u>Utility Water Act Group[12] definition in response to proposed rule:</u> Adverse environmental impact is a reduction in one or more representative indicator species that (1) creates an unacceptable risk to the population's ability to sustain itself, to support reasonably anticipated commercial or recreational harvests, or to perform its normal ecological function and (2) is attributable to the operation of the cooling water intake structure.
2001	Final rulemaking for new CWIS, 9 November 2001 (Federal Register Vol. 66, No. 243, pp. 65256-65345, 18 December 2001).	No definition. Default assumption that any entrainment or impingement is threat to aquatic resources.

1. Given the use of the phrase "adverse environmental impact" in Section 316(b) of the CWA and the extant disagreement over the meaning of the phrase, there should be a definition in regulation and/or guidance. Failure to do so would invite continued confusion and could lead to extended litigation among stakeholders regarding Section 316(b). Notwithstanding the lack of a definition in the final rulemaking for new facilities, there will be ample opportunity to resolve and define AEI as the 316(b)-rulemaking process continues.
2. Whereas much of the difficulty with the phrase "adverse environmental impact" has been with the word "adverse," we believe the word "environmental" has too often been ignored in attempts at definition of AEI. We believe Congress intended to minimize *environmental* impact and not impact at some finer level of biological organization. We interpret population impact – as embodied in most of the definitions reviewed above – as signaling the potential for AEI.
3. The concepts of public trust resources and AEI should be separated. They have been confused in the ongoing dialogue, and this, perhaps more than anything else, has hampered consensus on a definition of AEI. Public trust resources refer to resources held in trust for the benefit of the citizens of a political entity, usually a state. Strictly interpreted, the unauthorized taking of one fish would represent a loss of public trust resources. This is a matter of societal-based policy that has no relation to AEI.

Over the last 30 years, the scientific community has attempted to define AEI on a scientific basis, i.e., based on impacts at the population level. This is consistent with the clear intent of Section 316(b) to minimize *environmental* impact. Federal 316(b) guidance should define AEI on a scientific basis. This will not preclude a state from exercising its law and policy to protect its public trust resources.

REFERENCES

1. Sharma, R.K., Buffington, J.D., and McFadden, J.T., Eds. (1976) Proceedings of the Conference on the Biological Significance of Environmental Impacts. Argonne National Laboratory, sponsored by U.S. Nuclear Regulatory Commission. NTIS Rept. No. NR-CONF-002.
2. Buffington, J.D. (1976) A synthetic definition of biological significance. In Proceedings of the Conference on the Biological Significance of Environmental Impacts. Sharma, R.K., Buffington, J.D., and McFadden, J.T., Eds., pp. 319-327. Argonne National Laboratory, sponsored by U.S. Nuclear Regulatory Commission. NTIS Rept. No. NR-CONF-002.
3. U.S. Environmental Protection Agency (1977) Guidance for Evaluating the Adverse Impact of Cooling Water Intake Structures on the Aquatic Environment: Section 316 (b) P.L. 92-500. Draft. U.S. Environmental Protection Agency, Washington, D.C.
4. Fritz, E.S., Rago, P.J., and Murarka, I.P. (1980) Strategy for Assessing Impacts of Power Plants on Fish and Shellfish Populations. U.S. Fish and Wildlife Service, Biological Services Program, National Power Plant Team. Rept. No. FWS/OBS-80/34.
5. Voigtlander, C.T. (1981) If you can't measure an impact, there probably isn't an impact. In Issues Associated with Impact Assessment. Jensen, L.D., Ed. Proceedings of the Fifth National Workshop on Entrainment and Impingement, San Francisco, May 1980. Sponsored by Ecological Analysts, Inc. and Electric Power Research Institute. pp. 3–11.
6. Westman, W.E. (1985) *Ecology, Impact Assessment, and Environmental Planning*. John Wiley & Sons, New York.

7. Christensen, S.W., Van Winkle, W., Barnthouse, L.W., and Vaughan, D.S. (1981) Science and the law: confluence and conflict on the Hudson River. *EIA Rev.* **2**(1), 63–88.
8. Barnthouse, L.W., Klauda, R.J., Vaughan, D.S., and Kendall, R.L., Eds. (1988) Science, Law, and Hudson River Power Plants, a Case Study in Environmental Impact Assessment. American Fisheries Society Monograph 4, Bethesda, MD.
9. Plater, Z.J.B., Abrams, R.H., and Goldfarb, W. (1992) *Environmental Law and Policy: A Coursebook on Nature, Law, and Society*. West Publishing Co., St. Paul, MN.
10. Jensen, L.D., Ed. (1978) Fourth National Workshop on Entrainment and Impingement, Chicago, Dec. 1977. Sponsored by Ecological Analysts, Inc.
11. Power, J.P. (1995) Reinvigorating Natural Resource Damage Actions Through the Public Trust Doctrine. http://www.nyu.edu/pages/elj/issueArchive/vol4/2/4nyuelj418t.html. © New York University Environ- mental Law Journal 1995.
12. Bray, P.M. (2001) An Introduction to the Public Trust Doctrine. http://www.responsiblewildlifemanagement.org
13. Utility Water Act Group (2000) Comments of the Utility Water Act Group on EPA's Proposed § 316(b) Rule for New Facilities and ICR No. 1973.01. Submitted to the U.S. Environmental Protection Agency and the Office of Management and Budget, November 9, 2000. Docket No. W-00-03.
14. Jensen, L.D., Ed. (1981) Issues associated with impact assessment. Proceedings of the Fifth National Workshop on Entrainment and Impingement, San Francisco, May 1980.

Uncertainty and Conservatism in Assessing Environmental Impact under §316(b): Lessons from the Hudson River Case

John R. Young[1,*] and William P. Dey[2]
[1]ASA Analysis & Communication, 310 Goldfinch Drive, State College, PA 16801; [2]ASA Analsyis & Communication, 51 Old State Road, Wappingers Falls, NY 12590

Received November 15, 2001; Revised March 5, 2002; Accepted March 6, 2002; Published February, 2003

Initially, regulation of cooling water intakes under §316(b) was extremely conservative due to the rapid increase predicted for generating capacity, and to the uncertainty associated with our knowledge of the effects of entrainment and impingement. The uncertainty arose from four main sources: estimation of direct plant effects; understanding of population regulatory processes; measurement of population parameters; and predictability of future conditions. Over the last quarter-century, the uncertainty from the first three sources has been substan-tially reduced, and analytical techniques exist to deal with the fourth. In addition, the dire predictions initially made for some water bodies have not been realized, demonstrating that populations can successfully withstand power plant impacts. This reduced uncertainty has resulted in less conservative regulation in some, but not all venues. New York appears to be taking a more conservative approach to cooling water intakes. The conservative approach is not based on regulations, but in a philosophy that power plant mortality is an illegitimate use of the aquatic resources. This philosophy may simplify permitting decisions, but it does not further the development of a science-based definition of adverse environmental impact.

KEY WORDS: uncertainty, conservatism, entrainment, impingement, 316(b), power plant impact, environmental impact

DOMAINS: environmental management and policy, environmental modeling, environmental monitoring, water science and technology

---

> **Unless steps are taken to find alternate means of dispersing or utilizing this heat, there is a distinct possibility that all major rivers in the United States will reach the boiling point by 1980 and then evaporate entirely by 2010! –** Richard Wagner in *Environment and Man*, 1971[1]

* Corresponding author. Emails: jyoung@asaac.com; wdey@asaac.com

> **By the year 2000 the water flow through the condensers of power plants will exceed two million cubic feet per second, approximately 1.2 times the average freshwater discharge of the 48 contiguous States.** – C.P. Goodyear and B.L. Fodor in Ecological Implications of Anticipated Electric Power Development, 1977[2]

> **The staff analysis indicates that during June and July of most years from 30 to 50% of the striped bass larvae which migrate past Indian Point from upstream spawning areas are likely to be killed by entrainment. .... As a result, there is a high probability that there will be an initial 30 to 50% reduction in the striped bass fishery which depends upon the Hudson for recruitment.** – Atomic Energy Commission, Final Environmental Statement Related to Operation of Indian Point Nuclear Generating Plant Unit No. 2, 1972[3]

Although two of these quotes refer to the discharge of waste heat from power plant cooling systems and the need for cooling water, rather than to direct entrainment and impingement impacts, they nevertheless epitomize the attitude, prevalent at the time §316 was enacted, that once-through cooling systems would create huge environmental problems. These attitudes were fostered not only by a relatively rudimentary knowledge of the actual impacts of once-through cooling, but also by the projections for growth of electrical demand and especially nuclear power as a means of satisfying that demand. Projections were made that by 2000, the nationwide generating capacity would need to be 1,575,000 MW, nearly three times the capacity available in 1976[4].

Given the predictions for increasing electrical demand, the resultant need for cooling water, and the lack of information available on the effects of one-through cooling, it is not surprising that the new United States Environmental Protection Agency (USEPA) would take a conservative regulatory view, i.e., to err on the side of being over-protective regarding the use and discharge of cooling water. However, even in their conservatism, the agency focused on preventing effects at the population and ecosystem level. The guidance manuals provided by the agency clearly were directed at assessing and preventing impacts at the levels of populations and communities[5].

The conservative view to regulation was considered necessary because assessment of the impacts of power plant operations were highly uncertain. The uncertainty arose from four distinct sources. First, the direct effects on aquatic organisms were difficult to measure, and estimates were fraught with numerous untested assumptions. For instance, without any demonstration to the contrary, it seemed prudent to assume that all organisms entrained into the cooling system would be killed[6]. In addition, the calculation tools used to estimate numbers killed or a fraction of the population killed by power plants contained many parameters that were not amenable to empirical description with the data available at the time. Therefore, it was necessary to assess the sensitivity of the results to a range of assumed values for these parameters.

A second component to uncertainty was the incomplete knowledge of the processes that affect the population dynamics of the resident aquatic species. In the 1970s, the large ecological studies of power plant impacts (e.g., Hudson River, Delaware Bay, Niantic River) were just getting started. Many of these studies were conducted on estuarine systems. Although often very productive, estuaries are also highly variable, which makes it difficult, if not impossible, to understand population regulatory processes with only a few years of study. Assessments of impact conducted in the late 1970s typically had less than ten years of data available, therefore the understanding of the factors that influence the population dynamics of affected species was preliminary at best.

Sampling variability adds to the uncertainty in measuring population characteristics and the effects of power plants on these characteristics. Catches of fish in sampling programs are highly variable, thus estimates of abundance often have large confidence bounds. Life histories of many of the affected species are complex, involving only temporary occurrence near the power plants and/or long annual migrations, making them extremely difficult to sample for some parts of the life cycle. Invariably, all fish in a cohort do not follow the same life history pattern. For anadromous species, some individuals emigrate from the estuary at an earlier age than others, and similar variation exists for time and age at return. The length and timing of ocean migrations are also variable, as are growth, maturity, and fecundity.

Finally, uncertainty of future conditions also adds to the imprecision of our ability to predict impacts on future populations. Even if we had perfect knowledge of the direct impacts, the processes that regulate the population, present population characteristics, changes in climatic conditions, current patterns, habitat alterations, and commercial or recreational fishing mortality rates may occur in the future, which would then make our predictions of the future populations uncertain.

The result of these four sources of uncertainty was that regulation under 316(b) was initially very conservative and closed-cycle cooling was frequently mandated as the best available technology. During the 1970s the frequency of use of the various designs of cooling systems for new plants changed radically. For plants that began operating prior to 1970 and plants less than 500 MW prior to 1973, once-through cooling accounted for 75% of installed capacity with closed-cycle cooling comprising only about 10%. For plants completed after 1978, 80% of the capacity was cooled by closed-cycle systems, while once-through cooling was used at less than 5%[7].

Despite the clear trend toward closed-cycle cooling, some plants were able to reach agreement with USEPA and other regulatory agencies and find alternative measures to minimize adverse environmental impact; however, this was not easily accomplished. For example, the 1975 draft NPDES permits for the new Hudson River plants (Indian Point, Bowline Point, and Roseton) all contained conditions that would eliminate once-through cooling and greatly reduce the entrainment and impingement of fish. Finally, after lengthy legal proceedings, a settlement was achieved that reduced potential fish mortality through flow restrictions, appropriately timed outages, intake modifications, and mitigative stocking[8].

The key to reaching agreement on cooling system requirements lies in reducing the uncertainty of the assessment from as many of the four components as possible. In the Hudson River case, one of the key factors was the convergence of the estimates of direct power plant effects that was achieved as the technical experts from both sides met and discussed the impact models[9,10]. Part of this convergence was due to the clear demonstration that mortality of entrained organisms can be considerably less than 100% for particular species and life stages[11,12].

Uncertainty of the underlying ecological processes can also be reduced through long-term monitoring studies that provide a wider range of the conditions that affect the population in various ways and validate the predictions of the earlier methodologies. In the Hudson River, continuation of the environmental studies for nearly 30 years has provided the opportunity to observe both high and low abundance periods for striped bass and other species in response to fishing mortality rates, a wide range of climatic variation, and different levels of power plant mortality[13]. In addition, other human influences on the estuary have also changed dramatically over this time period. Untreated or inadequately treated sewage discharges to the estuary have been largely eliminated, with a concomitant improvement in water quality[14]. Chemical control of the invasive water chestnut (*Trapa natans*) was discontinued, resulting in a tremendous resurgence of the species in the freshwater regions of the estuary. In the early 1990s, zebra mussels (*Driessena polymorpha*) appeared in the freshwater portions of the estuary and caused a substantial alteration of the lower levels of the estuarine food web[15]. Long-term studies afford the opportunity to observe these ecologically important events, which offer unique opportunities for insights to population regulatory mechanisms.

It is impossible for any monitoring program to study all aspects of the environment that may be important in understanding the population dynamics of species subject to entrainment and impingement. It is critical to proper 316(b) evaluation to be aware of and facilitate other research efforts that could provide additional crucial information. In the Hudson River, there has been a great deal of other research conducted through funding provided by the Hudson River Foundation, by the New York Department of Environmental Conservation (NYSDEC) for fishery management purposes, and through other avenues. Through the years the owners of the Hudson River stations have attempted to promote these other research efforts through co-funding of projects, co-operating with researchers in collecting specimens, and by making the utility data available for legitimate research needs. These efforts have succeeded in assisting crucial pieces of scientific research that have helped el cidate some of the possible population regulatory mechanisms[16,17,18,19]. However, it must be remembered that monitoring studies provide no guarantee that they will uncover the primary regulatory processes[20], and will never be able to prove that particular mechanisms are the prime regulatory factors. They can, however, increase the confidence that the true regulatory processes are identified and understood.

Measurement uncertainty can also be reduced substantially with carefully designed and executed sampling programs. These programs need to consider inherent sampling variability and use sufficient sample sizes to provide suitably precise estimates. Data from the Hudson studies were used to determine how sample size and precision are related[21], knowledge which can be used to design an effective sampling program.

The always imperfect knowledge of future conditions may also be addressed in various ways. In choosing fisheries' harvest policies, the uncertainty is often ignored without substantially affecting the performance of the fishery; however, when mortality is high enough to permanently alter the health of the stock, explicit adjustment of policies for the uncertainty is preferable[22]. Explicit inclusion of uncertainty can be done through risk analysis if probabilities can be assigned to various possible future states[23,24]. Other techniques, such as fuzzy math[25], sensitivity analyses[26], and meta-analysis[27], can be used when information on probabilities is not available.

In some areas, fisheries management is moving toward the "precautionary approach" to setting management controls[28,29], and this approach may also be useful for 316(b) regulation. The precautionary approach explicitly recognizes the uncertainty of biological information and the imperfect ability of management policies to assure that biological targets are met. In recognition of this uncertainty, targets are set in a conservative manner so that the probability that numerical biological reference points, such as the minimum acceptable spawning stock biomass, are exceeded is acceptably low. The level of conservatism of the management policies varies directly with the level of uncertainty.

As a result of all the research and monitoring conducted since 316(b) was enacted, our understanding of the effects of entrainment and impingement in 2001, while still imperfect, is far better and less uncertain than it was in 1972. However, given that some uncertainty is still present, some will argue that conservative regulation, erring on the side of over-protection of aquatic species, is still the best policy for 316(b). If over-protection came at no cost, without trade-offs among other socially and ecologically beneficial attributes, then it would be difficult to argue against this position. After all, the technology exists to practically eliminate fish entrainment and impingement by using closed-cycle cooling. Unfortunately there are trade-offs to be made, and it is prudent to examine these trade-offs before settling on a final position on uncertainty and conservatism.

One of the trade-offs to be made is that elimination of entrainment and impingement by converting once-through power plants to closed-cycle cooling would be extremely expensive. In 1992 the estimated capital cost of converting all once-through plants to closed-cycle was $23 billion to $24 billion[30]. The extra electrical energy required to operate cooling towers and the reduced output from less efficient operation was estimated to cost an additional $13 billion to $24 billion[31], bringing the total cost to $36 billion to $48 billion. The prudence of the expenditure of this magnitude to eliminate entrainment and impingement losses when population level effects are not detectable is questionable.

Environmental impacts of other sorts are also a trade-off when once-through cooling is replaced by closed cycle. These impacts include destruction of vegetation and terrestrial habitat, noise, visual impacts, additional fuel use, increased air emissions, and construction-period impacts for any type of cooling tower. In addition, aerosol and saline drift, plumes, fogging, icing, discharge of chemicals and biocides, and evaporative water loss may be issues for wet towers.

Given the greater degree of certainty of assessment of effects that can be achieved in 2001 than was possible in 1972, it would seem logical that the degree of conservatism of regulatory approach could be reduced. In 1977, Van Winkle described the state of knowledge of assessing population-level power plant impacts from the viewpoint of an optimist, a pessimist, and a realist[32]. At that time, four aspects of population assessments needed improvement: estimating abundance, production, and mortality rates; monitoring programs and data analysis; compensation and stock-recruitment relationships; and use of population models. All four of these aspects have been explored diligently in the last 24 years, and many significant advances have been made. Although Van Winkle's optimist, who viewed these aspects as completely resolvable, has not been proven totally correct, his realist, who envisioned that significant improvements were possible, was probably not far off.

Have the reductions in uncertainty achieved over the last quarter-century been translated into reductions in conservatism in regulatory philosophy? Two east-coast states provide an interesting contrast in regulatory viewpoint. The state of Maryland appears to have adopted the "realist" viewpoint that population assessments remain uncertain, but data collected to date have shown that healthy populations and once-through cooling systems are not mutually exclusive. Maryland's regulations specifically exempt intakes of less than 10 million gallons per day (mgd)[33], presumably because intakes of this size would not be able to significantly harm the resident populations. Maryland also has a set formula for determining when costs and benefits of alternative technologies exceed the "wholly disproportionate" test.

The Maryland approach is in sharp contrast to that of the state of New York, which decidedly takes the pessimistic view. In a recent decision on best available technology for the proposed 1080 MW combined-cycle Athens Generating Station, the NYSDEC commissioner ruled that dry cooling was the best available technology for the plant, over the hearing examiner's recommendation that a hybrid wet-dry cooling tower, with wedge-wire screened intakes, and a fabric filter curtain would be sufficient. The commissioner found that the 4.2 mgd average flow with the hybrid towers and wedge-wire screens would kill 24,500 young-of-year American shad (0.2% of the population) and 1.8 million river herring (0.3% of the population), and would be unacceptable. In his view the hearing record did not support the additional application of a fabric filter curtain. Dry cooling would withdraw only 0.18 mgd and kill an estimated 1,000 young-of-year American shad and 76,500 young-of-year river herring annually. In the eyes of the commissioner, the incremental cost of $39 million for the dry cooling system over an assumed 20-year life of the plant was not "wholly disproportionate" to the environmental benefits to be gained[34]. The decision did not state what the benefits to be gained were, other than impact to aquatic organisms would be minimized. According to the decision, the applicant has the burden of proof to demonstrate that costs and benefits are disproportionate.

One might expect, given the highly conservative nature of the Athens decision, that New York had much more stringent regulations for cooling water intakes, but, in fact, the New York regulations simply parrot the language of 316(b). The state has not issued any formal guidance or regulations that support such a conservative interpretation. Like the federal government, New York State has not formally defined "adverse environmental impact." However, in comments to USEPA, one New York regulator proposed that adverse environmental impact was "any harmful, unfavorable, detrimental or injurious effect on *individual* (emphasis added) organisms of fish, wildlife or shellfish or their eggs or larvae; or the water, land or air resources of the U.S.....; or on human health, welfare, or safety; or on the human enjoyment of those resources"[35].

The reason given for proposing this simplistic definition is to avoid "analysis paralysis" that may result from a more complex standard. The New York regulator cited the Hudson River case as a prime example of this paralysis. After millions of dollars have been spent on environmental research for more than 25 years, "the state agency, regulated parties, and citizen conservation groups *still disagree with the interpretation,* despite probably the best data set on the planet, full agreement on sampling design, data collection, certain analysis techniques, and many aspects of modeling." This "paralysis" is used as an argument that a population-based standard is unworkable, yet the reality is that the paralysis occurs because there is no standard against which the data and analyses can be evaluated. If either USEPA or New York had adopted a workable population-based standard for adverse environmental impact, then it would be clear from the "best data set on the planet" whether the standard had been met. Certainly, if the 25+ years of Hudson River data are not sufficient to assess whether adverse environmental impact has occurred, then it is unlikely that any data set will prove adequate for the task.

Does a standard such as that being used in New York arise from a need to be conservative in the face of uncertainty, or from other considerations? In objecting to USEPA's proposal for cost-benefit analysis, the New York regulator stated, "EPA has no right to allocate State public trust resources to be killed in this manner." Clearly, New York has decided there are legitimate and illegitimate sources of fish mortality, and power plants fall into the latter category. Recreational and commercial fishing both are industries that derive income from the taking of fish, either by intent (legal sizes of target species) or by accident through the by-catch. However, New York's position is that these industries differ from power generation in that they have a historical and societal right to take fish. By categorizing industry-based mortality into legitimate and illegitimate sources, New York has no need to develop a logical, science-based approach to definition of adverse environmental impact.

After a quarter century of case-by-case decisions on 316(b) requirements, we still have plants using both once-through and closed-cycle cooling. Although we can't determine what would have happened had the plants with closed-cycle cooling not installed that technology, we can see, from those that have once-through systems, that local fish populations have not been decimated by entrainment or impingement[36]. There are no documented instances of populations being driven to the brink of collapse by power plant cooling systems. For systems that have been studied for long time periods, there is empirical evidence that, even with non-trivial levels of direct effects (conditional mortality rates on the order of 10% or more), fish populations continue to remain healthy[36,37,38]. If we have learned nothing else from the millions of dollars spent on studies and monitoring, we should have learned that there is not a one-size-fits-all solution to the best available technology requirement. Can we afford to be overly conservative on the cooling water intake issue when other environmental threats that appear more serious will also require resources to resolve?

We have now made it to the twenty-first century, so the accuracy of the quotes at the beginning of this paper is easily assessed. So far there have been no reports of any major rivers reaching the boiling point or entirely evaporating away as a result of heated discharges. In contrast to the 1.5 million megawatt demand envisioned for the end of the century, in 1999 the actual generation capacity in the United States was only 785,990 megawatts, about 50% of the prediction. In a similar vein, the dire prediction for the Hudson River striped bass population subject to entrainment and impingement has also not come to pass. It would seem logical that regulatory agencies would recognize the advances made in population assessments, and the empirical demonstrations of still healthy fish populations and communities, and adjust the conservatism of regulatory policies accordingly.

REFERENCES

1. Wagner, R.H. (1971) *Environment and Man*. W.W. Norton & Company, New York. 491 pp.
2. Goodyear, C.P. and Fodor, B.L. (1977) Ecological implications of anticipated electric power development. United States Fish and Wildlife Service. FWS/OBS-76/20.3.

3. United States Atomic Energy Commission (1972) Final Environmental Statement Related to Operation of Indian Point Nuclear Generating Plant, Unit No. 2.
4. U.S. Nuclear Regulatory Commission (1976) Nuclear energy center site survey – 1975. NUREG-0001.
5. United States Environmental Protection Agency(1973) Development Document for Proposed Best Technology Available for Minimizing Adverse Environmental Impact of Cooling Water Intake Structures.
6. United States Environmental Protection Agency (1977) Guidance for evaluating the adverse impact of cooling water intake structures on the aquatic environment: Section 316(b) P.L. 92-500 Draft http://www.epa.gov/waterscience/316b/1977AEIguid.pdf.
7. Reynolds, J.Z. (1980) Power plant cooling systems: policy alternatives. *Science* **207,** 367–372.
8. Barnthouse, L.W., Boreman, J., Englert, T.L., Kirk, W.L., and Horn, E.G. (1988) Hudson River Settlement Agreement: technical rationale and cost considerations. *Am. Fisheries Soc. Monogr.* **4**, 267 – 273.
9. Englert, T.L. and Boreman, J. (1988) Historical review of entrainment impact estimates and the factors influencing them. *Am. Fisheries Soc. Monogr.* **4**, 143–151.
10. Christensen, S.W., Van Winkle, W., Barnthouse, L.W., and Vaughan, D.S. (1981) Science and the law: confluence and conflict on the Hudson River. *Environ. Impact Assess.* **2,** 63–88.
11. Muessig, P.H., Young, J.R., Vaughan, D.S., and Smith, B.A. (1988) Advances in field and analytical methods for estimating entrainment mortality factors. *Am. Fisheries Soc. Monogr.* **4,** 124–132.
12. Electric Power Research Institute (2000) Review of Entrainment Survival Studies: 1970–2000.
13. Central Hudson Gas & Electric Corp., Consolidated Edison Company of New York, Inc., New York Power Authority, and Southern Energy New York (1999) Draft Environmental Impact Statement for State Pollutant Discharge Elimination System Permits for Bowline Point, Indian Point 2 and 3, and Roseton Steam Electric Generating Stations.
14. Brosnan, T.M. and O'Shea, M.L. (1996) Long-term improvements in water quality due to sewage abatement in the lower Hudson River. *Estuaries* **19**, 890–900.
15. Strayer, D.L., Caraco, N.F, Cole, J.J., Findlay, S., and Pace, M.L. (1999)Transformation of freshwater ecosystems by bivalves. *Bioscience* **49**, 9–27.
16. Hurst, T.P., Schultz, E.T., and Conover, D.O. (2000) Seasonal energy dynamics of young-of-the-year Hudson River striped bass. *T. Am. Fish. Soc.* **129**, 145–157.
17. Schultz, E.T., Cowen, R.K., Lwiza, K.M.M., Gospodarek, A.M. (2000) Explaining advection: do larval bay anchovy (*Anchoa mitchilli*) show selective tidal-stream transport? *ICES J. Mar. Sci.* **57**, 360–371.
18. Pace, M.L., Findlay, E.G., and Lints, D. (1992) Zooplankton in advective environments: the Hudson River community and a comparative analysis. *Can. J. Fish. Aquat. Sci.* **49**, 1060–1069.
19. Limburg , K.E., Pace, M.L., and Arend, K.K. (1998) Growth, mortality, and recruitment of larval *Morone* spp. in relation to food availability and temperature in the Hudson River. *Fish. Bull.* **97**, 80–91.
20. Chitty, D. (1996) Do Lemmings Commit Suicide? Beautiful Hypotheses and Ugly Facts. Oxford University Press, New York. 268 pp.
21. Cyr, H., Downing, J.A., Lalonde, S., Baines, S., and Pace, M.L. (1992) Sampling larval fish populations: choice of sample number and size. *T. Am. Fish. Soc.* **121**, 356–368.
22. Frederick, S.W. and Peterman, R.M. (1995) Choosing fisheries harvest policies: when does uncertainty matter? *Can. J. Fish. Aquat. Sci.* **52**, 291–306.
23. Dey, W.P., Jinks, S.M., and Lauer, G.J. (2000) The 316(b) assessment process: evolution towards a risk-based approach. *Env. Sci. Pol.* **3**(Suppl. 1), S15–S24.
24. Dunning, D., Ross, Q., Ginzburg, L. and Munch, S. (2001) Effects of measurement error on risk estimates for recruitment to the Hudson River stock of striped bass. In Defining and Assessing

Adverse Environmental Impact Symposium 2001. *TheScientificWorldJOURNAL*, **2(S1)**, in press. http://www.thescientificworld.com.

25. Saila, S.B., Lorda, E., Miller, J.D., Sher, R.A., and Howell, W.H. (1997) Equivalent adult estimates for losses of fish eggs, larvae, and juveniles at Seabrook Station with use of fuzzy logic to represent parametric uncertainty. *N. Am. J. Fish. Man.* **17(4),** 811–825.
26. Saila, S.B. and Lorda, E. (1977) Sensitivity analysis applied to a matrix model of the Hudson River striped bass population. In *Assessing the Effects of Power-Plant-Induced Mortality on Fish Populations*. Van Winkle, W. Ed. . Pergamon Press, New York. pp. 311–332.
27. Myers, R.A., Barrowman, N.J., Hilborn, R., and Kehler, D.G. (2002) Inferring Bayesian priors with limited direct data: applications to risk analysis. *N. Am. J. Fish. Man.* **22**, 351–364.
28. Restrepo, V.R, Thompson, G.G., Mace, P.M., Gabriel, W.L., Low, L.L., MacCall, A.D., Methot, R.D., Powers, J.E., Taylor, B.L., Wade, P.R., and Witzig, J.F. (1998) Technical guidance on the use of precautionary approaches to implementing National Standard 1 of the Magnuson-Stevens Fishery Conservation and Management Act. NOAA Technical Memorandum. 46 p. http://www.nmfs.noaa.gov/sfa/NSGtkgd.pdf.
29. Serchuk, F.M., Rivard, D., Casey, J., and Mayo, R.K. (1999) A conceptual framework for the implementation of the precautionary approach to fisheries management within the Northwest Atlantic Fisheries Organization (NAFO). NOAA Technical Memorandum NMFS-F/SPO-40. http://www.st.nmfs.gov/st2/nsaw5/serchuk.pdf.
30. Veil, J.A. 1993. Impact on the steam electric power industry of deleting Section 316(a) of the Clean Water Act: capital costs. Argonne National Laboratory. ANL/EAIS-4.
31. Veil, J., VanKuiken, J.C., Folga, S., and Gillette, J.L. (1993) Impact on the steam electric power industry of deleting Section 316(a) of the Clean Water Act: energy and environmental impacts. Argonne National Laboratory ANL/EAIS-5.
32. Van Winkle, W. (1977) Conclusions and recommendations for assessing the population-level effects of power plant exploitation: the optimist, the pessimist, and the realist. In *Assessing the Effects of Power-Plant-Induced Mortality on Fish populations*. Van Winkle, W. Ed. . Pergamon Press. New York. pp. 365–372.
33. McLean, R., Richkus, W., Schreiner, S.P. and Fluke, D. (2001) Maryland power plant cooling water intake regulations and their application in evaluation of adverse environ-mental impact. In Defining and Assessing Adverse Environmental Impact Symposim 2001. *TheScientificWorldJOURNAL*, **2(S1)**, 1–11. http://www.thescientificworld.com.
34. New York State Department of Environmental Conservation (2000) Interim Decision In the Matter of an Application for a State Pollutant Discharge Elimination System (SPDES) Permit pursuant to Environmental Conservation Law (ECL) Article 17 and Title 6 of the Official Compilation of Codes, Rules and Regulations of the State of New York (6NYCRR) Parts 750 et seq. by Athens Generating Company, LP. http://www.dec.state.ny.us/website/ohms/decis/athensid.htm.
35. William Sarbello, Letter to USEPA dated 11/9/2000.
36. Richkus, W.A. and McLean, R. (2000) Historical overview of the efficacy of two decades of power plant fisheries impact assessment activities in Chesapeake Bay. *Env. Sci. Policy*. **3**, 5283–5293.
37. Central Hudson Gas & Electric Corp., Consolidated Edison Company of New York, Inc., New York Power Authority, and Southern Energy New York (1999) Draft Environmental Impact Statement for State Pollutant Discharge Elimination System Permits for Bowline Point, Indian Point 2 & 3, and Roseton Steam Electric Generating Stations.
38. Barnthouse, L. W., Heimbuch, D.G., Anthony, V.C., Hilborn, R.L., and Myers, R.A. (2001) Indicators of AEI applied to the Delaware Estuary. In Defining and Assessing Adverse Environmental Impact Symposium 2001. *TheScientificWorldJOURNAL*, **2(S1),** in press. URL: http://www.thescientificworld.com.

A Holistic Look at Minimizing Adverse Environmental Impact Under Section 316(b) of the Clean Water Act

John A. Veil[1,*], Markus G. Puder[1], Debra J. Littleton[2], and Nancy Johnson[2]
[1]*Argonne National Laboratory, 955 L'Enfant Plaza, SW, Suite 6000, Washington, D.C. 20024;* [2]*U.S. Department of Energy, Office of Fossil Energy, 1000 Independence Avenue, SW, Washington, D.C. 20585*

Received November 1, 2001; Revised February 14, 2002; Accepted February 20, 2002; Published February, 2003

Section 316(b) of the Clean Water Act (CWA) requires that "the location, design, construction, and capacity of cooling water intake structures reflect the best technology available for minimizing adverse environmental impact." As the U.S. Environmental Protection Agency (EPA) develops new regulations to implement Section 316(b), much of the debate has centered on adverse impingement and entrainment impacts of cooling-water intake structures. Depending on the specific location and intake layout, once-through cooling systems withdrawing many millions of gallons of water per day can, to a varying degree, harm fish and other aquatic organisms in the water bodies from which the cooling water is withdrawn. Therefore, opponents of once-through cooling systems have encouraged the EPA to require wet or dry cooling tower systems as the best technology available (BTA), without considering site-specific conditions.

However, within the context of the broader scope of the CWA mandate, this focus seems too narrow. Therefore, this article examines the phrase "minimizing adverse environmental impact" in a holistic light. Emphasis is placed on the analysis of the terms "environmental" and "minimizing." Congress chose "environmental" in lieu of other more narrowly focused terms like "impingement and entrainment," "water quality," or "aquatic life." In this light, BTA for cooling-water intake structures must minimize the entire suite of environmental impacts, as opposed to just those associated with impingement and entrainment. Wet and dry cooling tower systems work well to minimize entrainment and impingement, but they introduce other equally important impacts because they impose an energy penalty on the power output of the generating unit. The energy penalty results from a reduction in plant operating efficiency and an increase in internal power consumption. As a consequence of the energy penalty, power companies must generate additional electricity to achieve the same net output. This added production leads to additional environmental impacts

* Corresponding author. Emails: jveil@anl.gov; puder@anl.gov; debra.littleton@hq.doe.gov; nancy.johnson@hq.doe.gov.

associated with extraction and processing of the fuel, air emissions from burning the fuel, and additional evaporation of freshwater supplies during the cooling process. Wet towers also require the use of toxic biocides that are subsequently discharged or disposed. The other term under consideration, "minimizing," does not equal "eliminating." Technologies may be available to minimize but not totally eliminate adverse environmental impacts.

KEY WORDS: cooling water, intake structure, adverse environmental impact, 316(b), entrainment, impingement

DOMAINS: freshwater systems, marine systems, ecosystems and communities, water science and technology, environmental technology, environmental management and policy, ecosystems management

---

INTRODUCTION

The U.S. Environmental Protection Agency's (EPA's) rationale for proposing rigorous new-facility intake structure requirements was based on the agency's desire to minimize the number of aquatic organisms that is trapped on an intake structure during cooling-water withdrawal (impinged) or carried by the cooling-water flow through the entire cooling system (entrained). While impingement and entrainment are real environmental impacts, some stakeholders in the regulatory process have viewed these impacts as the only basis for decision making[1,2]. Some of the alternative technologies to once-through cooling (e.g., wet and dry cooling towers) are extremely effective at minimizing impingement and entrainment impacts, but their use introduces other types of adverse environmental impacts (AEIs). This article develops a broader, more holistic concept of AEIs: impingement, entrainment, as well as several others.

Some stakeholders have postulated that cooling towers are not part of cooling-water intake structures and should therefore not even be considered as regulatory options under Section 316(b). The following discussion deals with minimizing AEIs rather than a full interpretation of Section 316(b). Therefore, the discussion does not enter into the debate about whether requiring cooling towers is an appropriate regulatory option.

Much of the discussion contained in the following sections was gleaned from the years of active debate surrounding the Section 316(b) issue. The authors have previously raised some of the points presented here, while others have been taken from the extensive public record that has been presented to the EPA during several public meetings and open comment periods.

POWER PLANT COOLING SYSTEMS

In 1999, more than 60% of the utility power-generating capacity in the U.S. (382,270 MW) utilized the steam electric process[3]. At nuclear and fossil-fuel power plants, electricity is produced by heating purified water to create high-pressure steam. The steam is expanded in turbines, which drive the generators that produce electricity. After leaving the turbines, the steam passes through a condenser with multiple tubes and a large surface area. A high volume of cool water circulates through the tubes, absorbing heat from the steam. As the steam cools and condenses, the temperature of the cooling water rises.

Most power plants use either once-through cooling or closed-cycle cooling. Once-through cooling systems withdraw large volumes of water, typically in the range of tens of millions to billions of gallons per day, from a river, lake, estuary, or ocean. The water is pumped through the condenser and finally returned to the same or a nearby water body. Closed-cycle cooling systems recirculate cooling water to a cooling tower and basin, cooling pond, or cooling lake before returning it to the condenser. Because evaporation and planned cooling-tower blowdown remove cooling water from the evaporative system, regular additions of "makeup" cooling water are needed. At many plants, the makeup water is withdrawn from surface water bodies. Makeup volumes are much lower than daily once-through volumes and may range from hundreds of thousands to millions of gallons per day.

The most commonly used type of closed-cycle cooling systems employs wet cooling towers, where water rejects heat to the atmosphere through evaporation and sensible heat transfer to the ambient air flowing through the tower. The air flow through the tower is maintained by fans (mechanical draft) or by convective currents created by the shape of the tower (natural draft).

Some stakeholders have advocated the dry cooling tower method because it requires even less makeup water than a wet tower. Dry towers remove heat to the atmosphere only by sensible heat transfer. They do not rely on evaporation and, therefore, require little makeup water. Few dry towers have been installed in power-plant-sized applications (typically such units have a capacity of several hundred megawatts) to date because of cost and practical thermodynamic heat transfer limitations.

LEGAL AND REGULATORY CONSIDERATIONS

Section 316(b) of the Clean Water Act

Steam electric power plants and other industries that withdraw cooling water from surface water bodies (e.g., pulp and paper, iron and steel, chemical, manufacturing, petroleum refineries, and offshore oil and gas production) must comply with the terms of Section 316(b) of the Clean Water Act (CWA) enacted by Congress in 1972:

> "Any standard established pursuant to Section 301 or Section 306 of this Act and applicable to a point source shall require that the location, design, construction, and capacity of cooling water intake structures reflect the best technology available for minimizing adverse environmental impact."

This single statutory sentence has spawned numerous biological studies and technological and operational plant modifications. The cost of these actions has been very high.

Implementation of Section 316(b)

In 1976, the EPA promulgated final Section 316(b) regulations (April 26, 1976; 41 FR 17387). However, these regulations were successfully challenged by a group of 58 utilities [*Appalachian Power Co. v. Train*, 10 ERC 1965 (4th Cir. 1977)]. In 1979, the EPA formally withdrew its Section 316(b) regulations (June 1979; 44 FR 32956). As a consequence of the vacuum created by the absence of more-detailed federal regulations, states implemented Section 316(b) in different ways. The broad statutory language facilitated widely differing interpretations by the states. Some adopted comprehensive programs, others imposed less rigorous requirements, and some never developed any formal regulations.

In the mid-1990s, a coalition of environmental groups, headed by the Hudson Riverkeeper, filed suit against the EPA over failure to repromulgate Section 316(b) regulations [*Cronin et al. v. Reilly*, 93 Civ. 0314 (AGS)]. On October 10, 1995, the U.S. District Court, Southern District of New York, entered a Consent Decree between the parties directing the EPA to regulate cooling-water intake structures. Under the Consent Decree, the EPA agreed to propose regulations by June 1999 and promulgate a final rule by 2001. The Consent Decree was modified on November 21, 2000, to: (a) take final action on new facility regulations by November 9, 2001; (b) propose existing source utility and nonutility power producer regulations by February 28, 2002, and take final action on those regulations by August 28, 2003; and (c) propose regulations for other existing facilities not covered in (b) above by June 15, 2003, and take final action on those regulations by December 15, 2004.

Legislative History

The language of Section 316(b) is quite short and easily subject to wide interpretation. Legislative history may prove to be a powerful tool for ascertaining the congressional intent behind Section 316(b). Legislative history consists of all legislative events that transpired throughout the process of enacting or defeating proposed legislation. The documentation generally includes the bill, the drafts, and the preceding markup documents; the testimony at hearings; the reports or studies commissioned by the legislature; the chronology of voting; the floor debates; and the message of the executive accompanying the signature or veto of the bill.

Little legislative history exists with respect to Section 316(b). A recent law journal article discusses this history[4]. The language appeared suddenly, without any further explanation, in the Federal Water Pollution Control Act (the law that was modified in 1972 to become the CWA) Conference Report under the heading "Regulation of Thermal Discharges" in "Title III – Standards and Enforcement." The Conference Report provided no additional commentary but merely included the language that was to become law. Neither the Senate nor House bill had included intake-structure language like the Section 316(b) language that the conference committee added.

Except for one contribution, subsequent floor debates failed to shed more light on the dynamics that led to the conference substitute. Representative Clausen, during the House consideration of the Conference Report, summarized Section 316(b) in his own words. In departure from the actual conference language, he used the word "any" in connection with "adverse environmental impact." He added that "the reference [in Section 316(b)] ... to 'best technology available' is intended to be interpreted to mean the best technology available at an economically practicable cost[5]."

Section 316(b) has never been amended. Although reform bills were periodically introduced over the years, no legislation has been passed. Thus, the minimalist language of Section 316(b) remains as initially enacted.

Interpretation of Section 316(b) Language

The following sections discuss and analyze two statutory terms that are central to Section 316(b): "adverse environmental impact" and "minimizing."

Adverse Environmental Impact

"Adverse" denotes undesirable attributes. In the context of Section 316(b), adverse has often been interpreted as relating to the extent of biological harm through impingement or entrainment. However, some agencies and other stakeholders have also considered increased quantities of an air or water pollutant or the amount of water lost through evaporation to represent adverse circumstances. Adverse is not an absolute but a subjective term that should be assessed in relation to the entire system or universe under review. For example, does the addition of 5 kg/day of nitrogen to a water body cause an effect that could be considered an adverse impact? If the water body were a small pond or slow-flowing stream, the nitrogen addition would likely cause an undesirable adverse effect. If the nitrogen entered a large river or an estuary with a total nutrient budget of thousands of kilograms of nitrogen, the consequences (if any) would be small. No adverse impacts would be expected in this latter case. Another example involves entrainment loss of fish larvae. In a small water body that hosts limited populations of important species, the entrainment of 1,000 larvae might have a critical adverse impact. However, in a large water body with extensive populations of species, the loss of 1,000 larvae would be insignificant to the species population, the food chain, and the overall ecosystem. In this light, the entrainment would not pose an adverse impact.

The other relevant word in Section 316(b) is "environmental." In a regulatory context, the term encompasses a wide range of attributes of the natural world, including air, water, land, noise, and their relationship to one another. The statutory definitions provided in major environmental laws support this broad reading. Section 101 of the Comprehensive Environmental Response, Compensation, and Liability Act defines the term "environment" as "surface water, ground water, drinking water supply, land surface or subsurface strata, or ambient air." Under Section 329 of the Emergency Planning and Community Right-to-Know Act and Section 3(5) of the Toxic Substances Control Act, "The term 'environment' includes water, air, and land and the interrelationship which exists among and between water, air, and land and all living things."

In the Section 316(b) debates of recent years, however, the EPA has confined "environment" to aquatic organisms and focused just on impingement and entrainment. These are very real impacts that are attributable to water intakes. However, the CWA mandates are not well-served by this narrow focus. In fact, a whole suite of environmental impacts is associated with any type of cooling-water intake or cooling-water system. For each intake or system, the relative importance of each type of impact will vary. A comprehensive impact determination should consider the composite and cumulative impacts, along with any benefit offered by a particular cooling-water intake system. This article discusses important environmental impacts and describes their relative weight for once-through cooling systems, wet cooling tower systems, and dry cooling tower systems.

Minimizing

"Minimize" and the related terms "minimum" and "minimal" appear throughout the CWA and its implementing regulations but are never defined. In the absence of a formal statutory or regulatory definition, two approaches can be used for ascertaining the meaning of the term. Consulting a dictionary provides a sense of the common usage of the word. Courts often use this approach to ascertain the plain meaning of terms. "Minimize" is defined in one dictionary as "to reduce to the smallest possible amount, extent, size, or degree[6]." Another dictionary defines the term as "to reduce or keep to a minimum" and defines "minimum" as "the least quantity assignable, admissible, or possible[7]." Neither definition implies complete elimination.

A second approach involves searching for the context and usage of the term in existing laws and regulations. The phrase "at a minimum" is commonly used in the CWA regulations and implies that the expected value for the parameter in question is a defined value greater than zero. Another example can be found on page 28869 of the EPA's recent Notice of Data Availability for the Section 316(b) regulations (May 25, 2001; 66 FR 28853). The relevant text states that "EPA would not define these technologies [closed-cycle cooling and extremely low approach velocities] as BTA for 'minimizing' adverse environmental impact but instead determine that they avoid adverse environmental impact altogether." In this example, the EPA emphasizes the difference between a technology that minimizes impacts and one that avoids or eliminates impacts entirely.

The concept of minimizing entails making a value judgment relative to the smallest amount of effect that is possible or acceptable. At each installation that falls under the purview of Section 316(b), it is necessary for accuracy and optimal decision making to make a site-specific evaluation of the relevant factors that may contribute to AEI.

During the past few years, the EPA has evaluated various types of technologies that can effectively reduce impingement and entrainment impacts (e.g., wedge wire screens, fine-mesh traveling screens with a fish return system, Gunderbooms). The degree of impact reduction achieved by these technologies varies at different locations, but in some instances they may perform well enough to be construed as effectively minimizing AEIs.

During the EPA's Section 316(b) public meetings, some stakeholders have offered that the threshold for minimization should be "one dead fish equals adverse environmental impact." The evidence presented in the previous paragraphs demonstrates that the loss of one fish does not equate to AEI.

TYPES OF ADVERSE ENVIRONMENTAL IMPACTS

Impingement and Entrainment

As a general rule, as more water is withdrawn from a surface water body, more aquatic organisms are likely to be impinged on the intake structure and entrained through the cooling-water system. Following this assumption, in most cases a cooling system that reduces the volume of water withdrawn will concomitantly reduce the number of organisms injured or killed by impingement and entrainment. Because this relationship is not linear (e.g., 10 times less flow does not necessarily result in 10 times less impingement and entrainment), the actual effects must be evaluated on a site-specific basis. Moreover, not all impinged and entrained organisms are killed or otherwise removed from the ecosystem. Nevertheless, the adverse impact associated with impingement and entrainment will usually be reduced when less cooling water is withdrawn. The highest impact in this category is associated with once-through cooling systems. Power plant once-through cooling systems typically withdraw in the range of tens of millions to billions of gallons per day. The volume of makeup water required for wet cooling towers is many times less than that used in once-through systems. The percentage reduction varies but generally falls in the range of 1% to more than 10% of the once-through flow volume. Dry towers use even less cooling water than wet cooling towers.

Cooling systems that rely on moving large volumes of air by fans – mechanical-draft wet towers and dry towers – may create their own form of impingement and entrainment. Insects and birds can be drawn into the intake plumes of large fans. Larger organisms can be trapped on the exterior of the fans or their intake coverings like insects caught on automobile radiators. Smaller organisms can be pulled through the moving fan and injured or killed. The authors are not aware of any published literature quantifying this impact, but the parallels to aquatic impingement and entrainment are obvious. Potentially, large batteries of fans may inflict harm to local populations of endangered insects or birds or important pollinator species. To place this in perspective, a dry tower installed to cool a power plant-sized unit might have banks of fans that cover several acres. For example, a dry tower system at a 40-MW geothermal power plant near Reno, Nevada, employs 240 fans covering a large surface area. Photographs of that facility are available at http://home.nvbell.net/sbgeo/steamboat.html.

Impingement and entrainment, when they result in death or harm to an organism, create an adverse impact to that organism. However, they do not necessarily create an adverse impact on the population or ecosystem at large. The principle of compensation – enhanced reproductive output by populations that have experienced loss of young members of the population – is well-documented in the literature. Compensation may serve to dampen the population-wide effects of impingement and entrainment. A thorough review of compensation in fish populations was submitted to the EPA by the Utility Water Act Group as part of its comments on the Section 316(b) new-facility proposed rule[8].

Energy Penalty

Steam condensers are designed to produce a vacuum at the outlet end of the turbine, thereby increasing the efficiency of the system. The temperature of the cooling water exiting the condenser affects the performance of the turbine: the cooler the temperature, the better the performance. As cooling-water temperatures decrease, a higher level of vacuum can be produced and additional energy can be extracted. On an annual average, once-through cooling water has a lower temperature than recirculated water from a cooling tower. As a result of switching a once-through cooling system to a cooling tower, less energy can be generated by the power plant from the same amount of fuel. In addition, cooling towers use more electricity for pumps and fans than once-through systems. The net output of a plant that has converted its cooling system from a once-through system to a cooling tower will be reduced through both of these mechanisms. This reduction in energy output is known as the "energy penalty."

A 1992 report funded by the U.S. Department of Energy (DOE) found that the majority of literature values for the energy penalty associated with retrofitting fossil-fueled plants using once-through cooling with wet cooling towers were clustered in a band between 1.5% and 2.5%[9]. This means that a plant now equipped with a wet tower will produce 1.5 to 2.5% less electricity on an annual average than previously with a once-through system, while burning the same amount of fuel. Results for nuclear power plants showed greater variability, ranging between 1 and 5.8%. The data points were not as clearly clustered in a narrow range when compared with the data for the fossil-fuel plants. The authors of that report selected a range of 2 to 3% for the decrease in net electrical power that could be experienced if existing nuclear power plants retrofit from a once-through to a wet cooling system.

A report being prepared by the DOE, the National Energy Technology Laboratory (NETL), and the Argonne National Laboratory, scheduled to be completed in 2002, calculates the energy penalties that result from converting plants with once-through cooling to wet towers and to dry towers[10]. In a working draft of that analysis, the energy penalties are estimated for the hottest time of the summer months when electricity demand would also be at its peak by modeling hypothetical 400-MW coal-fired plants in five regions of the country with an ASPEN simulator model. The preliminary results indicate that conversion of a plant to a wet tower could cause energy penalties ranging from 2.4 to 4.0%. Conversion to an indirect dry tower, where possible, could cause energy penalties ranging from 8.9 to 12.1% under conservative design assumptions and 12.7% to almost 16% under a more realistic set of design assumptions. Annual average energy penalties will be smaller than those calculated for the hottest period of the year, but must still be considered.

The implications of the energy penalty are quite important from an environmental standpoint. To make up lost energy, additional fuel would have to be burned at the affected plants, at other plants within the power grid, or at completely new units that would have to be constructed. In addition, extraction and combustion of the additional fuel would cause several new AEIs.

Fuel

In general, the most probable type of generating unit used to make up the lost energy would be a fossil-fueled unit. The process of extracting fossil fuels creates impacts with respect to water and air, solid-waste generation, and land use. Moving the incremental fuel to the power plants consumes additional fuel for pumps or engines in transportation equipment.

Air Emissions

As fossil fuels are combusted, they give off a range of air pollutants of concern, including sulfur dioxide, nitrogen oxides, mercury, particulate matter, and carbon dioxide. It has been estimated that a plant converting from a once-through cooling system to a wet cooling tower increases its carbon dioxide emissions by 22 tons/year/MW for oil and gas and by 58 tons/year/MW for coal[11]. The DOE/NETL/Argonne report will also include estimates of annual energy penalties and air emissions. In addition to affecting air quality, some of the airborne pollutants may be deposited into surface water bodies, where they can directly affect aquatic organisms.

Water Quality

Discharges of once-through cooling water and cooling tower blowdown to surface waters are subject to the conditions of National Pollutant Discharges Elimination System permits. Once-through cooling water may contain chlorine used as a biocide, but otherwise is unlikely to contain toxic chemicals added by the generating or the cooling processes. A wide range of toxic chemicals, including biocides (chlorine and other, more toxic chemicals), corrosion inhibitors, and scaling inhibitors, may be added to cooling towers[12]. Portions of the recirculating water in wet cooling tower systems are blown down periodically. These blowdown effluents contain residual levels of the toxic chemicals previously added to the towers. Dry towers will also have blowdown effluents, but the volume will be lower than for wet towers.

Heated discharges of once-through cooling water can have positive or negative effects. In temperate climates, warm discharges during the winter can enhance and sustain fish populations. At many locations, however, excessive heat can be damaging. Removal or addition of heat can yield different impacts on aquatic communities, depending on site-specific circumstances.

Water Quantity and Consumption

The first type of water-quantity impact is related to availability of sufficient water for the cooling needs without causing adverse impacts on the water body and its other uses. Historically, power plants were sited near large bodies of water that could supply large volumes of once-through cooling water. As the use of wet cooling towers increased, new plants were also sited in locations where smaller volumes of water were available. Plants with dry towers need very little water and, from a water-supply perspective, can be sited in a wide range of locations. Plants cannot be sited on water bodies where cooling-water withdrawal could reduce the water level in the water body to a point where significant habitat was lost or water-quality conditions became undesirable. Plants using once-through cooling systems withdraw large volumes of water, although they return nearly all of that water to the water body at the same or a nearby location. Plants using cooling towers must also consider water availability, particularly if they are situated on a water body that is too small to support a withdrawal of several million gallons per day of makeup water that is not returned to the water body.

The second type of water-quantity impact relates to evaporation. Many freshwater bodies face heavy demands on their water. Any cooling system use that removes freshwater from those systems will need to compete with other existing and future uses. Plants with once-through cooling systems do not directly evaporate water. Virtually all water used for cooling is returned to the surface water body. However, the returned water is warm and will raise the temperature of the receiving water body to an extent that may increase the rate of natural evaporation from the surface of the water body. The Nuclear Regulatory Commission estimates that water lost by evaporation from a once-through-cooled discharge is about 60% of the evaporation from cooling towers[13].

In contrast, wet cooling tower systems intentionally evaporate water as an intrinsic part of the cooling process. Anecdotal accounts suggest that wet towers may have twice the evaporation rate of once-through cooling systems. One study estimates that when plants are converted from once-through cooling to cooling towers, an additional 15 gal/min of freshwater would be evaporated for each converted megawatt of generating capacity. This estimate assumes a conservative energy penalty of just 1%[11]. For a higher percentage energy penalty, the impacts would be proportionally higher.

Dry cooling towers do not rely on evaporation for cooling. Therefore, the evaporative losses associated with dry towers are very low.

Solid Waste

Debris, trash, and other waterborne solids are captured by the bar racks at cooling-water intakes. The volume of solids corresponds to the volume of water that is passed through the intake structure. Therefore, once-through systems are likely to generate a larger volume of solid waste than closed-cycle systems. Wet towers accumulate sludge in the bottom of the towers that must be removed periodically. Dry towers collect some amount of airborne debris, including leaves and paper, on their intake screens. The relatively small solid-waste volumes attributable to all three types of systems are unlikely to cause AEIs.

Noise

Noise emissions pose AEIs to workers, nearby residents, and wildlife. Like many other major industrial facilities, power plants can be noisy. Once-through cooling systems do not add appreciably to the overall background noise at power plants. However, cooling towers can create considerable noise levels. Natural-draft towers are noisy because of water movement through the tower. Mechanical-draft wet towers use banks of fans that contribute to plant noise levels, particularly in the vicinity of the towers themselves. Because of the large surface area needed to house the towers (up to several acres) and the large number of fans, dry towers are likely to be even noisier than the other types of cooling systems. Many jurisdictions have regulations controlling noise levels. The issue of noise could restrict the ability of some facilities to add banks of fans for cooling.

Land Use/Habitat

It is difficult to assess what type of cooling system is likely to have the greatest adverse impact with respect to land use. Once-through systems need larger capacity intake structures that typically are built in the water body or on the shoreline. Some surface-water intakes require pipes that extend meters to kilometers offshore. Shore-side facilities for once-through systems are smaller than those for closed-cycle systems, however. The banks of towers, pumps, and piping used by closed-cycle systems may occupy significant land space.

Plants that convert from once-through systems to closed-cycle systems will consume more fuel. The process of extracting the fuel can disrupt terrestrial or aquatic habitats. As previously described, the energy penalty raises indirect land use/habitat concerns.

Plumes/Air Circulation

Once-through systems do not generate any significant plumes of water vapor and do not influence local air circulation. Wet towers release plumes that can contribute to deposition of undesirable particulates or salt onto surrounding land areas. Plants located in urban areas, near major highways, or in the flight paths of airports could create adverse impacts through fogging or icing conditions associated with plumes. These transportation-related impacts, though real, may not fall under the purview of "environmental impacts," depending on how broadly the term "environmental" is defined by a decision-making body. Such affects may be more correctly interpreted as safety impacts.

As a result of the large amount of air that must be moved through dry cooling systems, large banks of fans may affect local air circulation patterns. The authors are not aware of any research in this regard, but the issue could potentially affect windborne seed distribution and establish microclimates around the power plants. It is unknown whether these impacts would be adverse, positive, or neutral.

DISCUSSION

The language of Section 316(b) is brief, leaving appreciable room for alternative interpretations. The authors believe that a holistic approach to Section 316(b) leads to the most rational interpretation. The preceding paragraphs suggest that the language of Section 316(b) directs regulatory decision makers to consider a wide range of AEIs. Moreover, the decision makers should evaluate how to minimize those impacts in the context of the physical and environmental setting of the power plant and the nearby water bodies. The evaluation should consider the cumulative impacts posed by all facilities, pollutant inputs, and natural processes operating in the water body.

The types of adverse impacts and the authors' qualitative assessment of their relative magnitude (high, moderate, low, or none) for each type of cooling system are summarized in Table 1. These rankings are not absolute and will vary somewhat depending on site-specific factors. However, Table 1 provides a useful consolidated presentation of the multiple types of impacts and reflects the discussion in the preceding sections. The authors believe that decision makers should undertake a comprehensive evaluation of all the factors listed in Table 1. Because this evaluation is qualitative, the different types of impacts summarized in Table 1 should not necessarily be given the same weight. For example, impingement and entrainment impacts are probably of greater significance than noise, plume, or air circulation impacts.

EPA's Approach in the New-Facility Rules

The EPA proposed Section 316(b) rules for new sources on August 10, 2000 (65 FR 49060). Under the proposed rule, most new facilities would have had to install closed-cycle cooling systems (nearly all of these would use cooling towers). During public meetings and in written comments, the environmental community and several states have supported the use of dry cooling as the appropriate technology for addressing AEIs, while industry and several other federal agencies supported an approach involving site-specific evaluation.

The EPA published the final new-facility rules on December 18, 2001 (66 FR 65256). The new rules provide two alternative tracks. Under Track I, a fast track that does not consider actual site-specific environmental impacts, companies must meet a stringent set of rules, including installation of cooling towers or other equivalent flow-reduction measures, coupled with restrictions on the total flow volume, the ratio of intake flow volume to surface water body volume, and the intake velocity. In addition, plants are required to use other intake technologies. Since the process of obtaining permits becomes more predictable and timely, Track I may be attractive to many new facilities. Under Track II, companies must conduct intensive site-specific studies and projections at the facility in order to demonstrate to the permitting authority that an alternative set of intake controls

TABLE 1
Summary of Types of AEIs Associated with Cooling Systems

	Anticipated Magnitude of Impact*		
Type of AEI	**Once-Through Cooling**	**Wet Cooling Tower**	**Dry Cooling Tower**
Impingement	H	L	L
Entrainment	H	L	L
Energy penalty	N	M to H	H
Additional air emissions	N	M to H	H
Additional fuel usage	N	M to H	H
Water quality	Heat: + or – Biocides: L to M	Biocides: M to H	Biocides: L
Water consumption	L	M	M to H
Solid wastes	L	L	L
Noise	L	M	M to H
Land use/habitat	Aquatic: M to H Terrestrial: L	Aquatic: L Terrestrial: L to M	Aquatic: L Terrestrial: M to H
Plumes	N	M to H	N
Air circulation	N	L	Unknown, but potentially M to H

Note: Not all types of environmental impacts should be assigned the same weight.
*H = high, M = moderate, L = low, N = none.

will reduce impingement and entrainment to a level comparable to Track I controls. Approvals based on Track II would generally have a lower cost to applicants, but could make the permitting process more uncertain and lengthy.

The concept that AEI means more than just impingement and entrainment is captured in the EPA's requirements for Track II demonstrations at 40 CFR 125.84(d)(1)(i) and (ii):

> (i)This showing may include consideration of impacts other than impingement mortality and entrainment, including measures that will result in increases in fish and shellfish, but it must demonstrate comparable performance for species that the Director, in consultation with national, state or tribal fishery management agencies with responsibility for fisheries potentially affected by your cooling water intake structure, identifies as species of concern.
>
> (ii) In cases where air emissions and/or energy impacts that would result from meeting the requirements of paragraphs (b)(1) and (2) of this section would result in significant adverse impacts on local air quality, significant adverse impact on local water resources not addressed under paragraph (d)(1)(i) of this section, or significant adverse impact on local energy markets, you may request alternative requirements under Sec. 125.85.

EPA's Upcoming Rules for Existing Facilities

The EPA is in the process of developing cooling-water intake regulations for existing utility and nonutility power-generating facilities but has not yet revealed any details of its proposed rule. There is a potential that at least some of the requirements in the proposal for new sources may be carried over to the rule for existing facilities. About 44% of the U.S. steam electric power plants employ once-through cooling systems[14]. If the final existing facility rule requires many or all of these plants to install dry or wet cooling tower systems, serious impacts with respect to electricity costs and availability could arise. Moreover, such a decision could trigger other significant environmental impacts beyond impingement and entrainment, as described above.

CONCLUSIONS

Congress added Section 316(b) to the CWA to ensure that cooling-water intakes did not cause unnecessary AEIs on water bodies. Impingement and entrainment can adversely impact aquatic ecosystems. However, decisions with respect to cooling-water intakes and systems made solely on the basis of minimizing or eliminating impingement and entrainment do not meet the CWA's comprehensive mandate and do not necessarily provide the best environmental protection. A holistic approach to Section 316(b) allows for a more thorough and comprehensive evaluation of the suite of potential AEIs associated with a cooling-water intake structure. It also allows for a comprehensive evaluation of whether a plant's environmental setting is being adversely impacted by a cooling-water intake structure.

ACKNOWLEDGMENTS

Mr. Veil's and Dr. Puder's work was sponsored by the U.S. Department of Energy, Office of Fossil Energy, under contract W-31-109-Eng-38.

REFERENCES

1. Rabago, K.R. (1992) What comes out must go in: cooling water intakes and the Clean Water Acts. *Harv. Environ. Law Rev.* **16**, 429.
2. May, J.R. and van Rossum, M.K. (1995) The quick and the dead: fish entrainment, entrapment and the implementation and application of Section 316(b) of the Clean Water Act. *Vt. Law Rev.* **20**, 373.
3. Inventory of Electric Utility Power Plants in the United States 1999 (2000) DOE/EIA-0095(99)/1, Energy Information Administration, U.S. Department of Energy.
4. Anderson, W.A. and Gotting, E.P. (2001) Taken in over intake structures? Section 316(b) of the Clean Water Act.*Columbia J. Environ. Law* **26**, 1.
5. 118 Cong. Rec. 33,762 (1972), reprinted in A Legislative History of Water Pollution Control Act Amendments of 1972, at 264 (Jan. 1973).
6. *The American Heritage Dictionary of the English Language* (2000) Houghton Mifflin Company, 4th ed..
7. *Collegiate Dictionary*, Merriam-Webster Online, http://www.m-w.com/ .
8. Myers, R.A. (2000) Compensation in Fish: A Review. Submitted to Environmental Protection Agency by the Utility Water Act Group as part of the comments on the 316(b) new-facility proposal.
9. Veil, J.A., VanKuiken, J.C., Folga, S., and Gillette, J.L. (1993) Impact on the Steam Electric Power Industry of Deleting Section 316(a) of the Clean Water Act: Energy and Environmental Impacts. Report ANL/EAIS-5. Argonne National Laboratory.
10. U.S. Department of Energy, Office of Fossil Energy, National Energy Technology Laboratory, and Argonne National Laboratory (2002) Unpublished analyses.
11. Carter, D. (1991) Unpublished memorandum from Carter, U.S. Department of Energy, to James Gardner, Edison Electric Institute, Sept. 27.
12. Veil, J.A., Rice, J.K., and Raivel, M.E.S. (1997) Biocide Usage in Cooling Towers in the Electric Power and Petroleum Refining Industries. Report prepared for Office of Fossil Energy, Department of Energy; also published by National Petroleum Technology Office, Department of Energy as DOE/BC/W-31-109-ENG-38-3, DE98000455 (Nov. 1997).
13. Nuclear Regulatory Commission (1996) Generic Environmental Impact Statement for License Renewal of Nuclear Plants, NUREG-1437 (May 1996).
14. Environmental Directory of U.S. Powerplants (1996) Edison Electric Institute, Washington, D.C.

Modeling Possible Cooling-Water Intake System Impacts on Ohio River Fish Populations

Elgin Perry[1], Greg Seegert[2], Joe Vondruska[2], Timothy Lohner[3,*], and Randy Lewis[4]
[1]Consulting statistician, 2000 Kings Landing Rd., Huntington, MD 20639; Tel: (410) 535-2949; [2]EA Engineering, Science and Technology, Deerfield, IL 60015; Tel: (847) 945-8010; [3]American Electric Power, Columbus, OH 43215; Tel: (614) 223-1255; [4]Cinergy, Plainfield, IN 46168-1782; Tel: (317) 838-1723

Received November 2, 2001; Revised January 21, 2002; Accepted February 13, 2002; Published February, 2003

To assess the possible impacts caused by cooling-water intake system entrainment and impingement losses, populations of six target fish species near power plants on the Ohio River were modeled. A Leslie matrix model was constructed to allow an evaluation of bluegill, freshwater drum, emerald shiner, gizzard shad, sauger, and white bass populations within five river pools. Site-specific information on fish abundance and length-frequency distribution was obtained from long-term Ohio River Ecological Research Program and Ohio River Sanitation Commission (ORSANCO) electrofishing monitoring programs. Entrainment and impingement data were obtained from 316(b) demonstrations previously completed at eight Ohio River power plants. The model was first run under a scenario representative of current conditions, which included fish losses due to entrainment and impingement. The model was then rerun with these losses added back into the populations, representative of what would happen if all entrainment and impingement losses were eliminated. The model was run to represent a 50-year time period, which is a typical life span for an Ohio River coal-fired power plant. Percent changes between populations modeled with and without entrainment and impingement losses in each pool were compared to the mean inter-annual coefficient of variation (CV), a measure of normal fish population variability. In 6 of the 22 scenarios of fish species and river pools that were evaluated (6 species × 5 river pools, minus 8 species/river pool combinations that could not be evaluated due to insufficient fish data), the projected fish population change was greater than the expected variability of the existing fish population, indicating a possible adverse environmental impact. Given the number of other variables affecting fish populations and the conservative modeling approach, which assumed 100% mortality for all entrained fish and eggs, it was concluded that the likelihood of impact was by no means assured, even in these six cases. It was concluded that in most cases, current

* Corresponding author. E-mails: eperry@chesapeake.net; gls@eaest.com; jtv@eaest.com; twlohner@aep.com; randy.lewis@cinergy.com.

entrainment and impingement losses at six Ohio River power plants have little or no effect at the population level.

KEY WORDS: Clean Water Act 316(b), entrainment, fish, impingement, population modeling, Ohio River

DOMAINS: ecosystems management, freshwater systems, ecosystems and communities, environmental sciences, environmental management and policy, environmental technology, environmental modeling, environmental monitoring

INTRODUCTION

Cooling-water intake systems have the potential to adversely impact aquatic organisms through entrainment and impingement. Entrainment occurs when organisms (e.g., larval fish) pass through the intake traveling screens and into the power plant where they may suffer injury or death. Impingement occurs when organisms are drawn against intake trash racks or screens by the force of the incoming water current.

Section 316(b) of the Clean Water Act (CWA) requires that "the location, design, construction and capacity of cooling water intake structures reflect the best technology available for minimizing adverse environmental impact." Historically, the EPA has allowed section 316(b) of the Clean Water Act to be evaluated on a case-by-case basis, with individual plants performing 316(b) "demonstrations." These 316(b) demonstrations consisted of quantifying entrainment and impingement rates, then assessing whether the measured rates would affect populations of at-risk species. This has resulted in considerable variation in compliance requirements from plant to plant.

A key issue in the 316(b) rule development process has been how to define "adverse environmental impact" (AEI). The loss of a single fish could be considered an adverse impact and lead to an in-depth analysis of the number of fish killed and the cost of installing new intake technologies. However, the electric utility industry does not believe that Section 316(b) of the CWA was intended to address the loss of individual fish, but instead, was written to address the potential adverse impact on fish populations. The fact that an individual fish may die or suffer adverse physiological changes does not imply that the population will suffer a harmful decrease in number. In fact, the results of long-term monitoring studies in the Ohio River through 1985 have demonstrated that, within the permanent restrictions placed upon the river ecology by the lock and dam system, there is strong evidence of positive changes in the fish community due to improvements in water quality[1]. This has occurred in spite of the loss of millions of fish due to entrainment and impingement.

It is not possible to simply measure entrainment and impingement at a power plant and directly relate the results to population-level impacts. Rather, a suggested first step in assessing potential AEIs is to assess the condition of affected populations and to model the impact of various entrainment and impingement scenarios on those populations. This has the advantage of avoiding unnecessary studies and helping to focus actual field studies on those populations that are most vulnerable to adverse impact. Therefore, to assess whether Ohio River power plants may be adversely affecting fish populations, a Leslie matrix model was developed. For each fish species chosen, several life history parameters (age, growth, fecundity, and age-specific survival) were used and the population was projected forward for a specified time period (50 years). From the 316(b) perspective, the advantage of this approach is that it can be used to model the population as it currently is (i.e., with the power plant operational, inclusive of plant-specific entrainment and impingement losses) and it can be used to model the population assuming these losses were not occurring. By comparing populations with and without these losses, it can be judged whether the losses are of sufficient magnitude to significantly affect population size.

METHODS

Formulating the Population Model

Matrix models are widely used in ecology to investigate the structure and dynamics of natural populations[2,3,4,5]. The advantage of matrix models over other population models such as the logistic model, Ricker's spawner-recruit model, or the Beverton-Holt formulation, is that matrix models prescribe differing vital rates for different parts of the population while other models treat all individuals in the population as if they are identical. Because length data are available from the Ohio River electrofishing studies, it seems appropriate to partition the population into length categories and allow for the possibility that survival and fecundity rates may differ among size classes. One criticism of matrix models is that for populations with continuous reproduction, matrix models are not well suited. Matrix models use a time step that assumes that all births occur at the beginning of the time interval. For fish populations that typically have a relatively short spawning season, this assumption of a pulse of reproduction at the beginning of the time interval works well.

The basic form of the Leslie matrix model is described by Leslie[6,7]:

$$\begin{bmatrix} f_1 & f_2 & \cdots & f_k \\ s_1 & 0 & \cdots & 0 \\ . & . & \cdots & . \\ . & . & \cdots & . \\ 0 & 0 & s_{k-1} & 0 \end{bmatrix} \begin{bmatrix} n_1 \\ n_2 \\ . \\ . \\ n_k \end{bmatrix}_t = \begin{bmatrix} n_1 \\ n_2 \\ . \\ . \\ n_k \end{bmatrix}_{t+1} \quad (1)$$

where s_i is the probability that an individual will advance to the next age class, f_i is the recruitment rate for age class i, n_i is the number of individuals in size class i, t is the ordinal for time step (year), and k is the maximum number of size classes. In its basic form, the Leslie matrix model is deterministic. That is, given one set of estimates for the parameters on one initial population vector, it will always predict the same population trajectory. The current implementation of the Leslie matrix model uses the estimates described below as the mean of stochastic distributions to simulate random year-by-year variability in the s_i and f_i vital rates. The details of this simulation are as follows.

With each annual projection of the population, survival parameters (s_1 through s_k) were simulated from a Beta distribution with the range of the random number generator limited to the interval (0, 1). The means of these survival estimates were set equal to the survival estimates obtained from a length-frequency model. The variance of these survival estimates was restrained so that the coefficient of variation (CV) was 25% for survival in the interval (0.10 to 0.90) and 10% for survival outside this interval.

Only the fecundity component of the recruitment estimates was randomly simulated. The sex ratio, proportion of mature females, and larval survival components were held constant. The fecundity component was simulated using a Poisson distribution with the mean determined from fecundity estimates taken from literature values[8,9,10,11,12]. As described below, the larval survival parameter was tuned to yield long-term stability.

There are numerous strategies in the modeling literature for implementing compensation, such as the Ricker spawner-recruit model and the Beverton-Holt equations. In this model, compensation is implemented by the simple idea that each pool in the Ohio River has a carrying capacity for each age class of each species. That is to say, the population may be controlled by the population vital rates up to this threshold, but then some external factor, such as food supply, nesting sites, or habitat, limits the population. The carrying capacity for each age class was set equal to the maximum abundance observed for that age class for the period of record. If during simulations, the projected age-specific abundance exceeded the carrying capacity for the age class, the age-specific abundance (n_i) was set equal to the carrying capacity.

Data Sources

Since 1970, fish data have been collected near several power plants on the upper and middle Ohio River as part of the Ohio River Ecological Research Program (ORERP)[13]. Over this 30-year period, data have been collected primarily near six power plant locations:

Plant	River Mile	Pool	River Miles
WH Sammis	53.9	New Cumberland	31.7–54.4
Cardinal	76.7	Pike Island	54.4–84.4
Kyger Creek	260.0	Robert Byrd	237.5–279.2
Beckjord	453.0	Markland	436.2–531.5
Tanners Creek	494.0	Markland	436.2–531.5
Clifty Creek	560.0	McAlpine	531.5–605.0

While the ORERP was designed to gather information on the potential impacts of power plant operation on Ohio River biota, it did not require the collection of specific information on fecundity, recruitment, survival, or age and growth for individual Ohio River species[14]. This information was therefore obtained from the literature as noted in the following discussions. Only electrofishing data were used for the modeling effort as electrofishing is the single most effective sampling gear[15], and it has been used consistently as part of the ORERP. Since 1991, electrofishing has been conducted at night, rather than during the day, to be consistent with the approach recommended by Ohio EPA. Because catch rates and species composition changed in response to this change in sampling protocol, only electrofishing data from 1991 to 1998 were used in the model. Nighttime electrofishing data from 1992 to 1998 collected by ORSANCO were also used to supplement the ORERP electrofishing data.

Losses due to impingement and entrainment were based on data extracted from 316(b) study reports prepared for the above-listed power plants and for the Philip Sporn and Miami Fort power plants located in the Robert Byrd and Markland pools, respectively. Even though survival for some entrained species can be relatively high, the model conservatively assumed that all entrained fish were killed. Likewise, it was assumed that all impinged fish were killed.

Selecting Species and Estimating Vital Rates

Species were selected based on the availability of sufficient data to describe the population characteristics of that species, the availability of related data (e.g., fecundity), and the vulnerability of each species to entrainment and impingement. Based on the above criteria, six species were selected: bluegill, freshwater drum, emerald shiner, gizzard shad, sauger, and white bass. For three of these species (bluegill, gizzard shad, and sauger) sufficient data were available to allow populations to be estimated for all five pools studied, while data on other species (e.g., emerald shiner) restricted the analyses to as little as one pool. The Leslie matrix model used to estimate the Ohio River fish populations requires estimates of survival and fecundity. Fecundity estimates for each species were taken from relevant fisheries' literature[8,9,10,11,12]. These estimates were converted to recruitment estimates by estimating survival to midsummer for young-of-the-year (YOY) fish as described under the Recruitment Estimates section (below). Survival estimates between sequential age classes were obtained from length-frequency data using the length-frequency model also described below. Several intermediate steps were undertaken to obtain the length-frequency model, including estimation of a growth curve, adjusting the data for size-dependent electrofishing capture probability, and adjusting the data for within-season growth.

Recruitment Estimates

One major component of the Leslie matrix model is the recruitment component, which we define as the process of getting from eggs per female to YOY in midsummer after spawning. This is the component about which there exists the least amount of site-specific information. To obtain a reasonable estimate of the recruitment that is generated in each year, age-specific estimates of the number of eggs spawned per sexually mature female were compiled from the literature[8,9,10,11,12]. We assumed a sex ratio of 50:50. These estimates were used to set the mean of the stochastic distribution of the number of eggs spawned for the model. To obtain recruitment, the number of these eggs that survive to become YOY fish must be estimated. To model this survival, a parameter called larval survival (ls) was included in the model. While we call this the larval survival factor, it is the term that models the life history parameters: fraction of females that are mature and fraction of eggs that are actually spawned. Lacking information on all of these life history processes between number of eggs and YOY, larval survival was obtained by tuning the model to stability. Larval survival was set to the lowest level that would produce long-term stability for the population with the power plants operating. Stability was ascertained based on no observable increase or decrease in a graphical display of a 50-year projection using the stochastic version of the model. The rationale for this choice is that if the species continues to persist in the presence of the power plants, it must have achieved at least this minimal level of stability. Setting the parameter to this minimum level maximizes the effect of removing the power plants' influence on the long-term population level.

The data suggest that most species periodically have a good recruitment year. The model was designed so that approximately one year in five (randomly determined) is selected to be a good recruitment year in which YOY survival is ten times better than in a typical year. The parameter influencing recruitment is the high recruitment (HR) parameter. A Bernoulli random variable which takes on values of 1 or 0, was simulated to have probability 1/5 of being equal to 1. The

HR parameter was simulated once for each projection. If HR = 1, then the model estimates a high recruitment year by increasing larval survival (ls) by an order of magnitude (i.e., ls is replaced by ls × 10). The initial population vector was set equal to the mean population level for the period of record.

Growth Model

The von Bertalanffy growth equation[16,17] was used to model length as a function of age. This model follows the form:

$$l_a = l_{inf}(1 - e^{(-k*(a - ao))}) \qquad (2)$$

where l_a is the length at age a, l_{inf} is the maximum length, k is a growth rate coefficient, and a_0 is the initial length at age 0, usually set to 0. Estimates for the parameters for this nonlinear model were obtained using the nonlinear regression (NLIN) procedure of the SAS system[18]. The von Bertalanffy growth model served two purposes in the development of the length-frequency model. It was used to adjust the lengths of fishes caught at various times during the growing season to a common point in time. It was also used to determine distance between year class modes in the length-frequency model.

Length Adjustments

During each year of the electrofishing survey, collections were made in the late spring, midsummer, and early fall. The ideal data for parameterizing a Leslie matrix model would be an intensive snapshot of the population at a fixed date every year. In order to compose a length-frequency distribution at one point during the year, the lengths of the fish were adjusted by the von Bertalanffy model to estimate the length each fish would have been in midsummer (July 19). Using mean growth data obtained from the literature, the parameters of the von Bertalanffy model were estimated. Assuming an 8-month growing season, fish lengths were adjusted to the expected length for July 19. Lengths of fish captured in late spring were increased, while those of fish captured in the fall were decreased. Fish that were measured individually were assigned an adjusted length and then assigned to appropriate length groups. Fish that were measured in groups were prorated into adjacent length groups in proportion to the size of the adjustment.

Electrofishing Efficiency Adjustment

Because electrofishing is known to have a lower efficiency for collecting smaller fish, adjustments for electrofishing efficiency were made using a capture probability curve estimated for brown trout[19]. To adjust the data, the frequency for each length interval was multiplied by the inverse of the capture probability for the midpoint of the interval.

Length-Frequency (Mixed Beta) Model

The concept behind the methods employed to obtain estimates of survival rates is based on the idea that each age class of a species will have a length distribution. As the fish grow older the mean of the age specific length distribution will increase in a manner described by a growth function and the variance may also change. As fish age and approach the maximum size for the species in question, it is almost certain that age-specific distributions will overlap. By compositing the age-specific distributions, one can obtain the population (all ages) distribution of lengths. The length-frequency model employed in this study attempts to estimate the mean and variance of each age-specific length distribution by fitting the composite model to the population length-frequency histogram. After obtaining the parameters for the age-specific length distributions, it is then possible to compute survival estimates as the ratio of expected frequency at age $i + 1$ divided by the expected frequency at age i. The length-frequency model employed here was similar in structure to the length-frequency models used in the ELEFAN system[20]. In the Ohio River model, the length-frequency data are represented by a mixture of scaled beta density functions, whereas the ELEFAN system uses normal density functions. The classical beta density described in any mathematical statistics book is defined on a domain of {0, 1} and is defined as follows:

$$f(x; a,b) = \frac{x^{a-1}(1\text{-x})^{b-1}}{B(a,b)} \quad 0 < x < 1, a > 0, b < 0$$

$$\mathrm{B}(a,b) = \frac{\Gamma(a)\,\Gamma(b)}{\Gamma(a+b)} \tag{3}$$

where $\Gamma(b)$ is a complete gamma function and a and b are shape parameters for the beta density[21]. In order to use the beta density to define the probability that a randomly selected fish of a certain age falls in a particular length interval, the variable x must be rescaled from the interval {0, 1} to the interval {*minimum*, *maximum*} where *minimum* and *maximum* are defined as the minimum and maximum length for each year class of fish.

Survival Estimates

The conditional probability of survival from age class i to age class $i + 1$ is estimated by the ratio of the expected frequency for age class $i + 1$ to the expected frequency for age class i. The expected frequencies are determined from the length-frequency model. It is called the conditional probability because it is the probability of surviving to the next age class conditioned on having survived to the current age class. These conditional probabilities give us the survival terms of the Leslie matrix model. In the case where the population should have more age classes than are represented in the length-frequency data (as shown in the growth data found in the literature), the survival of additional age classes was set equal to that of the last class represented in the data.

Mean Population and Carrying Capacity

The mean age-structured population vector and the age-structured carrying capacity vector were respectively estimated by the mean and maximum of the age-structured population vectors for the period of record. After adjustments for capture probability and size, the mean number of fish per age class per hectare was computed assuming an effective shocking width of 4.25 m for the recorded length of the shocking effort. Multiple samples per year were averaged. If the data did not represent all the older age classes of the population vector for a species, these older age class vector components for the mean population vector and the maximum population vector were extrapolated based on the survival estimate obtained from the two oldest age classes represented in the data.

Power Plant Effects

Because the length-frequency data by which the model is calibrated were collected while the power plants were operating, the model up to this point estimates population behavior while under the influence of the power plants. To estimate the effects of entrainment and impingement, the model was modified to simply add back the entrainment and impingement harvesting. The total entrainment estimate for each pool is multiplied by larval survival (ls) and added to the YOY component of the population vector during each projection. Annual impingement estimates for each pool were summed by year class and these totals were added to the appropriate population cells in each projection of the population. In this configuration of the model, both the entrainment and the impingement effects are constants. The model projected the population 50 years into the future, and 500 simulations were run for both the power-plant-effect and no-power-plant-effect models.

RESULTS

Growth Equation

The first step of the length-frequency modeling process is to obtain parameters for the von Bertalanffy growth equation. To develop this growth model, species-specific age and length data from the literature for the Ohio River or other appropriate water bodies were used. The example von Bertalanffy growth curves for sauger and gizzard shad (Fig. 1) and the parameters for the equations describing these curves for the remaining species (Table 1) indicated that the fit of the model to the actual data was acceptable. The lowest growth rates (K in Table 1) were obtained

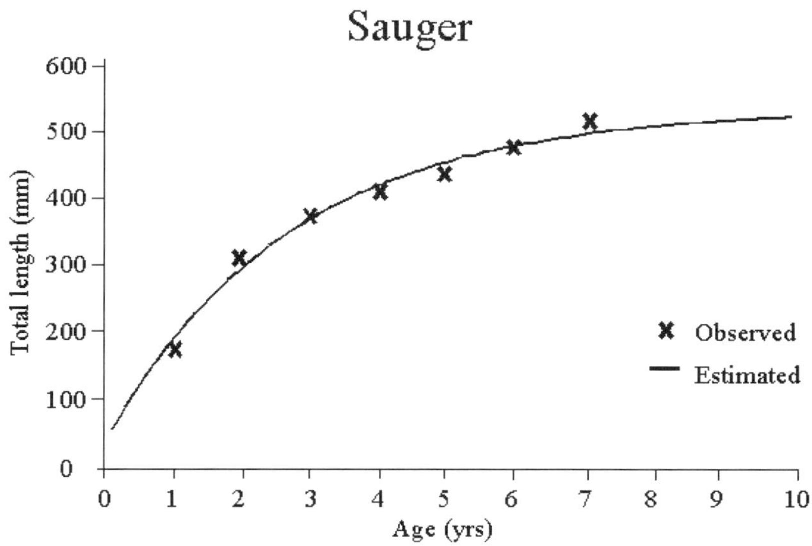

FIGURE 1A

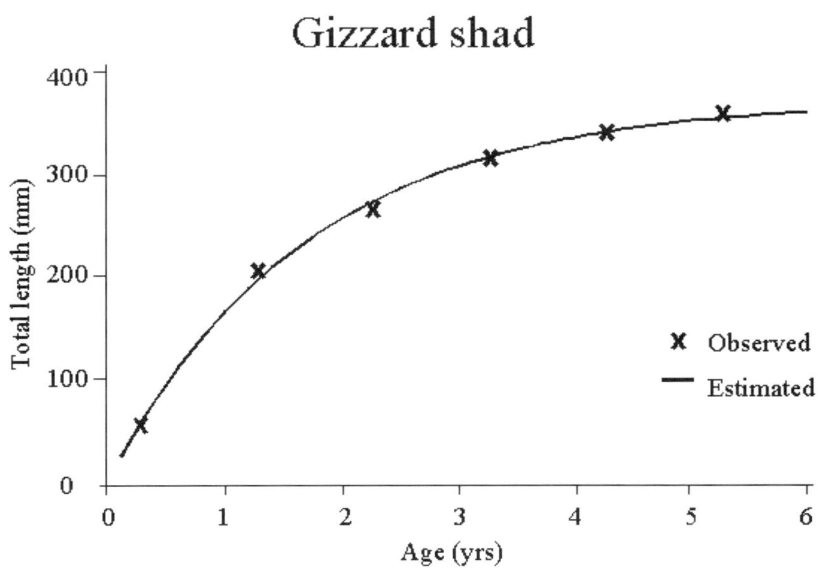

FIGURE 1B

FIGURE 1. Growth curves for sauger and gizzard shad showing observed data and estimated von Bertalanffy model.

for emerald shiner and bluegill. The highest growth rate was obtained for gizzard shad, which is reasonable because gizzard shad is a fast-growing species. The LINF parameter in the von Bertalanffy model is a theoretical maximum length (at age infinity); however one would not expect fish to approach that length unless they were extremely long lived.

Length-Frequency Model

Initially each species was modeled with as many age classes as were indicated in the growth data. When fitting the length-frequency model, it was common for the expected frequency of the last (youngest) age category to start converging toward zero. When this happened, the last age class would be eliminated under the

TABLE 1

Statistics for the Growth Curves by Species

Species	Parameter	Asymptotic Estimate	Std. Error
Sauger	LINF	533.3189072	34.335407684
	K	0.3711828	0.087560862
	TO	-0.1473548	0.273224912
Gizzard shad	LINF	370.1974256	10.034100589
	K	0.5841744	0.054676357
	TO	-0.0004830	0.051837130
Freshwater drum	LINF	458.0946917	12.177552434
	K	0.3298979	0.036129042
	TO	0.0071713	0.153225199
White bass	LINF	418.8457737	6.8258631136
	K	0.4490271	0.0375477085
	TO	0.0541263	0.1044621517
Bluegill	LINF	258.9146360	5.7377772322
	K	0.2568898	0.0198869626
	TO	-0.6382159	0.1263056293
Emerald shiner	LINF	164.5188804	0.00000000000*
	K	0.2460419	0.00068797160
	TO	-0.2357573	0.00142707420

Note: LINF is the theoretical maximum length (mm) of a fish under the von Bertalanffy model; K is the growth rate parameter; TO is a parameter to adjust the age (year) of the initial cohort.
* With only 3-year classes, the von Bertalanffy model obtains a perfect fit which leaves no information for estimating error.

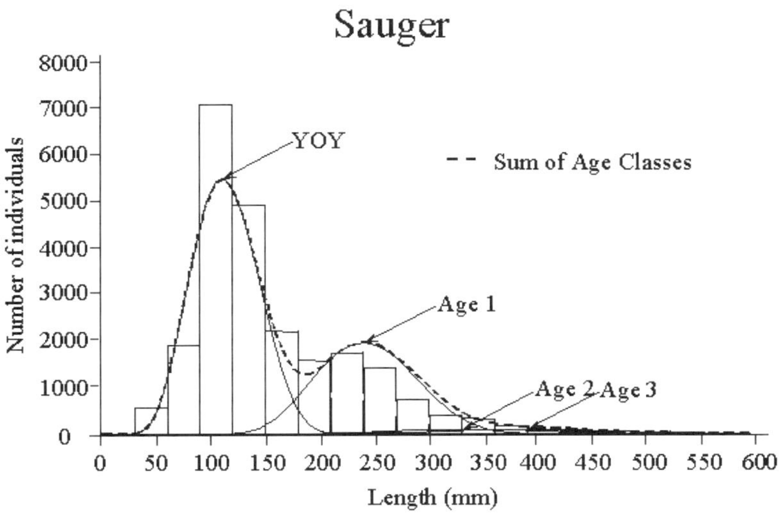

FIGURE 2A

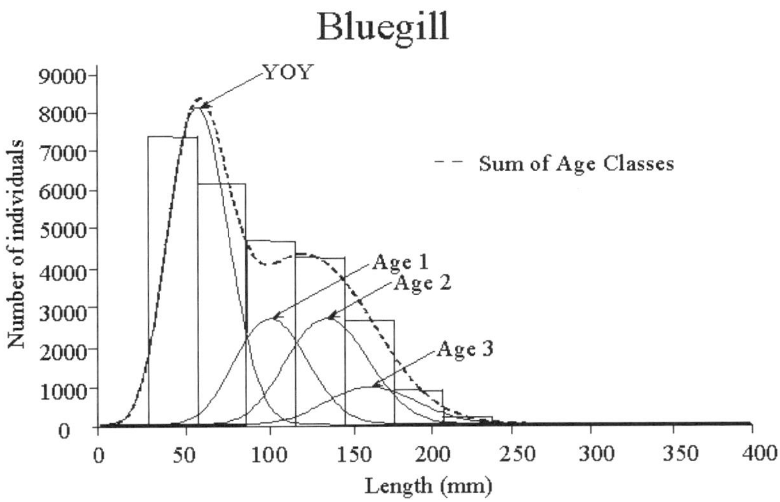

FIGURE 2B

FIGURE 2. The length frequency data (shown in 30-mm size groups) superimposed with the length frequency model for sauger and bluegill.

assumption that this age class (usually Age 0) was not adequately represented in the electrofishing data, and the fitting process was restarted with one fewer age category.

As a rule, the length-frequency model acceptably reproduced the observed length-frequency, as shown in the sauger example (Fig. 2A). One exception was the case of bluegill (Fig. 2B), where the second and third age classes (1- and 2-year-olds) were fitted with almost equal weight, suggesting that there is very little mortality between age 1 and 2. This anomalous result is probably an artifact of the fact that 30-mm length categories are too broad for bluegill, a fairly slow-growing and rather small species. Categorizing the bluegill into 30-mm size groups resulted in too low a resolution to accurately define the peaks associated with these two age classes.

The mean length for each age class is expressed by the von Bertalanffy growth equation. The 25th and 75th percentiles of length for each age class were computed from the beta density function parameterized with a, b, and m as defined previously (Table 2).

TABLE 2

The Mean, 25th Percentile, and 75th Percentile Lengths (mm) for Each Modeled Species (by Age Class, as Estimated by the Length-Frequency Model)

		Age				
Species	**Statistic**	**0**	**1**	**2**	**3**	**4**
Sauger	25th Percentile	71.02	197.78	285.23	345.57	387.20
	Mean	90.34	227.70	322.47	387.85	432.96
	75th Percentile	108.96	256.54	358.36	428.60	477.07
Gizzard shad	25th Percentile	77.02	191.25	254.94	290.45	310.25
	Mean	93.77	216.07	284.26	322.28	343.48
	75th Percentile	107.22	236.00	307.81	347.85	370.17
Freshwater drum	25th Percentile	54.49	156.43	229.72	282.42	320.31
	Mean	69.66	178.81	257.29	313.72	354.29
	75th Percentile	80.64	195.00	277.23	336.36	378.87
White bass	25th Percentile	59.45	162.23	227.83	269.70	296.43
	Mean	84.23	205.28	282.54	331.85	363.32
	75th Percentile	107.35	245.44	333.57	289.82	425.72
Bluegill	25th Percentile	19.79	68.29	105.81	134.83	157.27
	Mean	31.21	82.79	122.69	153.55	177.42
	75th Percentile	42.90	97.64	139.97	172.72	198.05
Emerald shiner	25th Percentile	10.74	42.47	64.28		
	Mean	19.04	50.77	75.58		
	75th Percentile	27.35	59.08	83.89		

Note: The means follow a von Bertalanffy growth model and the 25th and 75th percentiles are based on the beta distribution for each age category.

Population Simulations

The results of the 500 simulations of each 50-year population trajectory were summarized by four plots for each species and pool (e.g., sauger in Robert Byrd pool, Figs. 3 and 4). The first and second plots show results of simulations for the population under the influence of the power plants and the results with the power plant effects removed (i.e., no fish lost via entrainment or impingement or the no-power-plant-effects model). In these plots, the broad, solid line shows the mean of the 500 simulations for each of the 50 years. The dashed lines show the 5th and 95th percentiles of the 500 simulations for each year. The narrow, solid lines show the minimum and maximum of the simulated populations for each year.

The third plot (Fig. 4) is an overlay of the first two plots and illustrates how the population is predicted to respond if no fish were lost via impingement or entrainment. In these plots, the broad, dashed line is the estimate of the mean population if entrainment and impingement were eliminated (no-power-plants model). The broad, solid line is the estimated population under current conditions with the power plants operating and fish being lost due to impingement and entrainment. If the solid and dashed, broad lines follow a similar path (e.g., bluegill in Pike Island pool, Fig. 5), then no population level impacts would be predicted. However, if these two lines follow noticeably different trajectories (e.g., freshwater drum in Robert Byrd pool, Fig. 6), then impacts to that species in that pool may be occurring. The narrow, dashed line is the 5th percentile of the no-power-plants model and the narrow, solid line is the 5th percentile for the power plant model.

To help with the interpretation of these boundaries, a fourth plot was developed that illustrates the frequency with which the simulations fall below the 5th percentile for more than a specified number of years. For example, for sauger in the Robert Byrd pool (Fig. 4), it can be seen that the probability of falling below the 5th percentile for at least one of the 50 years is about 0.55, which is relatively high; the probability of falling below the 5th percentile for at least 10 of the 50 years is about 0.06, which is a fairly rare event; and the probability of falling below the 5th percentile for 25 of the 50 years is essentially zero.

The average change for each species and pool that would be expected if power plant entrainment and impingement could be removed is summarized in Table 3. This percent was calculated by finding the geometric mean of the abundance values from the 500 simulations for each year for both the power-plant and no-power-plant simulations. The average expected change for the 50 years is computed from these annual means. The change is calculated by the formula

$$\text{percent change} = 10^{(mn(nopp)\ -\ mn(pp))} * 100 - 100 \tag{4}$$

where $mn(nopp)$ is the mean of logarithms of total population of the no power plant simulations and $mn(pp)$ is the mean of logarithms of total population of the power plant simulations.

The projected changes range from a 3% decrease for white bass in Robert Byrd

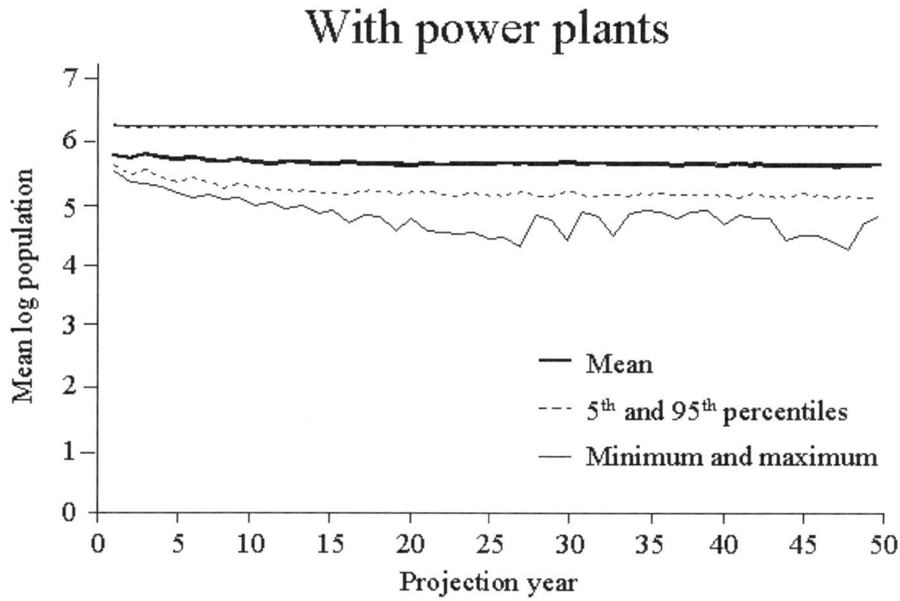

FIGURE 3A

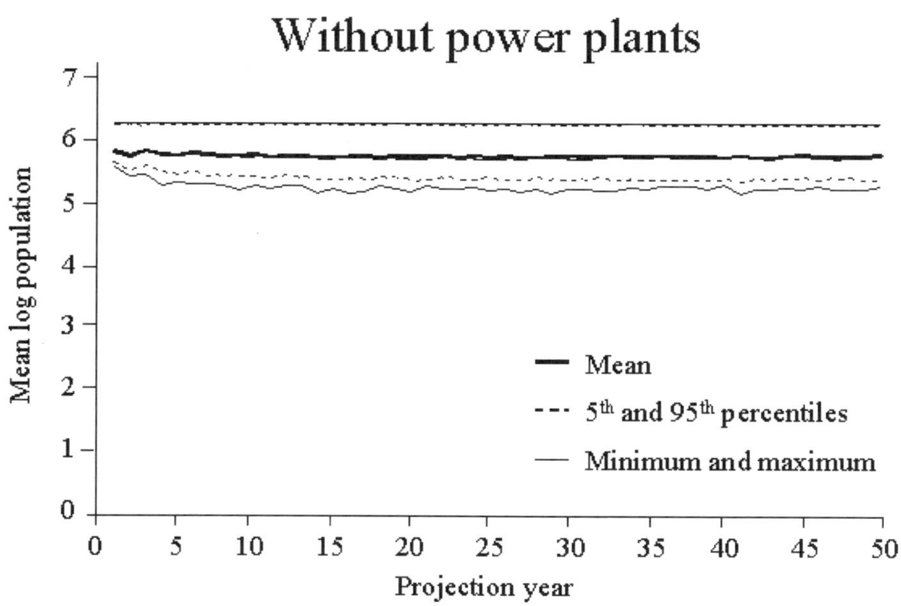

FIGURE 3B

FIGURE 3. Simulation results for sauger population in the Robert Byrd pool with and without the influence of power station entrainment and impingement.

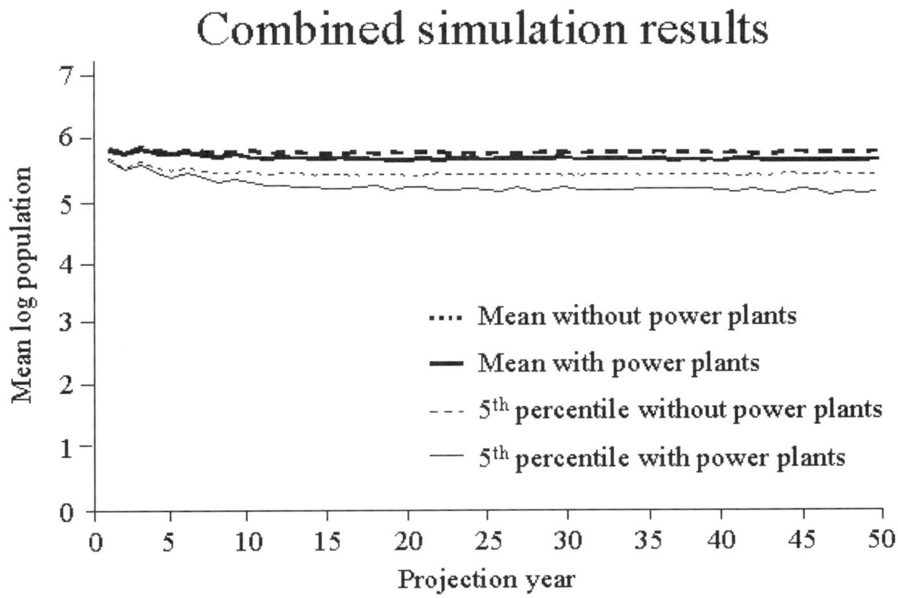

FIGURE 4A

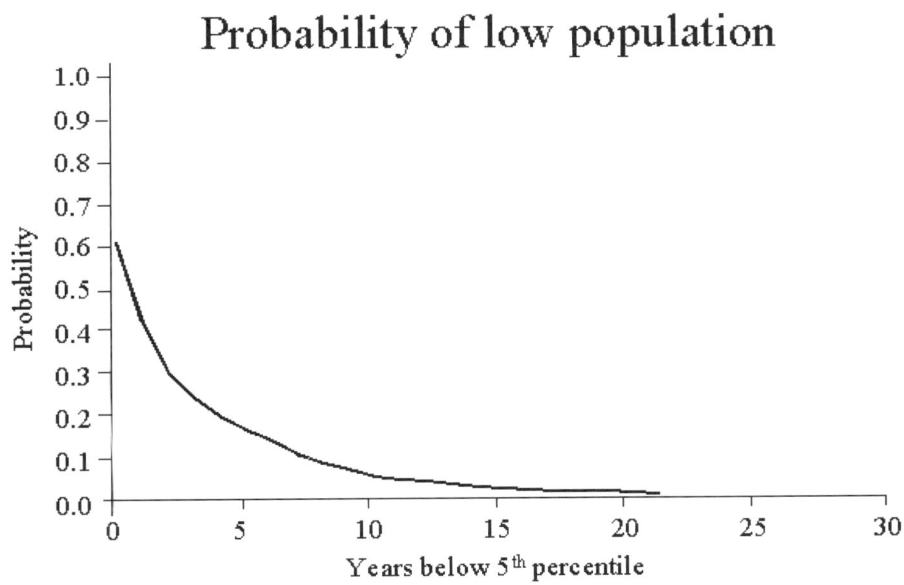

FIGURE 4B

FIGURE 4. Combined simulation results for sauger population in the Robert Byrd pool and the probability of falling below the 5th percentile for a specified number of years.

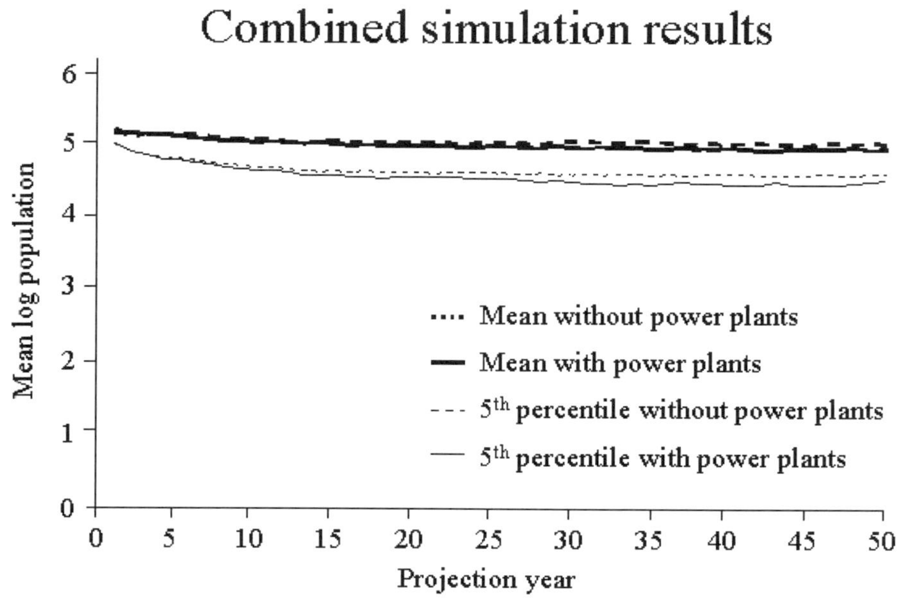

FIGURE 5

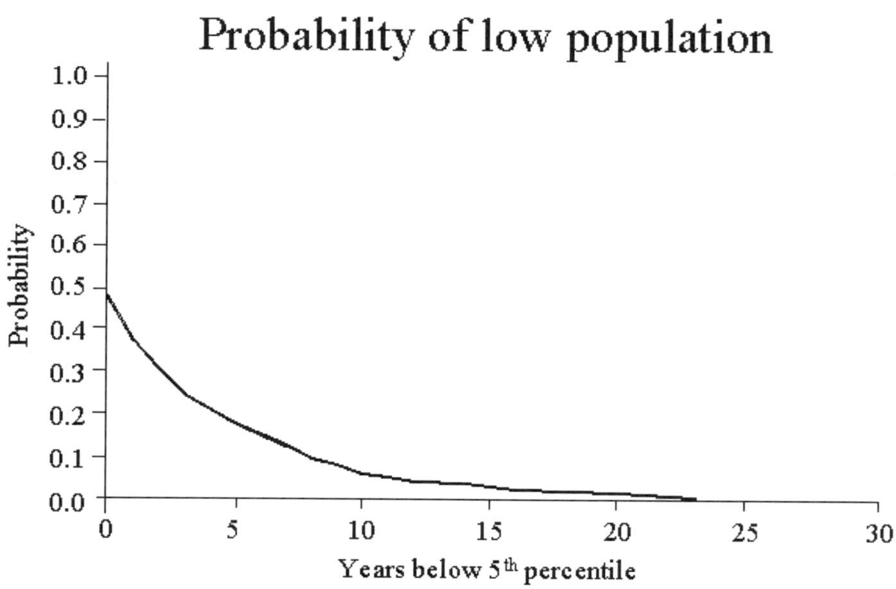

FIGURE 5B

FIGURE 5. Combined simulation results for bluegill population in the Pike Island pool and the probability of falling below the 5th percentile for a specified number of years.

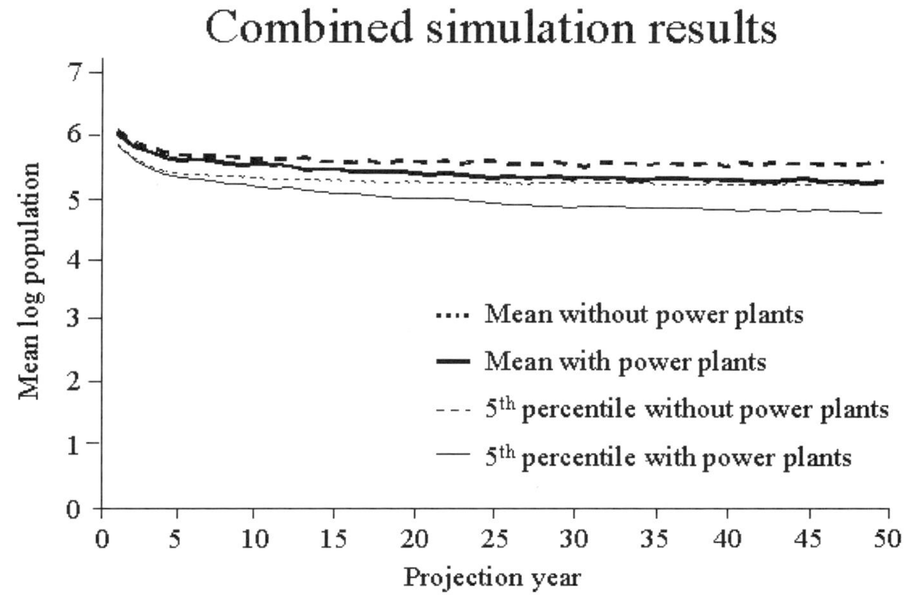

FIGURE 6A

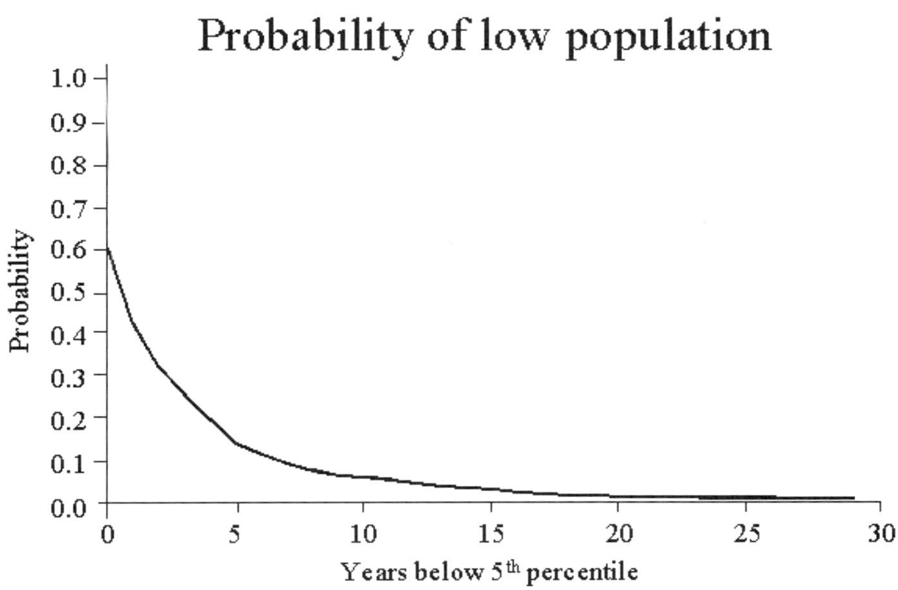

FIGURE 6B

FIGURE 6. Combined simulation results for freshwater drum population in the Robert Byrd pool and the probability of falling below the 5th percentile for a specified number of years.

TABLE 3

Estimated Percent Change in the 50-Year Average Population Size of Selected Ohio River Fishes if Impingement and Entrainment Losses are Eliminated through Mathematical Simulation

Species	Pool*	Mean Log with Power Plants	Mean Log without Power Plants	Percent Change	Mean** CV (%)	Effect** Level
Sauger	NC	4.88872	4.90693	4.28	62	17.5
	PI	5.57112	5.73589	46.14***		
	RB	5.60577	5.69861	23.84***		
	ML	6.34186	6.34501	0.73		
	MA	6.03326	6.05421	4.94		
Gizzard shad	NC	6.41621	6.53255	30.72	158	44.7
	PI	7.29137	7.29946	1.88		
	RB	7.06017	7.07516	3.51		
	ML	7.72982	7.73173	0.44		
	MA	6.85997	6.85992	-0.01		
Freshwater drum	PI	5.70266	5.75293	12.27	101	28.6
	RB	5.37173	5.56635	56.54***		
	ML	6.08379	6.11108	6.48		
	MA	5.94638	6.04799	26.36		
White bass	PI	5.05217	5.07317	4.95	98	27.7
	RB	5.27650	5.26525	-2.56		
Bluegill	NC	5.13175	5.15873	6.41	59	16.7
	PI	4.96286	5.01323	12.30		
	RB	4.93745	4.96330	6.13		
	ML	5.79204	5.89857	27.80***		
	MA	5.29899	5.38488	21.87***		
Emerald shiner	PI	6.44737	6.49330	11.16***	28	7.9

* NC = New Cumberland pool; PI = Pike Island pool; RB = Robert Byrd pool (formerly Gallipolis Pool); ML = Markland pool; MA = McAlpine pool.

** For comparison, the mean CV is an estimate of typical interannual percent deviation estimated from the observed data record. The effect level is an estimate of the 95th percentile of the percent deviation in the 50-year run. A change larger than this is unlikely to occur as a result of natural variation.

*** Projected percent change greater than would be expected from normal year-to-year population variation.

pool to a 57% increase for freshwater drum in Robert Byrd pool when the power plant effect is removed. The slight decreases predicted for gizzard shad and white bass are the result of random components of the model resulting in a larger population for the power plant simulations than for the no-power-plant simulations. Because the no-power-plant simulations add impingement and entrainment losses back into the populations, logic dictates that without random effects, the no-power-plant population should always be as large or larger.

DISCUSSION

Model Accuracy and Resolution

Despite the lack of site-specific data for some parameters (e.g., fecundity), the population model developed herein provides reasonable guidance as to which populations would most likely increase if entrainment and impingement losses were to be removed. To reduce the possibility of inappropriate decisions being made, a conservative approach was taken at each decision point in compiling the model. Until the model can be refined using additional site-specific data, interpretation of the model outputs should be done cautiously and should not focus on the absolute values for average expected change for a given species and pool. Rather, the focus should be on relative differences among species and pools in the values for average expected change. The resolving power of the model is currently limited by the lack of site-specific age/growth and fecundity data, as well as the inability of electrofishing to capture all age (size) classes equally. Even though an adjustment was made for electrofishing size-dependent efficiency, other unquantifiable factors undoubtedly also affect electrofishing results. For example, larger individuals of species such as sauger and freshwater drum are less shoreline oriented than the smaller individuals of these same species. Thus, the apparent absence of large sauger and freshwater drum in the electrofishing database is likely an artifact associated with the behavioral characteristics of these species. Examination of ORSANCO lock chamber rotenone data for sauger and channel catfish confirmed that larger individuals of these species were more abundant than the electrofishing data suggested[22].

Impact Predictions

With the above cautions in mind, the likelihood that populations might measurably increase in abundance if entrainment and impingement losses were eliminated was estimated by comparing the predicted percentage increase with the observed population variability shown by each species over the years of the electrofishing survey (Table 3). The interannual population CV was computed for each species and pool, and these estimates were averaged across pools for each species to estimate the typical year-to-year percent variation as it occurs in nature. The effect level was computed to correspond to a level of deviation from normal population levels that would be exceeded in only 5% of cases in a 50-year duration. The 50-year duration was chosen based on the life expectancy of an Ohio River power generating plant. The effect level was computed as $(2*CV/sqrt(50))$. When the model predicted a change greater than the effect level, it indicates that the projected population change exceeds what might be reasonably expected due to normal interannual variation.

Overall, the modeling results suggest that improvements at the population level

TABLE 4

Likelihood of Impacts to Fish Populations after 50 Years of Entrainment/ Impingement Losses based on Leslie Matrix Model Results*

	River Pool**				
Species	**NC**	**PI**	**RB**	**ML**	**MA**
Bluegill	No	No	No	Yes	Yes
Freshwater drum	–	No	Yes	No	No
Emerald shiner	–	Yes	–	–	–
Gizzard shad	No	No	No	No	No
Sauger	No	Yes	Yes	No	No
White bass	–	No	No	–	–

* No = No impact expected; Yes = impact may occur; Dash(–) = data insufficient to make comparisons.

** NC = New Cumberland; PI = Pike Island; RB = Robert Byrd (formerly Gallipolis); ML = Markland; MA = McAlpine.

are unlikely in 16 of the 22 cases (Table 4). In 9 of these 16 cases, the increases predicted by the model are small (<5%) and in all 16 cases the predicted increase was within the level of normal variability expected for populations of these species. In the other 6 of the 22 cases (bluegill in the Markland and McAlpine pools, drum in the Robert Byrd pool, emerald shiner in the Pike Island pool, and sauger in the Pike Island and Robert Byrd pools) the predicted increases (11–57%) were outside the range of expected variability. It is interesting to note that gizzard shad, despite being one of the most abundant fish species in the Ohio River[23] and dominating impingement catches at many power plants[24], is not predicted to increase in any of the pools if power plant entrainment and impingement were removed.

Although the increases predicted in these six cases are statistically real, there is no guarantee that these increases would actually occur, because the changes predicted by the model assume that populations of these species change only in response to entrainment and impingement losses and that none of the entrained or impinged fish survive. In reality, each population will respond to a combination of abiotic (e.g., river flow, water quality, habitat quality) and biotic factors (e.g., predation, exotic species, competition, etc.). In particular, spawning success and year class strength in fishes vary widely, depending on largely unknown interactions of these and other biotic and abiotic factors. Thus, it seems likely that the populations of these species 50 years from now are more likely to be affected by changes in factors such as habitat, water quality, and the occurrence of floods, droughts, and temperature extremes, rather than the minor (typically <10%) changes resulting from entrainment/impingement losses predicted by the model.

ACKNOWLEDGMENTS

Funding for this study was provided by American Electric Power, American Municipal Power, Buckeye Power, Cinergy, First Energy, Indiana-Kentucky Electric Corp., Ohio Valley Electric Corp., and Vectren. The authors thank Webb Van Winkle, Charles Coutant, and Gary Hickman for providing invaluable comments on an earlier draft of this manuscript.

REFERENCES

1. Van Hassel, J.H., Reash, R.J., Brown, H.W., and Thomas, J.L. (1988) Distribution of upper and middle Ohio River fishes, 1973-1985. I. Associations with water quality and ecological variables. *J. Freshwater Ecol.* **4(4),** 441–458.
2. Caswell, H. (1989) Matrix Populations Models: Construction, Analysis, and Interpretation. Sinauer Associates, Sunderland, MA.
3. Charlesworth, B. (1994) *Evolution in Age-Structured Populations.* 2nd ed. Cambridge University Press, New York.
4. Cullen, M.R. (1985) *Linear Models in Biology.* John Wiley & Sons, New York.
5. Cushing, J.M. (1998) An Introduction to Structured Population Dynamics. Society for Industrial and Applied Mathematics (SIAM), Philadelphia, PA.
6. Leslie, P.H. (1945) On the use of matrices in certain population mathematics. *Biometrika* **33,** 183–212.
7. Leslie, P.H. (1948) Some further notes on the use of matrices in population mathematics. *Biometrika* **35,** 213–245.
8. Bodola, A. (1966) Life history of the gizzard shad, *Dorosoma cepedianum* (LeSuer), in western Lake Erie. *U.S. Fish Wildlife Service Fish. Bull.* **65(2),** 391–425.
9. Carlander, K.D. (1977) *Handbook of Freshwater Fishery Biology.* Vol. 2. Iowa State University Press, Ames, IA. 431 pp.
10. Shaap, P.R.H. (1989) Ecology of the emerald shiner, *Notropis atherinoides* (Rafinesque) in Dauphin Lake, Manitoba [Thesis]. University of Manitoba, Manitoba. 178 pp.
11. LMS. (1993) Quad Cities Aquatic Program, 1992 Annual Report. Prepared for Commonwealth Edison Co. Lawler, Matusky & Skelly Engineers LLP, Pearl River, NY.
12. Carlander, K.D. (1997) *Handbook of Freshwater Fishery Biology.* Vol. 3. Iowa State University Press, Ames, IA. 397 pp.
13. Harding ESE, Inc. (2000) The 1999 Ohio River Ecological Research Program. Prepared for American Electric Power Service Corp., Indiana Kentucky Corp., Buckeye Power, Inc., and Cinergy. Harding ESE, Inc., St. Louis, MO.
14. Lohner, T.W., Seegert, G., Vondruska, J., and Perry, E. (2000) Assessment of 316(b) impacts on Ohio River fish populations. *Environ. Sci. Policy.* **3,** S249–259.
15. Ohio Environmental Protection Agency. (1987) Biological Criteria for the Protection of Aquatic Life. Vol. II. Users Manual for Biological Field Assessment of Ohio Surface Waters. Division of Water Quality Monitoring and Assessment, Surface Water Section. OEPA, Columbus, OH.
16. Everhart, W.H., Eipper, A.W., and Youngs, W.D. (1975) *Principles of Fishery Science.* Cornell University Press, Ithaca, NY.
17. Sanders, M.J. (1987) A simple method for estimating the von Bertalanffy growth constants for determining length from age and age from length. In *Length Based Methods in Fisheries Research.* Pauly, D. and Morgan, G.R., Eds. International Center for Living Aquatic Resources Management, Manila, Philippines, and Kuwait Institute for Scientific Research, Safat, Kuwait.
18. SAS Institute. (1989) SAS/IML Software: Usage and Reference, Version 6, 1st ed. SAS Institute, Cary, NC.
19. Anderson, C.S. (1995) Measuring and correcting for size selection in electrofishing mark-recapture experiments. *Trans. Am. Fish. Soc.* **124,** 663–676.
20. Pauly, D. (1987) A review of the ELEFAN system for analysis of length-frequency data in

fish and aquatic invertebrates. In *Length Based Methods in Fisheries Research.* Pauly, D. and Morgan, G.R., Eds. International Center for Living Aquatic Resources Management, Manila, Philippines, and Kuwait Institute for Scientific Research, Safat, Kuwait.
21. Patel, J.K., Kapadia, C.H., and Owen, D.B. (1976) *Handbook of Statistical Distributions.* Marcel Dekker, New York.
22. Ohio River Valley Water Sanitation Commission (ORSANCO). (1994) Ohio River Water Quality Fact Book. ORSANCO, Cincinnati, OH.
23. Pearson, W. and Krumholz, L. (1984) *Distribution and Status of Ohio River Fishes.* Water Resources Laboratory, University of Louisville, Louisville, KY.
24. EA Engineering, Science, and Technology. (1987) Clifty Creek Station Impingement Study and Impact Assessment. Report prepared for Indiana-Kentucky Electric Corporation, Piketon, OH.

A Process for Evaluating Adverse Environmental Impacts by Cooling-Water System Entrainment at a California Power Plant

C.P. Ehrler[1,*], J.R. Steinbeck[1], E.A. Laman[1], J.B. Hedgepeth[1], J.R. Skalski[2], and D.L. Mayer[3]

[1]Tenera Environmental, 225 Prado Road, Suite D, San Luis Obispo, CA, 93401; [2]Columbia Basin Research, 1325 Fourth Ave., Suite 1850, Seattle, WA, 98101-2509; [3]Tenera Environmental, 100 Bush Street, Suite 850; San Francisco, CA, 94104

Received November 15, 2001; Revised February 19, 2002; Accepted February 20, 2002; Published February, 2003

A study to determine the effects of entrainment by the Diablo Canyon Power Plant (DCPP) was conducted between 1996 and 1999 as required under Section 316(b) of the Clean Water Act. The goal of this study was to present the U.S. Environmental Protection Agency (EPA) and Central Coast Regional Water Quality Control Board (CCRWQCB) with results that could be used to determine if any adverse environmental impacts (AEIs) were caused by the operation of the plant's cooling-water intake structure (CWIS). To this end we chose, under guidance of the CCRWQCB and their entrainment technical working group, a unique approach combining three different models for estimating power plant effects: fecundity hindcasting (FH), adult equivalent loss (AEL), and the empirical transport model (ETM). Comparisons of the results from these three approaches provided us a relative measure of confidence in our estimates of effects. A total of 14 target larval fish taxa were assessed as part of the DCPP 316(b). Example results are presented here for the kelp, gopher, and black-and-yellow (KGB) rockfish complex and clinid kelpfish. Estimates of larval entrainment losses for KGB rockfish were in close agreement (FH ≈ 550 adult females per year, AEL ≈ 1,000 adults [male and female] per year, and ETM = larval mortality as high as 5% which could be interpreted as ca. 2,600 1 kg adult fish). The similar results from the three models provided confidence in the estimated effects for this group. Due to lack of life history information needed to parameterize the FH and AEL models, effects on clinid kelpfish could only be assessed using the ETM model. Results from this model plus ancillary information about local populations of adult kelpfish suggest that the CWIS might be causing an AEI in the vicinity of DCPP.

* Corresponding author. Emails: cehrler@tenera.com; jsteinbeck@tenera.com; nlaman@tenera.com; jhedgepeth@tenera.com; jrs@fish.washington.edu; docfish@ix.netcom.com

KEY WORDS: adverse environmental impact, AEI, 316(b), entrainment, cooling-water intake structure, larval mortality, fecundity hindcasting, adult equivalent loss, empirical transport model, rockfish, nearshore fish

DOMAINS: marine systems, ecosystems and communities, environmental sciences, environmental management and policy, environmental modeling, environmental monitoring

---

INTRODUCTION

Section 316(b) of the 1972 Federal Water Pollution Control Act (Clean Water Act [CWA]) requires that "the location, design, construction, and capacity of cooling water intake structures reflect the best technology available [BTA] for minimizing adverse environmental impact [AEI]." However, the CWA does not define AEI. This has caused much concern, debate, and financial hardship for industries using water for cooling and for electric utilities in particular.

Most of the studies describing the effects of cooling-water withdrawals by electric utilities were completed in the late 1970s and early 1980s. The case of the Hudson River power plants is one of the best documented from this period[1]. After many years of debate, the case was settled out of court with the utilities contending that the intake technologies did minimize AEI even though a definition was never developed[2]. Englert and Boreman[3] stated that two points assisted in finalizing the Hudson River case: first, that converging estimates of the effects yielded increased confidence in their "realness," and second, focusing on conditional mortality instead of long-term impacts and on "defining the relative importance of each component to the analysis" was a beneficial approach.

Growing demands for new power production and a court-ordered consent decree (*Cronin v. Browner*, U.S. District Court for the southern District of New York, 93 Civ. 034), required the EPA to develop regulations for minimizing AEI caused by cooling-water intake structures (CWIS). This has kept alive the debate over the development of a clear and concise definition of AEI. Several potential definitions of AEI were presented in the proposed rules for cooling-water structures at new facilities (*Federal Register* Vol. 65, No. 155, pp. 49060–49121, August 10, 2000), but until the rule is finalized, it is not certain which, if any, of them will be used.

In an effort to evaluate the level of entrainment impact caused by the CWIS at the Pacific Gas and Electric (PG&E) Diablo Canyon Power Plant (DCPP) located in central California, we estimated entrainment effects using three mathematical models[4]. Although the plant began operation in 1985, a final 316(b) demonstration was not completed at DCPP until 1999. The DCPP 316(b) demonstration was completed under the direction of the Central Coast Regional Water Quality Control Board (CCRWQCB).

The CCRWQCB assembled a team of scientists, consultants, and industry representatives to assist their staff in the design and implementation of all aspects of the study. This Entrainment Technical Work Group (ETWG) consisted of CCRWQCB staff and their consultants, U.S. Environmental Protection Agency

(EPA) staff, PG&E and their consultants, and a consultant for an intervener group. The ETWG met every 1 to 2 months during the study to review interim reports and to discuss aspects of the study design, implementation, and analytical methods used to assess results. The CCRWQCB also convened two workshops with a larger group of state, federal, and academic fishery experts to discuss assessment approaches with the ETWG.

The ETWG determined that the 316(b) study at DCPP would only address CWIS entrainment effects because previous studies[5] had demonstrated low potential for impingement losses. The ETWG, in consultation with state, federal, and academic fishery experts, determined that using multiple approaches to assess entrainment effects would produce results that could be used to identify whether environmental impacts were adverse for a broad range of target organisms. Convergence of the results of the multiple models would provide a relative measure of confidence in our estimates of effects. However, many of the fish entrained by the DCPP CWIS were small, nearshore species with little or no reported life history information. Thus there was no way to assess impacts for many of the taxa using models that require demographic information (e.g., adult equivalent loss[6]).

In the recent 316(b) demonstration at DCPP, two demographic models, fecundity hindcasting (FH: Alec MacCall, NOAA/NMFS, Tiburon Laboratory, personal communication; [7]) and adult equivalent loss (AEL[6,8]), were used to analyze impacts on adult populations where life history information was available. A third approach, the empirical transport model (ETM[9,10]) was used on all target organisms.

Similar to the Hudson River case[1], the DCPP 316(b) was settled before a site-specific definition of AEI could be determined. Despite this, we remain hopeful that the approach we employed at DCPP could have yielded at least a site-specific definition. By combining the three assessment approaches with ancillary local adult abundance information and harvest data, we began to converge on estimates of losses due to entrainment. The next logical step would have been to determine if these losses represented an AEI.

METHODS

Site Description

The DCPP is a 2,200-MW, two-unit, nuclear-powered, steam-turbine plant owned and operated by PG&E. Units 1 and 2 began commercial operation in May 1985 and March 1986, respectively. Diablo Canyon is located on a coastal terrace about midway between the communities of Morro Bay and Avila Beach in San Luis Obispo County on the central coast of California (Fig. 1).

The plant's cooling-water intake is a shoreline structure consisting of bar racks, vertical traveling screens, auxiliary cooling-water systems, and four main circulating water pumps (Fig. 2). There are seven vertical traveling screens per unit that are designed to trap and remove debris that passes through the bar racks.

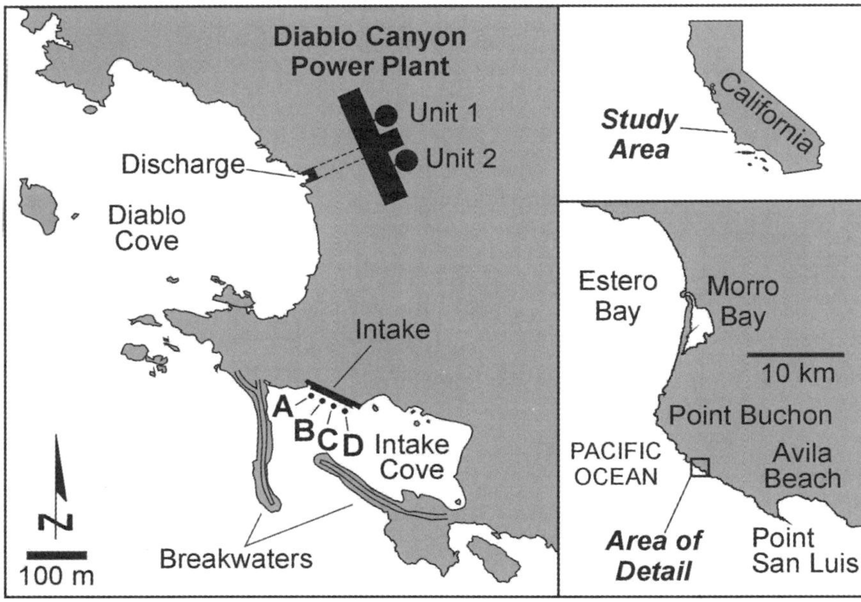

FIGURE 1. Diablo Canyon Power Plant, intake cove, and entrainment sampling locations (A, B, C, and D).

The screens extend from the upper deck of the intake structure to the bottom of the intake cove at a depth of approximately 10 m below sea level. The traveling screen baskets are covered with 0.95-cm mesh designed to prevent material from entering the conduits and clogging the 2.5-cm diameter condenser tubes.

The manufacturer's rated average flow rate for each of the four cooling-water pumps (CWP) at DCPP is 1,641 m^3/min (433,500 gallons/min)[5]. The total daily intake volume is 9.45 million m^3/day (2.5 billion gallons/day) when all four CWPs (two per unit) are operating. The combined flow rate of the two pumps that feed seawater to the auxiliary plant systems is 240,000 m^3/day (63.4 million gallons/day). The cooling-water volume withdrawn can vary daily due to changes in tidal and swell height as well as resistance caused by occlusion of the bar racks, traveling screens, or condenser tubes.

Sampling and Processing Methods

Weekly entrainment samples were collected from a survey vessel between October 1996 and June 1999 at four permanent sampling stations (Fig. 1). Entrainment sampling took place over one 24-h period each week, with each sampling period divided into eight 3-h cycles. The four stations were sampled in random order during each cycle. Samples were collected from a boat moored to buoys located approximately 10 m from the intake and used to mark the permanent stations (Fig. 2). A 0.71-m diameter standard CalCOFI (California Cooperative Oceanic

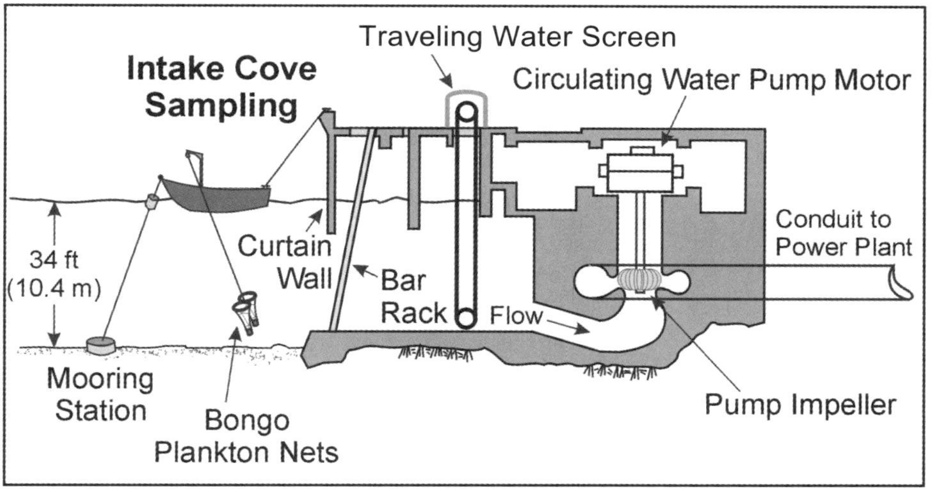

FIGURE 2. Cross-section view of the DCPP intake structure illustrating the location of the sampling boat and bongo nets.

Fisheries Investigation) style bongo frame[11,12] with two 1.8-m long nets was used for these collections. Each net mouth was fitted with a calibrated flow meter to measure the volume of water filtered. The majority of samples collected during this study employed 335 μm Nitex™ mesh nets. To achieve an adequate volume filtered, the frame and nets were fished from the surface to the bottom and then back to the surface a maximum of eight times. The net was turned at the surface and within ca. 10–30 cm of the bottom. The sinking speed of the net (0.3–0.45 m/s) was determined primarily by gravity and drag resistance on the frame and nets, while the retrieval speed (0.3 m/s) was controlled by an electric winch. The material collected in each net for all the samples collected during this study were preserved separately in either 5% buffered formalin or 70–80% ethanol. Formalin preserved samples were transferred to ethanol before laboratory processing. A total of eight subsamples (four samples) were collected per 3-h cycle for a total of 64 subsamples (32 samples) during each 24-h sampling period.

Calculation of proportional entrainment for the ETM requires an estimate of larval abundance in the source population. A survey grid centered on the DCPP intake cove was established and sampled to characterize larval abundance in the source water body (Fig. 3). The grid consisted of 64 cells set up in a symmetric eight-by eight-cell pattern. The grid extended along the coastline approximately 14 km and offshore about 3 km. The boundaries of the grid were Point Buchon to the north and Point San Luis to the south. Most areas inshore of the grid were too shallow to safely conduct boat operations in and were not sampled.

The study grid was sampled monthly from July 1997 through June 1999. Each of the 72-h study grid surveys was scheduled to bracket a 24-h entrainment survey, overlapping the day before and the day after entrainment sample collection. This

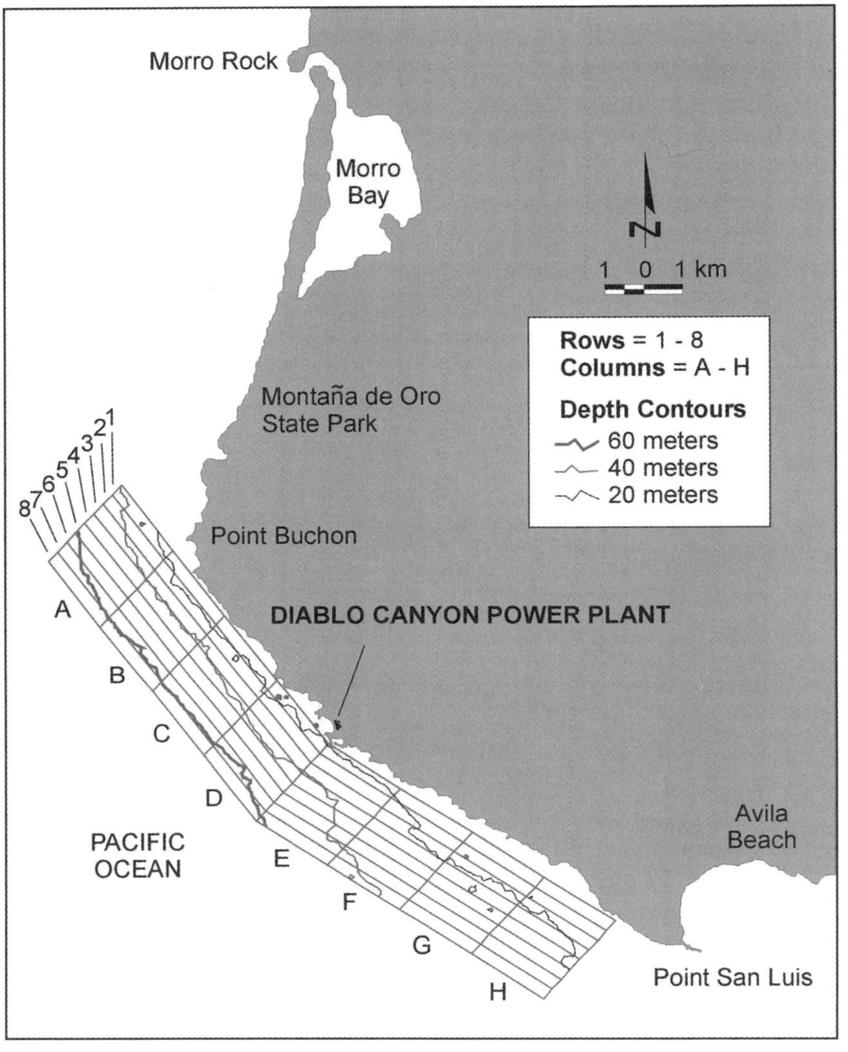

FIGURE 3. DCPP study grid and depth contours.

was done to minimize temporal variation between the entrainment and study grid sampling. During each grid survey two randomly selected locations within each cell were sampled with a bongo frame using two 3.3-m long, 335-μm mesh nets. The nets were fished through the water column in an oblique manner following CalCOFI protocol[12]. The nets were lowered through the water column to within approximately 3 m of the bottom and then retrieved to the surface. Net speed through the water column was similar to that used for the entrainment sampling. A calibrated flowmeter in each net mouth measured the volume of water filtered.

In addition to the entrainment and source water samples collected during this study, data for comparison were also available from a long-term plankton sampling study conducted from 1990–1998 in the DCPP intake cove. Three samples were collected near the surface at dawn once per week by towing a 0.5-m diameter, 335-μm mesh net from the intake structure to approximately the outer end of the west breakwater. A calibrated flowmeter in the net mouth measured the volume of water filtered by the net.

Fourteen larval fish taxa and megalopal stages of all species of *Cancer* crabs were the organisms chosen by the ETWG for assessment based on ten criteria[4]. Laboratory processing consisted of removing all larval fish and *Cancer* spp. megalopae from the entrainment subsamples and from the formalin preserved grid subsamples (two per cell). A quality control program verified the removal of the target organisms from the processed samples. Larval fish and crab megalopae were identified to the lowest possible taxonomic level. A quality control program verified the identification of the larvae and megalopae. Some of the larval fish could only be identified to the familial or generic level, due to the fact that the larval stages of many fish are poorly known or undescribed.

Notochord length of most individuals of the target fish taxa was measured in the laboratory using a computer imaging system. The average length of each fish taxon per entrainment survey was used with information on larval growth found in the scientific literature to estimate the number of days the larvae had been in the plankton before being entrained.

Ancillary Field Studies

Adult and juvenile fish populations were counted along permanent benthic subtidal transects in the vicinity of DCPP as part of the plant's Receiving Water Monitoring Program[13]. All fish observed by SCUBA divers within 2 m of either side and 1 m above the 50-m-long transect line were identified and logged onto datasheets. Two divers swam each transect but from opposite directions, with all fish being identified to the species level whenever possible. The resulting survey data were the combined species counts for both divers, divided by two, yielding an average count per 50-m transect. One area sampled by this method was located approximately 700 m to the south of the intake cove, in an area not influenced by the plant's thermal plume. The three transects in this control area range from approximately 3–10 m in depth.

Analytical Methods

The density of the fish in the entrainment samples was used to estimate the total annual entrainment of each larval taxon (E_T). A daily entrainment estimate (number of organisms/m^3) and its variance were calculated for each 24-h entrainment survey[7]. An estimate of the number entrained during the survey was determined by multiplying the density of each taxon by the intake water flow measured during the survey. A 100% mortality was assumed for all entrained organisms. Entrainment estimates for the period between surveys (usually 7 days) were determined by summing the product of the entrainment estimate and the daily intake volumes for the survey period. These estimates and their associated variances were then summed to obtain estimates of annual entrainment and variance using the following formulae:

$$E_T = \sum_{i=1}^{52}\left(\frac{V_i}{v_i}\right)\cdot E_i$$

and

$$Var(E_T) = \sum_{i=1}^{52}\left(\frac{V_i}{v_i}\right)^2 \cdot Var(E_i)$$

where V_i is the intake volume on the survey day of the ith survey period (i = 1,...,52), V_i is the total intake volume for the ith survey period (i = 1,...,52), and E_i is the estimate of daily entrainment during the entrainment survey of the ith survey period.

The estimate of annual entrainment at DCPP was adjusted to better represent long-term trends for each taxon by using the longer-term intake cove plankton tow data set. These data were used to provide an index of annual trends in larval abundance for the period 1990–1998. The estimated total annual entrainment was multiplied by the quotient of the average index value from the intake cove plankton tows (1990–1998) and the index value from the surface tows during the ith year, thus adjusting annual DCPP entrainment by the annualized long-term average.

$$E_{adj-T} = E_T\cdot\left(\frac{\bar{I}}{I_i}\right)$$

and

$$Var(E_{Adj-T}) = Var(E_T)\cdot\left(\frac{\bar{I}}{I_i}\right)^2$$

where E_{Adj-T} is the adjusted estimate of total annual entrainment adjusted to the long-term average for 1990–1998, I_i is the index value from intake cove plankton tows in the ith year, and $\bar{I}$ is the average index value from intake cove surface tows, 1990–1998. The variance of E_{adj-T} does not include the between-day, within-stratum variance, interannual variance, nor the variance associated with the indices used in the adjustment. So the actual variance is higher than what would be calculated by the above formula.

The fecundity hindcast (FH) model estimates the amount of potential female reproductive output eliminated using entrainment losses combined with estimates of female fecundity and demography (A. MacCall, NOAA/NMFS, personal communication; [7]). The number of larvae entrained by DCPP was used, along with mortality schedules from the egg stage up to the age at entrainment, to hindcast the number of females whose reproductive output could have been effectively removed from the population. This method has the advantage of needing to estimate survivorship over only a relatively short time period (i.e., egg to age at entrainment). To be extrapolated to adult losses, however, FH does require age-specific mortality rates and total lifetime fecundities that are largely unreported for species affected at DCPP. In addition, adult population estimates, typically unavailable for unfished taxa, are required to estimate population-level effects.

Estimates of the annual rate of entrainment for larval fish and subsequent FH and AEL calculations were determined for the following two analysis periods:

Period 1 – October 1996 through September 1997,
Period 2 – October 1997 through September 1998.

The plankton samples collected at the surface in the DCPP intake cove were analyzed only for the months of December through June, as this was the peak period of larval fish abundance for most of the species in this area. These data were used to estimate the long-term average abundance of each taxon that was then used to adjust the estimated annual number of larvae entrained.
The estimated total annual entrainment of each taxon (E_T) was used to estimate the number of breeding females whose fecundity was potentially lost using the following formula:

$$FH = \frac{1}{F_T} \sum_{i=1}^{w} \frac{E_{Ti}}{S_i}$$

where F_T is the average total lifetime fecundity for females, equivalent to the average number of eggs spawned per female over their reproductive years, w is the number of weeks the larvae are vulnerable to entrainment, E_{Ti} is the estimated total entrainment for the ith weekly survey period ($i = 1,\ldots,w$), and S_i is the survival rate from the fertilized eggs to larvae of the stage present in the ith weekly survey period.

This equation was based on the simple case of a single synchronized spawning for a given taxon. For most taxa with overlapping or continuous spawning, larval abundance would have to be specified by week and age class. At DCPP, we used the mean size of the larvae entrained to estimate a representative larval age using daily growth rates, and then estimated a survival rate to that age. The age of the average-sized larvae in the entrainment samples was determined from length measurements and growth rates available from the scientific literature.

Assuming average rates of survival were the same between years, the adjusted annual entrainment (*Eadj–T*) was used in the FH approach, using the following formula:

$$FH = \frac{E_{adj-T}}{\prod_{j=1}^{n} S_j \cdot F_T}$$

where S_j is the age specific survival of eggs and larvae for the jth age class (j = 1,…,n), and F_T is the expected number of eggs produced in a reproductive lifetime.

The expected total lifetime fecundity was approximated by the equation:

$$Fr = (\text{average eggs/year}) \cdot (\text{average number of reproductive years}).$$

The midpoint between the ages of maturation and longevity was used as the average number of reproductive years. This was based on an assumption of linear survivorship (uniform survival) between the ages of maturation and longevity. It was assumed that for exploited species, such as northern anchovy and Pacific sardine, the expected number of years of reproductive life could be less, so the estimated longevity was based on the oldest individuals caught in the fishery.
The variance of FH was approximated using the Delta method[14] in the following formula:

$$Var(FH) = FH^2\left(CV^2(E_{adj-T}) + \sum_{j=1}^{n} CV^2(S_j) + CV^2(F) + \left(\frac{Var(A_L)+Var(A_M)}{(A_L - A_M)^2}\right)\right)$$

where $CV(E_{adj-T})$ is the coefficient of variation of the adjusted entrainment estimate, $CV(S_j)$ is the CV of the estimated survival of eggs and larvae up to entrainment, $CV(F)$ is the CV of the estimated average annual fecundity, A_M is the age at maturation, and A_L is the age at maturity.

The following additional assumptions were made for the calculation of FH at the DCPP:

- Values of parameters from the scientific literature represent the population parameters for the years and location of this study and are constant for the population of inference;
- Reported values of egg mass are lifetime averages to calculate an unbiased estimate of lifetime fecundity;
- Reproductive life expectancy can be accurately calculated by assuming that time of death is uniformly distributed between age at maturation and age of longevity;
- Egg and larval survival rates are constant over time;
- No population reserve or compensation counters the entrainment mortality;
- The loss of the reproductive potential of one female is equivalent to the loss of an adult female; and
- A CV of 30% was assumed when no estimates of variance were available from the literature.

The AEL model estimates the loss of an equivalent number of adults (male, female, or both) based on the estimated number of entrained larvae and species-specific mortality schedules[6,8]. Survival estimates from the age of entrainment to adulthood are required for these calculations. These age-specific survivorship rates are generally not well known, except for the adults of some commercial species. For species where age-specific survival rates from larvae to adults have been estimated, AEL was calculated based on the average age of the larvae entrained. This age was determined as described for FH.

To calculate two annual estimates of larval mortality from the ETM, the monthly grid and the paired entrainment surveys were divided into the following two analysis periods:

Period 3 – July 1997 through June 1998,
Period 4 – July 1998 through June 1999.

Survivorship to adulthood (recruitment) was separated into several age stages, and AEL was calculated using the entrainment estimates adjusted to the long-term average using the following formula:

$$AEL = E_{adj-T} \prod_{j=1}^{n} S_j$$

where n is the number of age classes from entrainment to recruitment, and S_j is the survival rate from the beginning to end of the jth age class.

The variance of AEL was estimated using a Taylor series approximation (Delta method[13]) as follows:

$$Var(AEL) = AEL^2 \left(CV^2(E_{adj-T}) + \sum_{j=1}^{n} CV^2(S_j) \right)$$

In cases where survival estimates from larval entrainment to adulthood were unavailable, the fecundity hindcasting estimates could be generated as $AEL \equiv 2FH$. This treatment assumes that two animals would survive to the age to generate the average number of eggs produced in a lifespan, calculated as follows:

$$S_{\text{larvae to adult}} = \frac{2}{(S_{\text{egg + larvae}}) \cdot (\text{Fecundity}_{\text{lifetime}})}$$

where both AEL and FH can be calculated independently they offer an indication of the confidence in the accuracy of the estimate.

The following assumptions were made for the calculation of AEL:

- Literature-based life history parameters represent the fish populations during the years and at the location of the DCPP study;
- If survivorship values from the literature are limited to a single observation,

they are assumed constant over time or representative of the mean;
- Survival rates used in the calculation represent the life stages of fish in the DCPP area;
- No population reserve or compensation counters the entrainment mortality; and
- A CV of 30% was assumed when no estimates of variance were available from the literature.

In some instances, survival rates were not available for the individual target species, but values for similar species were found. In these instances, an additional assumption was made for both FH and AEL:

- survival values for both species were the same.

The ETM was used to generate an estimate of the probability of larval mortality caused by entrainment (P_M). This model uses an estimate of the daily entrainment mortality (proportional entrainment, or PE) for each taxon based on each monthly survey. Such mortality has been referred to as conditional mortality[15]. Conditional mortality was calculated by compounding daily survival for the estimated duration that larvae would be susceptible to entrainment. The adjusted entrainment values used in the FH and AEL models were not used in the ETM results because this calculation relies on a PE ratio that uses larval abundance values from the paired entrainment and study grid surveys.

The general equation to estimate the ith day's PE values is:

$$PE_i = \frac{N_{Ei}}{N_{Gi}}$$

where N_E is an estimate of the number of larvae entrained and N_G is the estimate of the number of larvae in the study grid. To estimate the PE values, a daily entrainment estimate was paired with a corresponding estimate for the study grid survey collected over 72 h. N_G was calculated using the following formula:

$$N_{Gi} = \sum_{k=1}^{64} A_{Gk} \cdot D_k \rho_{ik}$$

where A_{Gk} is the area of grid cell k, D_k is the average depth of the kth grid cell, and ρ_{ik} is the density (#/m^3) of larvae in the kth grid cell during survey i.

The area inshore of study grid row 1 was too shallow to safely collect samples (Fig. 3). Since adults of many of the taxa entrained in high numbers at DCPP were likely to reside in these areas, we developed a method to include the unsampled areas in the estimates of PE[7]. The volumes of inshore areas were estimated and multiplied by the larval density in the adjacent cell to yield an estimated number of larvae in the unsampled area. The exceptions to this adjustment were cells A1, D1, and E1. Cell A1 was further offshore than the other row 1 cells due to a bend in the coastline at Point Buchon, so no adjustment was made for this cell. Cells D1 and E1 were directly off of the DCPP intake cove, so the ETWG decided that the number of larvae in the area between the grid and the intake structure would be best represented by the entrained density of each larval taxon.

The boundaries of each taxon's population could range from local (a portion of the grid) to regional (i.e., fishery management units). Boreman et al.[10] point out that if any members of the population were located outside of the area studied (the study grid at DCPP), then the ETM would overestimate the conditional[15] entrainment mortality for the entire population. The fraction of the larvae being entrained from the population of inference on a given day is then the product

$$(PE_i)\cdot(P_S)$$

where

$$P_S = \frac{N_G}{N_P};$$

the proportion of the larval population of inference (N_P) that is represented by the larval population within the study grid (N_G). The "proportion of the parental stock"[15], or P_S, can also be calculated using an estimate of the adult population in the study area. Assuming that the distribution in the larger area is uniform, the value of P_S could be approximated as a ratio based on the size of the two areas. At DCPP, P_S was estimated using the distance the larvae could have traveled based on the number of days it was subject to entrainment and the current velocities and patterns measured during that period. Measurements were collected at a single current meter suspended at a depth of ca. 6 m, approximately 1 km from shore. For taxa dispersed throughout the grid, both alongshore and onshore current was used in P_S calculations as

$$P_S = \frac{L_G}{L_P}$$

where A_G is the area of the grid and A_P is the area of the population calculated from the alongshore and onshore current excursions. For taxa whose larvae were concentrated in the nearshore portions of the grid, P_S was calculated as

$$P_m = 1 - \sum_{i=1}^{12} f_i \cdot \left(1 - PE_i \cdot P_s\right)^{\text{days}}$$

where L_G is the length of the grid and L_P is the estimated alongshore current movement through the grid which estimates the population at risk.

The daily conditional survival is the value 1 - PE_i. An estimate of the larval population surviving entrainment during the ith survey period was generated by applying the number of days the larvae are subject to entrainment ($[1 - PE_i]^{days}$). In an attempt to provide a relevant range of survivorship estimates, the number of days that the larvae were subject to entrainment was calculated using both the average and maximum larval ages at entrainment. This provided both an average and minimum (maximum exposure to entrainment mortality) estimate of survivorship.

The monthly estimates of PE were weighted by the monthly survey fraction (f_i) of the source water population at risk. This was obtained from the monthly fraction of the total annual entrainment for the source water survey periods. The weighted estimates of survivorship for each survey period was then summed to provide a final estimate of P_M using the following formula:

$$P_m = 1 - \sum_{i=1}^{12} f_i \cdot \left(1 - PE_i \cdot P_s\right)^{\text{days}}$$

The following assumptions were made in the P_m estimations:

- Larval lengths and growth rates accurately estimate larval duration for the taxa studied;
- The estimates of conditional PE are constant within monthly survey periods;
- The monthly estimates of larval abundance represent a proportion of total annual larval production during that month; and
- P_S accurately characterizes the fraction of the population of inference represented by the sampling grid.

Our intent in using three approaches to estimate the effects of larval entrainment at DCPP (i.e., FH, AEL, and ETM) was to provide several methods for determining the magnitude and quality of resulting population level impacts and as an aid to determining what constituted an AEI. While it is true that none of the three approaches is completely independent of the others, their combination still allowed us to estimate possible effects using three different methods of calculation.

RESULTS AND DISCUSSION

There were 169,440 larval fish identified and enumerated from the processed samples (Table 1). They represented a total of 193 different taxonomic categories, ranging from the ordinal (6 taxa), family (28 taxa), genus (30 taxa), and species level (129 species). We also had a category for unidentifiable or damaged larvae and also larval fragments. From the different categories, the ETWG chose 14 fish taxa for detailed assessment using FH, AEL, and ETM.

TABLE 1

Collection Period, Number of Subsamples Laboratory Processed, and Number of Larval Fish Found During the DCPP 316(b) Demonstration

	Sample Collection Dates	# Subsamples Processed	# Larval Fish in Subsamples Processed
Entrainment samples	Oct. 1996–June 1999	4,693	98,593
Study grid samples	July 1997–June 1999	3,163	43,785
Intake cove surface tows	1990–1998	660	27,062

We present results for two of these taxa as a demonstration of our assessment approach using three models. Our first example is a grouping of rockfish that we nominally refer to as the kelp, gopher, and black-and-yellow rockfish (KGB) complex, and our second example is a grouping of clinid kelpfish. These two were selected for presentation here due to their high abundance in entrainment samples and because they represented varying levels of available life history information. A more detailed presentation of the results of these and the other 12 taxa can be found in the final DCPP 316(b) demonstration report[4].

KGB Rockfish Complex

Rockfish (*Sebastes* spp.) comprise a large marine commercial and recreational fishery along the California coast and are caught from nearshore coastal habitats out onto the continental shelf and slope. Lea et al.[16] report that there are 59 species of *Sebastes* in the coastal waters of California. Although *Sebastes* are an economically important genus, larval, juvenile, and adult life history parameters are not well known for many of the species in the group.

Larval *Sebastes* are very difficult to visually identify to the species level[17,18,19,20,21,22]. Perhaps 5 or 6 of the 59 rockfish species expected to occur in the vicinity of DCPP can be identified at the early larval stage to the species level[22]: aurora rockfish (*S. aurora*), shortbelly rockfish (*S. jordani*), cowcod (*S. levis*), blue rockfish (*S. mystinus*), bocaccio (*S. paucispinis*), and stripetail rockfish (*S. saxicola*). We placed the other larval *Sebastes* into one of eight broad subgeneric groupings based on larval pigment patterns[22,23]. The most abundant *Sebastes* pigment group collected in the DCPP plankton samples was the nominal KGB complex. Based on available descriptions of larvae from identified females, species in the KGB complex (Table 2) have a common pigment pattern that distinguish them from the other larval rockfish occurring in the DCPP vicinity. Genetic analysis of a subset of larvae verified the visual identification of the KGB complex in the DCPP samples[24].

Age at maturation is approximately 5 years, and longevity is about 15 years for the species in the KGB complex[16,25,26,27, R. Larson, San Francisco State Uni-

TABLE 2

Larval *Sebastes* Species Assigned to the KGB Complex

Sebastes atrovirens	Kelp
S. auriculatus	Brown
S. carnatus	Gopher
S. caurinus	Copper
S. chrysomelas	Black-and-yellow
S. dalli	Calico
S. maliger	Quillback
S. nebulosus	China
S. rastrelliger	Grass
S. semicinctus	Halfbanded

versity, personal communication]. KGB rockfish are generally thought to spawn once per year, with an estimated average annual fecundity of 213,158 eggs per female[28,29,30]. Female rockfish are viviparous with internal fertilization[31] and internal development of the larvae[27]. Newly released larval *Sebastes* can reside in the plankton for a period of 1 to 3 months[32,33,34].

Presence of KGB larvae in our samples was seasonal (Fig. 4a). Using estimates of weekly entrainment densities, the estimated numbers of KGB rockfish complex larvae entrained annually for the two periods, adjusted to the long-term average intake cove surface plankton tow index, were

October 1996 through September 1997 – 275,000,000 (SE = 24,700,000) larvae,

and

October 1997 through September 1998 – 222,000,000 (SE = 28,900,000) larvae.

The FH calculations require estimates of the mortality rate and age at entrainment in addition to the estimated number of larvae entrained. The only mortality rate estimate available for very young larval rockfish is 0.14/day for blue rockfish (M. Yoklavich, NOAA/NMFS/PFEG, unpublished data). Despite the fact that blue rockfish are not included in the KGB complex, this value was presumed to be representative of the genus and used in FH calculations. It was estimated that the average age of entrained KGB complex larvae at DCPP was 6.2 days based on the mean length of the larvae in this group (4.2 mm) and an estimate of the daily larval growth rate from brown rockfish of

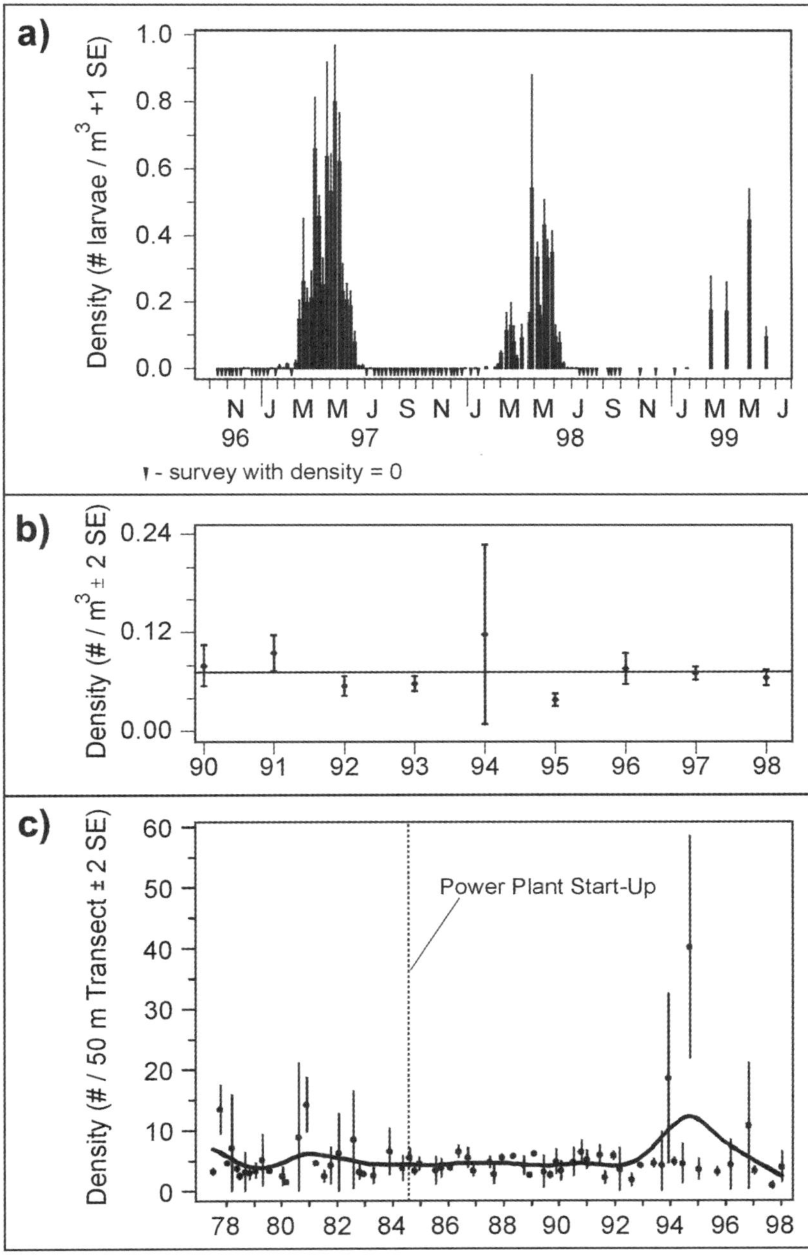

FIGURE 4. **(a)** Weekly mean density of larval KGB rockfish (#/m^3 + 1SE) at the DCPP intake. **(b)** Annual mean density ± 2 SE of larval KGB rockfish (vertical lines) and grand mean density for all years combined (horizontal line) for the intake cove surface plankton tows. **(c)** Mean density of juvenile and adult KGB rockfish (#/50 m transect ± 2 SE) estimated from SCUBA surveys in an area 700 m south of the DCPP intake cove. Spline smoothing algorithm used to draw the curve through the points.

0.14 mm/day[30,31]. Using these values in FH calculations, the estimated number of adult female KGB rockfish whose reproductive output was potentially lost due to larval entrainment was 617 adult females for the period 1996–1997 and 497 adult females for the 1997–1998 period.

The AEL model requires survivorship estimates from the time of larval entrainment through adulthood. No estimates of KGB complex larval, juvenile, or adult survivorship were available, but survivorship for these life stages of blue rockfish had been described[22]. Early blue rockfish mortality estimates through year one were provided by M. Yoklavich (NOAA/NMFS/PFEG, Pacific Grove, CA., personal communication) and annual instantaneous mortality was assumed as 0.2/year after 1 year (Table 3). Using these survival values, the estimated number of adult equivalents (male and female) lost due to entrainment and based on the adjusted annual larval entrainment was 1,120 for the 1996–1997 period and 905 for the 1997–1998 period.

The monthly PE estimates used in calculating ETM for KGB larvae ranged from 0 to a maximum of 0.587 ± 0.297 (± 1 SE (PE)) for the 2 years studied. The highest value was calculated for March 1998, a period of peak parturition for many species in the KGB complex[33]. Due to the wide distribution of the KGB larvae throughout the grid, P_S and P_M were calculated using both alongshore and onshore current movements as well as average maximum estimates of larval duration. The values of P_M varied from a low of 0.005 to a maximum of 0.05 depending on larval duration and current speed and direction.

Additional larval and adult abundance information collected in the vicinity of

TABLE 3

3-Year Survival for the KGB Rockfish Complex Based on Blue Rockfish Data

Day (start)	Day (end)	Instantaneous Natural Mortality (Z)	Survival ($\hat{S}$)
0	6.21	0.14	0.419
6.21	20	0.14	0.145
20	60	0.08	0.041
60	180	0.04	0.008
180	365	0.0112	0.126
365	1,095	0.0006	0.670

Note: Survival was estimated from release as $\hat{S} = e^{(-z)\ (Day(end)-Day(start))}$. Daily instantaneous mortality rates (Z) up to 1 year of blue rockfish, S. mystinus, larvae that were used to calculate KGB larval survivorship were provided by M. Yoklavich (NOAA/NMFS/PFEG, Pacific Grove, CA, personal communication). Annual instantaneous mortality was assumed as 0.2/year after 1 year. Average age of entrainment was estimated as 6.21 days based on average size at entrainment and a growth rate of 0.14 mm/day[31].

the DCPP implies a low entrainment impact on KGB rockfish complex larvae. The annual mean density of KGB complex larvae in the DCPP intake cove plankton tows appears similar among years (Fig. 4b). In addition, abundance data from a combination of juvenile and adult KGB rockfish observed by SCUBA divers along permanent transects between 1978 and 1998 in an area 700 m south of the intake cove showed much intra- and interyear variation but no apparent declines in abundance over time (Fig. 4c).

Catch data from the port of Morro Bay (reported in the Pacific States Marine Fishery Council's online Pacific Coast Fisheries Information Network database were also used to provide some context for interpreting results from the three models. KGB rockfish were mainly landed as part of the live-fish fishery, and had an average price per kilogram of $7.65 in 1999 (PacFIN database). Assuming an average weight of 1 kg for a 3-year-old KGB rockfish, 100% catchability of the adult equivalents, and no compensatory mortality, the annual average estimate of 977 KGB rockfish translate to a value of about $7,500. The estimate of P_M from this study for the area fished from Port San Luis (south of DCPP) to Morro Bay was between 4 and 5%. Based on the dollar value for KGB landings at Morro Bay in 1999, the proportional reduction caused by entrainment translated to a value of about $20,000 or about 2,600 1-kg adult rockfish.

The results of the three impact assessment approaches, in conjunction with additional adult abundance data, show that KGB rockfish in the vicinity of DCPP are not adversely impacted by power plant entrainment. The close concurrence of the three model results (i.e., FH - ca. 550 adult females annually; AEL - ca. 1,000 adults annually [500 adult females] worth approximately $7,500; and ETM ca. 5% or $20,000 of the local catch) provides us high confidence in our results and the conclusion that potential impacts are relatively small. Combining these results with the adult fish observations indicating a fairly stable population size confirms the conclusion of no AEI for KGB rockfish.

Clinid Kelpfish

There are four species of adult clinid kelpfish in the DCPP area, three species of *Gibbonsia* and the giant kelpfish *Heterostichus rostratus*. The *Gibbonsia* larvae collected at the DCPP were not identifiable to the species level, so they were analyzed as a group (*Gibbonsia* spp.); *H. rostratus* were uncommon in the samples. Very little information is available about the adult, juvenile, or larval stages of *Gibbonsia* or *Heterostichus*. *G. elegans* was reported to have a fecundity of about 2,300 eggs/female[35]. Fitch and Lavenberg[36] stated that *Gibbonsia* spp. first spawn at 2 years of age, might spawn more than once per year, and have a life expectancy of about 7 years. No survivorship information was available for either genus of kelpfish, so no FH and AEL estimates could be calculated. Daily growth rates of *Gibbonsia* spp. were also unavailable, but estimates for lab-reared larval *H. rostratus*[37] were determined using linear regression as 0.25 mm/day ± 0.013 mm/day (slope ± 1 SE). This growth rate, although not for the correct genus, was substituted for *Gibbonsia* spp. to allow calculation

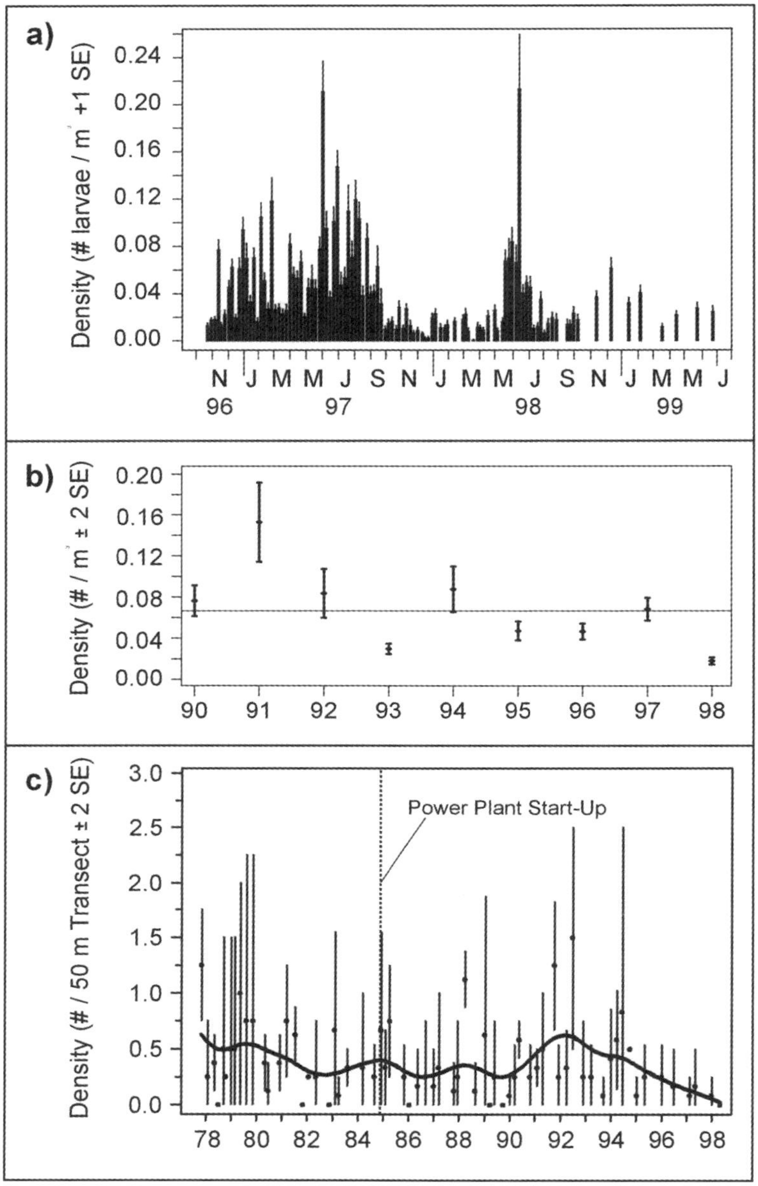

FIGURE 5. (**a**) Weekly mean density of larval kelpfish (#/m^3 + 1SE) at the DCPP intake. (**b**) Annual mean density ± 2 SE of larval kelpfish (vertical lines) and grand mean density for all years combined (horizontal line) for the intake cove surface plankton tows. (**c**) Mean density of kelpfish (#/50 m transect ± 2 SE) estimated from SCUBA surveys in an area 700 m south of the DCPP intake cove. Spline smoothing algorithm used to draw the curve through the points.

of the ETM for kelpfish.
Kelpfish larvae were present throughout the year in entrainment samples (Fig. 5a). Using estimates of weekly entrainment densities, the estimated numbers of larval kelpfish entrained annually for the two periods, after adjustment to the long-term average intake cove surface plankton tow index, were

October 1996 through September 1997 – 181,000,000 (SE = 4,610,000) larvae,

and

October 1997 through September 1998 – 308,000,000 (SE = 15,300,000) larvae.

The monthly PE used in ETM calculations ranged from 0.001 ± 0.002 (± 1 SE (PE)) to a maximum of 0.346 ± 0.189. These larvae were mainly collected in the nearshore area of the grid, and therefore P_S was calculated using only alongshore current movements and not onshore movement as was done for the KGB complex larvae. The values of P_M from both years based on the average larval age at entrainment ranged from 0.294–0.318, and from 0.395–0.410 for the maximum age at entrainment.

Gibbonsia spp. are small and cryptic, not commercially or recreationally sought, and almost nothing is known of their trophic role in the coastal ecosystem where they occur. The calculated P_M values cannot be converted into an estimate of adult equivalent loss because nothing is known about the population size or adult density of kelpfish. Thus, we must turn to other sources of information to determine whether entrainment losses constitute an AEI for this taxon. Data from the intake cove surface plankton tows indicate a decline in larval kelpfish abundance from 1995–1998 (Fig. 5b), and the local adult kelpfish abundance appears to be declining from 1993–1998 (Fig. 5c). These local declines combined with ETM results showing up to a 40% reduction of the larvae from an area ca. six to seven times the length of the study grid indicate that the effects on this taxon could be significant and represent a population decline in the vicinity of DCPP.

CONCLUSION

Three unique assessment models were used to determine the effects of the DCPP cooling-water system on local larval and adult populations. Although AEI was not defined, comparison of the model results in combination with ancillary information on local larval and adult populations of KGB rockfish and clinid kelpfish was helpful in defining the level of impact caused by entrainment at DCPP. The similar results from the three models and stable local populations provide us with high confidence in our determination of no localized impact for this taxa. In the case of clinid kelpfish, withdrawal of about 40% of the available larvae appears to have led to a measurable decrease in the local adult population. It was estimated that the operation of the CWIS at San Onofre Power Plant in California reduced the adult recruitment and adult standing stock in the Southern California Bight by 13% for queenfish and 6% for white croakers[38]. An entrainment rate of 23% by the Wabash River Generating Station was felt to possibly be high enough to impact year-class strength of certain species, yet follow-up studies detected no short-term adverse impacts to the fish community[39].

The DCPP study was unique in having long-term data on abundances of larval and adult fish populations in the vicinity of the plant. The larval data collected from 1990–1998 allowed us to adjust annual entrainment estimates to the long-term average for a species. Entrainment studies are typically done for a period of 1 to 2 years and have an implicit assumption that the data for those years are representative. By adjusting the entrainment estimates to the average larval abundance over a 9-year period, we were able to address the question of sampling in a representative year. The long-term data on adult populations provided context for interpreting the results of our modeling. In the cases of the small, nearshore species that have not been extensively studied, it was the only data available.

Ultimately, the 316(b) demonstration at DCPP did not progress to a formal determination of which effects, if any, could be designated AEIs. Thus, while our approach to defining AEI remains untested, it still shows promise as a way to qualitatively decide if an effect is important and whether it might be considered an adverse environmental effect. To determine this we would have to arbitrarily define a cutoff for AEI (e.g., 40% reduction of larval population) and then combine the interpretation of results from the three approaches as a measure of confidence that the "adverse" effect was either significant or not. If results from the three approaches agreed with each other, then confidence would be high and vice versa. Nevertheless, this definition would likely have been site- or species-specific since much of the context for qualitatively assessing the value of the effects would have to rely on local landings, economics, and population sizes.

ACKNOWLEDGEMENTS

Grateful thanks go to Pacific Gas and Electric Co. for funding this project, and to its project manager, Ms. Anne Jackson, for her support during this entire project. We also thank the members of the ETWG for their technical guidance and support during this study: Michael Thomas (CCRWQCB), Dr. Greg Cailliet (Moss Landing Marine Laboratories), Drs. Allan Stewart-Oaten and Roger Nisbet (University of California at Santa Barbara), Dr. Pete Raimondi (University of California at Santa Cruz), Deborah Johnston (California Department of Fish and Game), and Anne Jackson and Pat Eckhardt (Pacific Gas and Electric Company). Finally, we acknowledge the staff of Tenera Environmental for all of their time and hard work on the field collection and laboratory processing of the samples. We would also like to thank Mary Nishimoto and an anonymous reviewer for comments that assisted the authors in clarification of certain aspects of this document.

REFERENCES

1. Barnthouse, L.W., Klauda, R.J., Vaughan, D.S., and Kendall, R.L. (1988) Science, law, and Hudson River power plants: a case study in environmental impact assessment. *Am. Fish. Soc. Spec. Monogr.* **4**, 346 pp.
2. Yost, T.B. (1988) Science in the courtroom. *Am. Fish. Soc. Spec. Monogr.* **4,** 294–301.
3. Englert, T.L. and Boreman, J. (1988) Historical review of entrainment impact estimates and the factors influencing them. *Am. Fish. Soc. Spec. Monogr.* **4,** 143–151.
4. Tenera Environmental, Inc. (2000) Diablo Canyon Power Plant 316(b) Demonstration Report. Doc. No. E9-055.0. Prepared for Pacific Gas and Electric Co., San Francisco, CA.
5. Pacific Gas and Electric Co. (PG&E). (1988) Diablo Canyon Power Plant. Cooling Water Intake Structure: 316(b) Demonstration. April 28, 1988. Prepared by Tenera Environmental Services. San Francisco, CA.
6. Goodyear, C.P. (1978) Entrainment impact estimates using the equivalent adult approach. U.S. Fish and Wildlife Service. FWS/OBS-78/65. Ann Arbor, MI.
7. Tenera Environmental, Inc. (1998) Diablo Canyon Power Plant 316(b) Demonstration Study: Phase 3-Sampling Plan and Modelling Evaluation. Doc. No. E7-205.10. Prepared for PG&E, San Francisco, CA. 119 p.
8. Horst, T.J. (1975) The assessment of impact due to entrainment of ichthyoplankton. In *Fisheries and Energy Production: A Symposium.* Saila, S.B., Ed. Lexington Books, DC Heath and Company, Lexington, MA. pp. 107–118.
9. Boreman, J., Goodyear, C.P., and Christensen, S.W. (1978) An Empirical Transport Model for Evaluating Entrainment of Aquatic Organism by Power Plants. U.S. Fish and Wildlife Service. FWS/OBS-78/90. Ann Arbor, MI.
10. Boreman, J., Goodyear, C.P., and Christensen, S.W. (1981) An empirical methodology for estimating entrainment losses at power plants sited on estuaries. *Trans. Am. Fish. Soc.* **110,** 253–260.
11. McGowen, J.A. and Brown, D.M. (1966) A new opening-closing paired zooplankton net. *Oceanogr. Res.* **66,** 1–56.
12. Smith, P.E. and Richardson, S.L. (1977) Standard techniques for pelagic fish egg and larva surveys. *FAO Fish. Tech. Paper* **175,** 1–100.
13. Tenera, Inc. (1999) Receiving Water Monitoring Program-1995–1997 Progress Report. PG&E, San Francisco, CA.
14. Seber, G.A.F. (1982) *The Estimation of Animal Abundance and Related Parameters.* McMillan, London. 654 p.
15. Ricker, W.E. (1975) Computation and interpretation of biological statistics of fish populations. *Fish. Res. Board Can. Bull.* **91**. 382 p.
16. Lea, R.N., McAllister, R.D., and VenTresca, D.A. (1999) Biological Aspects of Nearshore Rockfishes of the Genus *Sebastes* from Central California. California Department of Fish and Game. Bulletin 177. 107 p.
17. Moser, H.G., Ahlstrom, E.H., and Sandknop, E.M. (1977) Guide to the Identification of Scorpionfish Larvae (Family Scorpaenidae) in the Eastern Pacific with Comparative Notes on Species of *Sebastes* and *Helicolenus* from Other Oceans. NOAA Technical Report NMFS Circ. 402. 71 p.
18. Moser, H.G. and Ahlstrom, E.H. (1978) Larvae and pelagic juveniles of blackgill rockfish, *Sebastes melanostomus*, taken in midwater trawls off southern California and Baja California. *J. Fish. Res. Board Can.* **35(7),** 981–996.
19. Baruskov, V.V. (1981) A brief review of the subfamily Sebastinae. *J. Ichthyol.* **21,** 1–26.
20. Kendall, Jr., A.W. and Lenarz, W.H. (1987) Status of early life history studies of northeast Pacific

rockfishes. Proc. Int. Rockfish Symp. Univ. Alaska Sea Grant Report 87-2. pp. 99–128.

21. Moreno, G. (1993) Description of Early Larvae of Four Northern California Species of Rockfishes (Scorpaenidae: *Sebastes*) from Rearing Studies. NOAA Technical Report NMFS 116. 18 p.
22. Nishimoto, M.M. (2002) Identification of *Sebastes* larvae, in preparation.
23. Ehrler, C.P., Vetter, R., Nishimoto, M.M., Stannard, J., and Laman, E.A. (2000) A Simple Method of Grouping Preflexion *Sebastes* Larvae on the Basis of Pigment Patterns and its Validation by Molecular Genetics. Poster presented at CalCOFI Conference 2000, Lake Arrowhead, CA.
24. Vetter, R.D. and Stannard, J. (1999) Molecular Identification of Rockfish (*Sebastes* spp.) Larvae. National Marine Fisheries Service, NOAA, Southwest Fishery Science Center. La Jolla, CA.
25. Burge, R.T. and Schultz, S.A. (1973) The Marine Environment in the Vicinity of Diablo Cove with Special Reference to Abalones and Bony Fishes. Marine Research Technical Report No. 19. California Department of Fish and Game. 433 p.
26. Wyllie-Echeverria, T. (1987) Thirty-four species of California rockfishes: maturity and seasonality of reproduction. *Fish. Bull.* **85(2),** 229–250.
27. Bloeser, J.A. (1999) Diminishing Returns: The Status of West Coast Rockfish. Pacific Marine Conservation Council, Astoria, OR. 94 p.
28. DeLacy, A.C., Hitz, C.R., and Dryfoos, R.L. (1964) Maturation, Gestation, and Birth of Rockfish (*Sebastodes*) from Washington and Adjacent Waters. *Wash. Dep. Fish. Fish. Res. Pap.* **2(3),** 51–67.
29. MacGregor, J.S. (1970) Fecundity, Multiple Spawning and Description of Gonads in *Sebastodes*. U.S. Fish and Wildlife Services Special Science Report, Fisheries No. 596. 12 p.
30. Love, M.S. and Johnson, K. (1999) Aspects of the life histories of grass rockfish, *Sebastes rastrelliger*, and brown rockfish, *S. auriculatus*, from southern California. *Fish. Bull.* **97(1),** 100–109.
31. Yoklavich, M.M., Loeb, V.J., Nishimoto, M., and Daly, B. (1996). Nearshore assemblages of larval rockfishes and their physical environment off central California during an extended El Niño event, 1991–1993. *Fish. Bull.* **94,** 766–782.
32. Matarese, A.C., Kendall, Jr., A.W., Blood, D.M., and Vintner, B.M. (1989) Laboratory Guide to Early Life History Stages of Northeast Pacific Fishes. NOAA Technical Report NMFS 80. 652 p.
33. Moser, H.G., Ed. (1996) The Early Stages of Fishes in the California Current Region. California Cooperative Oceanic Fisheries Investigations, Atlas No. 33. National Marine Fisheries Service, La Jolla, CA. 1505 p.
34. Cailliet, G.M., Burton, E., Cope, J., Kerr, L., and Wright, N. (2000) Marine Species Life History, Version 1. Database on compact disc developed at Moss Landing Marine Laboratories for California Department of Fish and Game.
35. Bane, G.W. and Bane, A.W. (1971) *Bay Fishes of Northern California*. Mariscos Publications, Hampton Bays, NY. 143 p.
36. Fitch, J.E. and Lavenberg, R.J. (1975) *Tidepool and Nearshore Fishes of California.* University of California Press, Berkeley. 156 p.
37. Stepien, C.A. (1986) Life history and larval development of the giant kelpfish, *Heterostichus rostratus* Girard, 1854. *Fish. Bull.* **84(4),** 809–826.
38. Murdoch, W.W., Fay, R.C., and Mechalas, B.J. (1989) Final Report of the Marine Review Committee to the California Coastal Commission. MRC Document No. 89-02.
39. Lewis, R. and Seegert, G. (1988) Entrainment and Impingement Studies at Two Power Plants on the Wabash River in Indiana. EPRI Clean Water Act Section 316(b) Technical Workshop (Coolfont). September 1988.

Comparing Clean Water Act Section 316(b) Policy Options

John Kadvany
Environmental Consultant, Policy and Decision Science, 1070 College Avenue, Menlo Park, CA 94025

Received November 16, 2001; Revised February 13, 2002; Accepted February 19, 2002; Published February, 2003

This paper develops a comparative framework for policy proposals involving fish protection and Section 316(b) of the Clean Water Act (CWA). Section 316(b) addresses the impingement and entrainment of fish by cooling-water intake structures used principally by steam electric power plants. The framework is motivated by examining the role of adverse environmental impacts (AEIs) in the context of Section 316(b) decision making. AEI is mentioned in Section 316(b), but not defined. While various AEI options have been proposed over the years, none has been formalized through environmental regulations nor universally accepted. Using a multiple values approach from decision analysis, AEIs are characterized as measurement criteria for ecological impacts. Criteria for evaluating AEI options are identified, including modeling and assessment issues, the characterization of ecological value, regulatory implementation, and the treatment of uncertainty. Motivated by the difficulties in defining AEI once and for all, a framework is introduced to compare options for 316(b) decision making. Three simplified policy options are considered, each with a different implicit or explicit AEI approach: (1) a technology-driven rule based on a strict reading of the 316(b) regulatory text, and for which any impingement and entrainment count as AEI, (2) a complementary, open-ended risk-assessment process for estimating population effects with AEI characterized on a site-specific basis, and (3) an intermediate position based on proxy measures such as specially constructed definitions of littoral zone, sensitive habitat, or water body type. The first two proposals correspond roughly to responses provided, respectively, by the Riverkeeper environmental organization and the Utility Water Act Group to the U.S. Environmental Protection Agency (EPA)'s proposed 316(b) new facilities rule of August 2000; the third example is a simplified form of the EPA's proposed August 2000 new facilities rule itself. The simplified policy positions are compared using the three dimensions of the comparative policy framework: (1) the role of CWA philosophy or vision, such as the use of technology-forcing rules, (2) regulatory policy implementation, and (3) the role for scientific information and the knowledge base. Strengths and weaknesses of all three 316(b) policy approaches are identified. The U.S. EPA's final new facilities rule of November 2001 is briefly characterized using the comparative policy framework and used to further illustrate the approach.

* Corresponding author. Email: kadvany@aol.com.

KEY WORDS: adverse environmental impact, aquatic ecology, Clean Water Act, cooling-water intake systems, decision analysis, entrainment, environmental policy, fish populations, fisheries, impingement, multiple values, power plants, regulatory affairs, risk, Riverkeeper, stakeholders, tradeoffs, U.S. Environmental Protection Agency, Utility Water Act Group

DOMAINS: freshwater systems, marine systems, ecosystems and communities, water science and technology, environmental technology, environmental management and policy, ecosystems management, decision analysis, environmental modeling

CLEAN WATER ACT SECTION 316(b) AND DEFINING ADVERSE ENVIRONMENTAL IMPACT

Section 316(b) of the Clean Water Act (CWA) is a remarkably brief and controversial piece of environmental law regulating cooling-water intake structures (CWIS). A CWIS is the structure used by steam-electric power plants (and some manufacturing facilities) to obtain water from a nearby river, lake, ocean, or estuary to cool purified steamwater that rotates the turbines generating electricity. Recooled steamwater is recirculated, while the cooling water from a "once-through" cooling system is returned to the source water body. Millions or even billions of gallons of water per day may be conveyed through a CWIS in this way. Consequently, adult or juvenile fish, or fish eggs and larvae, may be injured or killed, either by being *impinged* on external screens or other barriers, or by being *entrained* by the cooling water as it passes through the cooling system proper. Actual mortality or injury is not always 100%, and can depend on factors such as temperature gradients, species-specific survival strength, CWIS and cooling system design, intake flow velocities and gradients around the CWIS, types of protective mechanisms, and water body characteristics.[1]

Section 316(b) is intended to protect fish (or fish populations) from the potential hazards created by a CWIS. In full, Section 316(b) states: "Any standard established pursuant to section 301 [regulating effluent limitations] or section 306 [describing effluent performance standards] of this Act and applicable to a point source shall require that the location, design, construction, and capacity of cooling water intake structures reflect the best technology available [BTA] for minimizing adverse environmental impact [AEI]."

Does the key 316(b) term "adverse environmental impact" (AEI) need to be defined explicitly to advance 316(b) policy? The CWA provides little or no direct guidance on what AEI should mean, and no further explicit characterization has been provided to date by the U.S. Environmental Protection agency (EPA). Advocates on different sides of the 316(b) debate hold a variety of views. The electric power industry, at least as represented by its advocacy group the Utility Water Act Group (UWAG), and the Electric Power Research Institute (EPRI), have said that, without a characterization of AEI, it is not possible to know what impact is to be minimized and therefore how to select BTA. They have both also asserted that AEI should at the least only refer to the health and fecundity of fish populations as a whole, and not the acute impacts caused directly by impingement or entrainment. The Riverkeeper environmental organization, which successfully brought suit against the EPA in 1995, after the EPA failed for decades to promulgate 316(b) regulations, takes AEI to include *any* mortality to fish or other aquatic life caused by a CWIS. They also consider "closed-cycle" dry cooling to be BTA.[2] (A "dry" closed-cycle system is like a radiator in which evaporative loss is minimized, while a wet closed-cycle system, such as a cooling pond or open tower, allows evaporative loss. Both systems recirculate cooling water instead of returning it directly to its source. An evaporative system may count as a consumptive water use, unlike a once-through system, and hence may be subject to regulatory constraints governing in-stream flows.)

The Riverkeeper position may be supported by a straightforward (but perhaps overly-literal) reading of the short 316(b) text, notwithstanding the ambiguity associated with AEI and BTA. But a consequence could be the retrofitting of many existing CWISs at high costs (in the tens to hundreds of millions of dollars per plant) for the industry. For new facilities, a strict 316(b) reading could imply using cooling alternatives, such as recycling cooling towers, that have other environmental impacts. UWAG and others contend that the ecological benefits that would be obtained by eliminating an existing CWIS or using closed-cycle cooling are almost always, if not always, minimal or zero. That while thousands or tens to hundreds of thousands of individual fish may be killed yearly by a single CWIS, and even billions or trillions of fish eggs and larvae, such mortality usually has no important impact on a fish population as a whole, especially in the context of commercial and recreational fishing, which has a far greater influence on fish population size and health. Regardless of technology cost, the electric power industry also contrasts the Riverkeeper position with numerous examples of natural resource management (fisheries, forests, and livestock) for which only populations are relevant.

In any case, no existing CWIS has ever been replaced (though many have been modified) because of 316(b). While regulatory motivations are a matter of conjecture, the reasons likely include a combination of the high cost of retrofitting (cooling system design is fundamental to plant efficiency and hence cannot easily be compensated for if changed); the near impossibility of eliminating *entrainment* in once-through systems (by contrast, impingement, while often challenging to control, is more manageable and less costly); and, most of all, the difficulty in establishing that the ecological disturbance change by a CWIS indeed constitutes ecological harm. The Riverkeeper asserts that such large numerical impacts are of obvious ecological significance, while the industry considers that judgment to be a superficial risk perception at odds with fisheries science and the management of fish resources generally. Possible adverse environmental outcomes, Section 316(b)'s "AEI", therefore are part of a complicated decision often involving unique ecological circumstances, a shortage of useful remediation options, and unclear policy about ecological value. The question of AEI is in part a scientific judgment regarding just what happens to fish in the water. It is also a nonscientific value judgment regarding whether an ecological change is to be deemed adverse, while 316(b) language gives no particular guidance about the adverse impacts to be "minimized."

Hence it is reasonable to return to the 316(b) text and ask: "What is an AEI"? Why could the EPA not create a substantive and workable AEI definition (not to mention actual regulations) over the decades since the CWA was enacted and the suit brought by the Riverkeeper? Without providing details of this fascinating and important story in environmental regulation, and with due respect to the Riverkeeper's position that any form of CWIS-induced fish mortality constitutes AEI, the following contrasts should suggest sources for the controversy and the EPA's difficulties in grappling with CWIS decisions:

- Fish are a consumed natural resource *and* an ecological resource to be protected.
- The large numbers often associated with fish impingement and entrainment may be of *prima facie* concern *and* of little evident concern from the perspective of fisheries management, while EPA's regulatory scope does not include fisheries management per se.
- Judgments of ecological quality or health depend on science for their prediction *and* an implicit or explicit social value judgment for their importance.
- Fish are valued differently when they are created by a stocking program in a reservoir, or are essential to the productivity of a precious estuary, or are a nuisance fish to be eliminated, or are so fecund that huge croppings may be accepted with no concern.
- While risk assessment does not distinguish fish mortality caused by fisherman or other predators from mortality caused by a CWIS, the inconsistency, if there is one, of so doing can nonetheless be upheld by a society and codified as law.
- The ecological changes caused by impingement or entrainment can be hard and costly to predict with few or no generalizable indicators across a variety of water body types(e.g. rivers, estuaries, oceans, lakes and reservoirs), and with the causal importance of various anthropogenic and nonanthropogenic influences also being hard to disentangle.

Defining AEI, therefore, has embedded in it several factors contributing to making 316(b) risk controversial and resistant to obvious risk management solutions. It is an important example because it primarily involves ecological and not human health risk where such controversies most often arise.

UWAG has recently proposed an AEI definition, but it amounts to site-specific risk assessment of fish population impacts due to an existing or future CWIS.[3] The definition is broad enough to include any possible stakeholder concern, but provides no specific criteria for levels of acceptable population decrease; that choice would be part of a site-specific risk and values assessment. It is a reasonable guess that UWAG arrived at their proposal for site-specific risk assessment after recognizing that useful and meaningful general rules (e.g., "x% reduction in regional population size" or "probability of at least p that population falls below critical level *x* for *n* years") for characterizing adverse ecological impacts are almost impossible to define in the CWIS context. The industry, seeing also that dozens of 316(b) decisions have been made over the years through the National Pollutant Discharge Elimination System (NPDES) permitting process *without* detailed regulatory guidance, also feel confident proposing a more flexible process in which AEI is effectively defined locally by regulators and stakeholders. The only real constraint on that would be limiting ecological impacts to the population level (except for species covered by the Endangered Species Act).

Thus the Riverkeeper and UWAG have staked out polar positions on AEI, with the EPA still not opting to define AEI. Instead, the EPA proposed in August 2000 to regulate new (vs. existing) CWISs by water body type (estuary, river, marine, lake) along with a set of supporting proxy measures, such as location of CWIS relative to a defined "littoral zone," and CWIS intake flow velocity. These proxy measures are intended, one presumes, to protect aquatic life from AEIs, though the latter is again not characterized. A rough summary of the positions staked out by the Riverkeeper and UWAG in response to the EPA's August 2000 proposed new facilities rule is contained in the Appendix, and similar issues may be raised through debate over existing facilities rules. For the purposes of this paper, a policy goal is assumed of some intermediate position between an open-ended site-specific risk assessment and stakeholder process, like UWAG's, and the Riverkeeper's strict reading of the brief 316(b) text. Just what to take and reject from these is part of the policy decision faced by the EPA. This policy goal is assumed here as a means of exploring the main proposals put forward, and because the 316(b) policies finally adopted by the EPA may be the result of multiple compromises or policy tradeoffs.

To compare 316(b) policy options, the paper begins by asking whether defining AEI is a good starting point for organizing the environmental, stakeholder, scientific, and regulatory issues involved in articulating a coherent 316(b) policy. The answers provided are Yes and No. Yes, because the problem of characterizing AEI leads to a broader set issues which should be addressed by any defensible 316(b) policy; No, because these issues cannot be answered only through an *a priori* or general AEI definition. AEI is important because it stands for environmental and stakeholder consequences or outcomes generally; thus it is a necessary piece of environmental regulation. However, it is not sufficient, and much of the 316(b) controversy can be understood by looking for other institutional, regulatory, and judgmental factors underlying 316(b) policy design. It is important to understand why this approach is being taken. From a broad values-based stakeholder perspective, in which 316(b) decisions incorporate whatever ecological, social, and financial are considered relevant, AEI should just reflect appropriate local or societal value judgments about impacts on biological health. Indeed, the position of many in the electric power industry has been close to simply taking 316(b) to be a site-specific risk assessment and decision process that allows just that. What should be made of opposing views to what is arguably a widely held approach to environmental decisions? What are rationales, if any, for their alternative views?

The goal here is not to defend any particular view, but to understand the logic behind some complicated combinations of policy choices.

This paper's approach, therefore, is to show how evaluating pros and cons of AEI options leads to other important 316(b) policy choices for regulators and stakeholders. These additional questions about 316(b) policy arise regardless of whether a definition of AEI is central to, or explicit in, a 316(b) policy proposal. The questions define the policy space, so to speak, in which various 316(b) policies can be defined. EPA's first attempt, in 2000, to forge 316(b) policy for new CWISs through synthetic proxy definitions can be seen as a classic institutional response to the messy but tractable reality of an environmental decision burdened by poor regulatory history, considerable stakeholder interests, and unclear scientific and social directives. The Riverkeeper's and UWAG's options are characterized as strong on some dimensions, but weak on others. Different options that build on these three positions are possible. The reader can decide which, if any, are desirable, including the EPA's November 2001 final new facilities rule[4]. In any case, one goal of this paper is to show that comparing existing 316(b) policy options, and defining new ones, can be simplified and made considerably more transparent. By starting with the problem of defining AEI, challenges for risk assessment and public policy are raised by two roles for the Clean Water Act: as enabling legislation for 316(b) regulation and as law that has to be practically implemented with respect to substantive but imperfect science and competing stakeholder values.

The next section first locates AEI in the context of 316(b) decision making. That perspective will suggest other policy questions whose answers help evaluate the merits of AEI proposals. Tradeoffs in satisfying all 316(b) policy needs are raised, and these tradeoffs are used to define comparisons for 316(b) policy options.

WHAT IS AN AEI?

What is an AEI generically thought of as a component of environmental decision making? That is, what role does AEI play in actual 316(b) choices, analogous to other environmental decisions? The answer is that an AEI is a measure or criterion for regulators and stakeholders to evaluate the ecological or other benefits and costs of making 316(b) choices. Examples may include:

- Any acute mortality to adult fish, larvae, or eggs; similar acute mortality, but above some fraction of the total species population size, or above some absolute number;
- Acute mortality to adult fish only;
- A decrease in fish population size threatening its long-term local or regional viability, but not acute impacts per se;
- A probability of fish population decline greater than some critical value;
- A similar probability of decline for multiple species;

- Any of the previous, but ignoring invasive or nuisance species;
- Estimated economic impacts to commercial fisheries and recreational losses;
- Ecological productivity losses to aquatic populations beyond those immediately affected by the CWIS.

316(b) *choices* here means selecting among impingement or entrainment reduction technologies or approaches of any kind, including barrier or diversion devices, CWIS operational, location, or design changes, use of dry or wet closed-cycle cooling; and, for completeness, mitigation or enhancement options, such as fish stocking or ecological restoration projects. The latter may be chosen if it is decided that CWIS modification itself has low benefits, but that a compensating action should be taken nonetheless (as may occur when CWIS technology costs are extremely high). AEI measures may be taken to include any required definitions (e.g. fish population geographical ranges, relevant species) involving ecological scale and biological function (e.g., reproductive success, predator-prey relations, energy transfers), and the judgmental, empirical, and mathematical means used to make such definitions operational. AEI for practical purposes includes *how* AEI is determined as well as *what* it is or should be.

This characterization of AEI follows from identifying generic categories of stakeholder *objectives and values* associated with 316(b) decisions as a whole. These stakeholder values include:

- The ecological consequences to fish and other aquatic (or even terrestrial[4]) species, at any level of ecological scale;
- The direct capital and operating marginal cost of the technology choice (e.g., barrier installation and maintenance, flow reductions, intake relocation, etc.);
- Energy production changes associated with a 316(b) choice, e.g. comparative efficiency losses due to cooling towers;
- Economic impacts on relevant commercial fisheries;
- Changes to recreational fishing, possibly including a mix of noneconomic and economic factors;
- Land use or aesthetic issues associated with the use of cooling ponds or towers;
- Ecological changes associated with mitigation options possibly chosen in lieu of impingement or entrainment reduction; and
- Possible environmental side-effects such as water quantity use and changed air emissions, due for example to the use of cooling towers in place of a CWIS.

Figure 1 graphically summarizes these 316(b) stakeholder values. Benefits and costs here have their broadest possible meaning, and are not limited to market-valued resources.[5]

Whether defined by the CWA or elsewhere, these are value categories that are relevant to 316(b) decisions. In particular 316(b) decisions, different subsets may assume greater or less importance and their measures may be operationalized

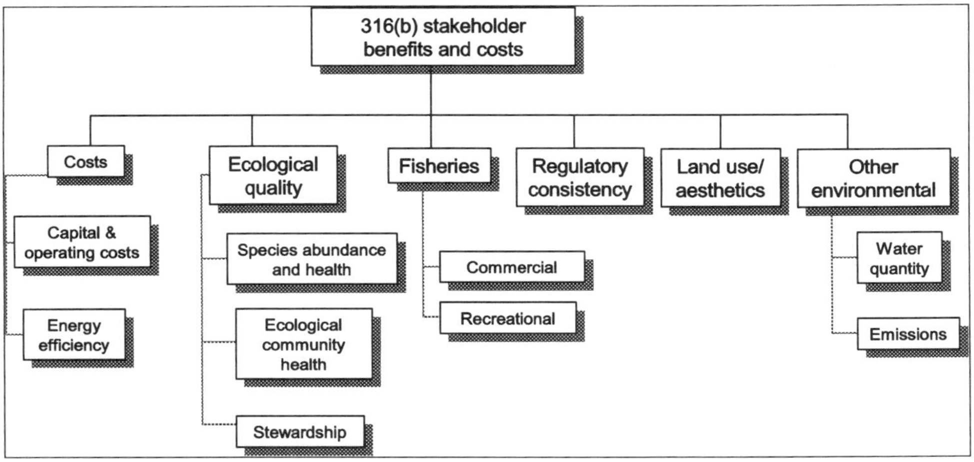

FIGURE 1. Stakeholder values relevant to most 316(b) decisions. Graphical organization does not indicate any priority. Measurement criteria are required for all categories and may differ for different 316(b) decisions.

differently. But however a 316(b) decision is actually made, it implies changes for some of these stakeholder values. These changes are measured somehow, and the measures associated with biological or ecological change effectively define AEI, or its absence, for that decision.[6] For example, suppose a 316(b) choice is to install a barrier net to reduce impingement and to carry out a fish stocking program to mitigate entrainment. That choice might be labeled the "best technology available" (BTA) selection, as provided by the 316(b) text. But its benefits and costs depend on ecological quality changes; a fisheries benefit due to impingement reduction; plus benefit to be achieved through the stocking program (including the possible continuing negative effect, if any, of entrainment); and all technology and enhancement program costs. AEIs in this way are determined by the quantitative or qualitative measurement criteria used to evaluate stakeholder interests or make 316(b) choices, even if not as explicitly organized as in Fig. 1. That perspective applied even to choices assuming the most conservative technology standard, such as limiting BTA to dry cooling.[7] For even if a technology standard is not intended as a direct measure of environmental change, it implies changes in benefits and costs, and thus becomes a proxy measure in practice. In such a situation, regulators or stakeholders will effectively back-calculate, as best they can, the consequences of interest as shown in Fig. 1. Section 316(b)'s BTA language makes it sound like a technology-based standard, along the lines of much of the CWA. But in practice, BTA decisions, including broadly defined technology rules, incorporate measures effectively defining AEI. Thus, it matters in the end not whether a biological change is labeled "adverse." Rather, what matters is simply what 316(b) choice (including no action), if any, is made, and how outcomes associated with that choice differ from the status quo or other benchmarks.

Thus we have a first organizing principle for AEIs. Starting with a values-focused view of 316(b) decisions, AEIs are implied by the many options available for measuring, and then comparing, the benefits and costs for stakeholder interests. This purely consequentialist perspective does not depend on limited or out-of-date views on economic cost-benefit, or the choices for how tradeoffs between competing values (e.g., dollar cost and ecological change) are carried out.[8] It is a ubiquitous feature of 316(b) decisions. So the problem of defining AEI, whether uniformly or on a site-specific basis, is first that of designing and implementing such measurement choices, and second, deciding which measured levels constitute "adverse." More generally, one may evaluate the tradeoffs accepted among various 316(b) choices for all relevant consequences or outcomes. The multiple values perspective of Fig. 1 shows that ecological AEI only is only one piece of the 316(b) decision-making process.

In principle, everything needed to compare a set of 316(b) choices for a new or existing CWIS is contained in Fig. 1 plus criteria for measuring each value category.[9] First, one or more proposed 316(b) entrainment or impingement reduction technologies, CWIS operational changes, mitigation proposals, or other options, are evaluated along the dimensions in Fig. 1. Then, regulators or stakeholders (implicitly or explicitly) rank or compare options based on their value judgments (e.g., comparative value of ecological or fisheries change and technology costs) and 316(b) regulatory policy constraining choices (e.g. whether mitigation is allowed, and then what kinds). Such evaluations may integrate uncertainty about various outcomes, some combination of impingement and entrainment counts, forecasts of fish population changes, cost estimates, and so on. There also may or may not be a formal process for how 316(b) options are compared and ranked. Nonetheless, once we back up from AEI to the implied stakeholder value hierarchy in Fig. 1 and see what measurement criteria and tools are actually used, AEI has been effectively defined at least on a case-specific basis.

So 316(b) decisions, like all environmental decisions, will always effectively use some notion of AEI, whether AEI has been formally defined or not, and whether or not in advance of an individual 316(b) decision. That sounds like defining AEI should then be central to 316(b) policy. As mentioned above, the electric power industry has effectively framed a general approach to site-specific tradeoffs as an AEI definition, with the only major constraint being to focus only on fish populations, not acute impingement and entrainment. But as is argued next, while that consequentialist approach (or even one also allowing acute impacts) is correct in principle, the many measurement or AEI options possible make it difficult to provide much substantive guidance without further goals for what 316(b) policy should achieve and how that is to be accomplished. The many AEI alternatives may in this way be faced with decision-making constraints. This paper's main thesis is that 316(b) policy options are largely determined by how these decision-making constraints are interpreted and addressed. In the language of decision analysis, 316(b) policy alternatives revolve around major disagreements of the 316(b) decision *frame*; it is the decision "dog" wagged by the AEI "tail."

EVALUATING AEI OPTIONS

This paper began by asking how important defining AEI was for formulating 316(b) policy. That question can now be revised.

Are the stakeholder values of Fig. 1 plus a set of measures, or AEI proposed definitions, adequate for formulating a substantive 316(b) policy perspective? To answer that question means evaluating various AEI options, including the option of allowing site-specific approaches for different rivers, estuaries, lakes, or marine environments. That is, imagine all the ways that AEI, whether explicitly or not, can or has been defined in a 316(b) decision. How are the choices or implementations of measures in an actual decision to be compared? Following are some of the evaluative questions that can be raised, with no answer intended as right or best. Importantly, these are not abstract questions, but ones that have occurred in actual 316(b) decisions (and are cited in the notes). Thus, even if AEI is not formally defined, these or similar questions about AEI adequacy will appear anyway, and imply real choices to be made by regulators or stakeholders.

What are the roles for individual fish, fish populations, and/or ecological community measures? Which fish population or acute mortality changes are measured? What additional ecological or trophic level changes may be measured, and over what temporal and spatial scales? How do such outcomes change depending on 316(b) technology choices? How are these selected for the aquatic life in a given water body, and what is the role of current resource management practices?[10]

What is the role of proxies? Technology standards, such as CWIS intake velocity criteria or biocriteria (i.e., measures integrating a suite of relevant ecological measures and judgments), can be seen as proxies for outcomes of genuine interest. For example, an intake velocity is not valued per se; it is a stand-in for achieving some genuine environmental objective. How well or poorly have 316(b) decision-making objectives been measured, given the use of such proxy measures? Do they work well for some water bodies but not others, as has been argued for biocriteria?[11] Are they as useful as other measurement approaches?[12]

What modeling approaches, simplifications, background assumptions, calibration techniques, and data collection methods are to be used? Of the many approaches for modeling fish populations, which are used and why? What is the quality of data used to estimate model parameters? What methods are used and over what time period? What types of judgment are needed to calibrate a model and ensure that it gives realistic or credible results? What background assumptions such as projected future hydrological flows and temperature regimes are assumed? How does the scientific process of theory proposal and testing fit in to 316(b), given the challenges of characterizing local ecological change?[13] Even if technological standards are used as proxies for site-specific modeling, one nonetheless makes some ecological assumptions, and uses some data and some conceptual model of a water body and its aquatic life.[14]

How are constructed measures used? Constructed measures are those created or designed specifically for a particular decision, and not necessarily applicable to others. These occur throughout environmental decision making, and examples from 316(b) could include estimates of trophic level changes specific to a particular water body or possible changes associated with a site-specific mitigation or enhancement project. What professional judgments and modeling or empirical considerations lay behind such a measure?[15]

How are priorities set for species of concern? In some situations, there will be too many fish species to track all of them, and typically several species dominate a 316(b) decision. How are these selected and related to one another? What is judged when or if, say, just a single species of several is doing poorly?[16]

How are model outputs aggregated, and uncertainty and causality addressed? In using fish population forecasting models, a typical output could be a population size *trajectory*, including confidence intervals, over multiple years and for multiple species. How are such stochastic trajectories aggregated over time and compared for 316(b) decision making? How are modeling and data uncertainty distinguished from natural variability, and how are probabilistic modeling outputs interpreted? If necessary, how are CWIS causal influences disentangled from nonanthropogenic influences such as water temperature and water quality, or overfishing, as another potentially important anthropogenic cause?[17]

How is ecological value defined? By what means are ecological changes compared to other stakeholder values, including costs? What do such comparisons mean? Are they explicit or implicit in the 316(b) decision-making process? To what extent may they be compared to other environmental decisions?[18] Do water body characteristics play a role in determining ecological value, such as differences between productive estuaries, managed reservoirs, or impacted water bodies? What is the role for considering changes to fish populations as opposed to acute mortality?

What kind of stakeholder and risk assessment process is followed? To the extent that any choices are made for assessing fish protection benefits, or comparing various stakeholder values, what process is to be followed? For example, are the suggestions made in the EPA's Ecological Risk Assessment guidelines (as yet not recommended by EPA for 316(b) but supported, for example, by EPRI) followed, or other risk assessment and stakeholder paradigms?[19]

How are predictive accuracy, environmental benefit, and regulatory effort balanced? What, in practice, are realistic options for reducing uncertainty for understanding fish population impacts, with or without various 316(b) choices?

Regulators and others have limited resources to expend on studying a 316(b) issue. How much effort should be expended on data collection and modeling tasks, given options for deciding with or without more information, and relative to ecological and other consequences, including costs, associated with a particular decision?[20]

What is characteristic of the 316(b) policy debate is that the variety of water body types and uses, fish populations, and resource management settings makes a great range of reasonable and realistic answers possible to these questions. Indeed, because permanent 316(b) regulations (as opposed to the original CWA legislation) were effectively never created, the historical implementation of 316(b) has in practice led to a variety of site-specific answers. It would indeed be extraordinarily complex to enumerate all options and relevant conditions associated with various water body types, impingement and entrainment scenarios, ecological settings, and so on. It is also doubtful that they can be easily sorted out by starting with the question of defining AEI. Nor is there a generic 316(b) decision or paradigm that appears representative of all.

There are two immediate answers in response to the complexity associated with subtle environmental effects, challenging assessments, and widely varying ecological significance. One is to avoid AEI implementation issues altogether through a simplifying, uniform technology standard. Another is site-specific risk assessment through which AEI implementation is carried out case-by-case. However, each approach still needs some consideration of how it would be implemented to achieve CWA goals.

This implies that some broader context than AEI measures alone (or AEI plus stakeholders and their multiple values) is needed to compare policy options for the 316(b) debate. The next section proposes one such approach.

ORGANIZING AEI EVALUATION

The first step in understanding the role of AEIs in 316(b) decision making was to recognize that AEIs appear as implicit or explicit ecological measurement criteria for 316(b) stakeholder values.[21] In the previous section, key questions about AEI measurement were raised, with the observation that there appears to be no satisfactory generic approach to defining AEI once and for all (or at least one different from counting any acute impact as significant). Following are three categories for organizing these questions and characterizing the adequacy of possible AEI measures; these evaluative categories are implicit in any 316(b) policy approach and are not answered only by the comparison of stakeholder benefits and costs as illustrated in Fig. 1:

1. What is 316(b) trying to achieve in the context of the CWA?
2. How is that to be accomplished in practice?
3. What knowledge base informs these goals?

Figure 2 graphically summarizes the approach, followed by further explanation of how the questions help compare 316(b) policy proposals. Answers to these questions help characterize the decision frame, or fundamental constraints and assumptions, through which a 316(b) decision may take place.

Clean Water Act 316(b) Policy Goals and Vision. AEI proposals are tools for implementing a vision for what 316(b) policy is intended to be and what, in broader social, political, and regulatory terms, CWA legislation is, or could be, intended to make happen. For example, should 316(b) be organized along the lines of technology standards, as are many CWA wastewater rules?[22] That can be construed narrowly as one of legal interpretation, but that is not the intention here. Rather, for 316(b), are technology standards of various kinds a good idea for achieving CWA goals? Is it appropriate to "force" strong technology standards to improve environmental quality, as was done for waste- and stormwater through the CWA? A positive answer would be supportive of the Riverkeeper's position, and implies that aggregate "errors" (e.g., overprotection) made through a uniform rule are small compared to overall benefit. A negative answer implies other justification is needed for uniform standards; alternatively, a negative answer could mean that no feasible technology standard is justified by the benefits associated with 316(b) outcomes.

Given that fish protection and harvesting is also managed and regulated outside of EPA, just what is EPA's role? One position is that "a fish is a fish," that a resource should be managed uniformly across all agencies. But that itself is a policy choice, so it is possible that one agency could decide that it has enabling powers not shared by others, and not regulate a resource in exactly the same

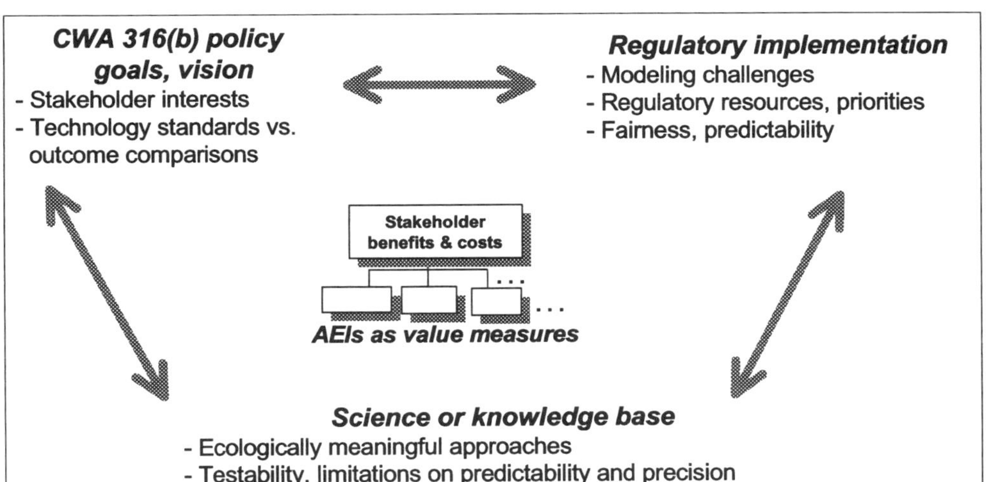

FIGURE 2. Comparative framework for evaluating AEI and 316(b) policy options.

way as others.[23] It may be that, for whatever reasons, we choose to protect a fish resource in one way and manage its consumption differently (e.g., because U.S. fisheries policy has many weaknesses), or we may not. Importantly, there are broad questions of equity and fairness between industries as consumers of the fish resource. That is, why should the electric power industry not be able to treat fish mortality as a resource harvest? Why should fish killed for food differ from fish killed to produce societally important electric power? It surely does not matter from the perspective of the fish.[24] These questions are not necessarily that difficult to answer, but some answer should be forthcoming. No obvious social values appear to decide the answer, as many people are likely unaware of the numbers of fish normally impinged and entrained as part of the power production process, and risk perceptions of how fish are killed may, for better or worse, significantly influence judgments of those outside the 316(b) debate. Framers of 316(b) language may not have anticipated the equity and resource consumption problem, either. Entrainment and impingement affect different phylogenetic levels, namely eggs and larvae vs. adult fish, about which societal values may be ambiguous, also. To make a stark comparison, if instead of fish, adult dogs or cats were killed, albeit with no significant population impacts, someone would surely take notice; whether that is "obviously" not true for fish appears unclear. This is not to argue that the Riverkeeper's position against almost all forms of fish mortality is correct, only that the basis for the ecological value choice is not transparent.

Other CWA issues include the role of technology costs, which are not mentioned in 316(b), but play a role elsewhere in the CWA. Relevant also to CWA philosophy is the role of cumulative risk in attaining the CWA goals of "swimmable, fishable" waters. Are 316(b) impacts to be regarded as individually small, but regulated as one of multiple aquatic hazards? The overall risk aversion or degree to which more certain or timely outcomes are preferred may also be a reason for supporting technology standards as opposed to allowing site-specific determinations.[25]

Finally, the BTA and AEI language of 316(b) likely would not be created today because of (at least for some) its evident ambiguity and brevity. The simplicity of 316(b) language is comparable in its logic to the now-repealed Delaney Clause of the Food and Drug Act prohibiting absolutely any amount of a carcinogen (e.g., from a pesticide used on a vegetable crop) to persist in processed food, regardless of the outcome to human health.[26] Similarly, 316(b) almost prima facie disallows any de minimus CWIS-caused impact, regardless of its ecological significance, and regardless of seeming inconsistencies with the management of other harvested resources. Much of the Riverkeeper's policy position depends on just such a strict reading. Therefore, it may be asked, though no major stakeholder group to date has done so, to what extent is a "rewrite" of 316(b) in order using more up-to-date concepts of risk assessment or stakeholder involvement? A negative answer implies sticking to traditional CWA approaches, while a positive answer implies consideration of broader risk management and stakeholder involvement options.

Various coherent 316(b) policy positions can be created by different answers to these CWA questions. The questions have in common that they should be addressed uniformly as part of 316(b) policy, and not reinvented locally when particular 316(b) decisions are made.

Regulatory Implementation. Implementation refers to the decision-making process required by regulators or resource managers, and involved stakeholders, to make 316(b) technology choices (including no action and mitigation); of particular importance is the *effort* needed to evaluate 316(b) choices compared to the *accuracy* of such assessments. .[27]

If modeling or empirical study are required, what kinds of requirements are useful for ensuring useful results? Both the Riverkeeper and the EPA have commented on the practical need to simplify 316(b) implementation, reacting in part to the challenges in assessing and modeling fish population changes due to a CWIS. However, mere technical difficulty is not a reason to impose a conservative technology standard. If it is judged that 316(b) problems generally do not pose especially significant ecological risks, then it is consistent to allow more leeway in local regulatory decisions, with the belief that the consequences of possible errors may be small. Conversely, if 316(b) represents a more important ecological concern, that may justify allocating or shifting more resources toward it and away from other water problems. But, if resources required for extensive study are great and leave uncertainty unresolved, then more uniform technology rules may be justified depending on their cost. Overall judgments of comparative risk and their tradeoffs thus are closely tied to regulatory implementation priorities.

In any case, regulators need to place 316(b) on their priority list for allocating their own and stakeholders' resources. How important, for example, is 316(b) either as a watershed or fish protection issue compared to, say, nonpoint source runoff, combined storm-sewage water overflows, or other priority water quality and quantity issues? Or are regulators to set such priorities locally? More generally, what guidance is there on making ecological value judgments related to 316(b)?[28]

Finally, how predictable and fair is a 316(b) decision process, either for regulators or the regulated community? Are perceived fairness and predictability largely provided via technology standards? If technology standards are not in place, what expectations can the regulated community have for predictability and fairness of the decision-making process? For example, in historic 316(b) cases where technology options have been ruled out for cost or effectiveness reasons, mitigation projects have been allowed as a substitute, with costs ranging from hundreds of thousands to many tens of millions of dollars.[29] What fairness considerations are relevant here?

Science or Knowledge Base. Even given a CWA policy vision and an approach to how it should be implemented, any 316(b) policy or AEI proposal should relate *somehow* to knowledge and the science of fish lives, populations, ecosystems, and options for affecting ecological outcomes. For example, there are significant ecological differences between impingement of adult fish and the entrainment of eggs and larvae. The natural mortality rates of the latter can be above 90%, which, while not justifying entrainment, and depending on other causal influences, may mean that fish populations may be only weakly influenced by CWIS mortality at that life stage. That important biological difference in fish life stages also is not reflected in 316(b) language. Fish population declines should also account for other causes of fish mortality or population size changes including (over-) fishing, water quality degradation, and natural variability in water temperature and fish habitat. An important piece of policy guidance therefore is recognizing the influences on fish populations other than CWISs. Also important are the judgmental considerations relevant to assessing their importance, as it is often difficult to unambiguously model or assess causal strength for ecological changes associated with CWISs.

Policy guidance for 316(b) science can also be provided for different types of modeling, modeling outputs, data collection, and the simplifications often needed to calibrate and implement fish population models for practical use. In particular, if technology standards, biocriteria, or other regulatory proxies are to be used, then what are reasonable compromises intended for policy purposes? Finally, what is the role for fish science in valuing fish mortality at the individual versus population level, a major point of disagreement in the 316(b) debate? This is a value judgment and *not* a scientific question.[30] But since biologists are accustomed to making implicit value judgments about ecological health, policy advice is needed to distinguish neutral fish mortality outcomes from judgments of biological or social value. Biologists can participate in those judgments, but policy guidance is needed to distinguish, as best as can be done, roles for value-laden and technical judgments.

That completes the description of the three main areas of 316(b) policy evaluation. Most of these issues raised have been argued by others involved in the 316(b) debate. The organization provided here shows, first, that many questions that can be asked of 316(b) policy can be usefully arranged under the three broad categories described. Second, while any 316(b) policy implies an explicit or implicit role for AEI, one can ask many critical questions about how 316(b) policy could or should work without making an AEI definition axiomatic. That does not mean ignoring AEI, but putting it in the decision-making context created by the CWA, the need for pragmatic regulatory guidance, and the practical use of scientific knowledge.

Therefore these three evaluative dimensions, plus the values-focused perspective of Fig. 1, provide a simple functional model of 316(b) policy. AEI or their implied 316(b) policy proposals can be examined by seeing how well they perform along each dimension. What is important so far is that each policy component raises important issues, with no right or wrong choice in isolation, and whose answers should cohere with those provided from the other policy components. Any 316(b) approach needs to address each component to work well, or at least to be immune to some immediate objections.

COMPARING SOME 316(B) POLICY STRATEGIES

Three simplified 316(b) policy strategies are now used to examine how the categories of the previous section (or Fig. 2) work in practice. While illustrative only, they indicate issues relevant to proposals from important 316(b) stakeholder groups.[31] The three simplified approaches are:

- *"Strict" 316(b) interpretation.* Here, acute impingement and entrainment are sufficient metrics, and are to be minimized without respect to cost considerations. The 316(b) choice is largely or entirely technology driven, based on the most literal interpretation of minimizing BTA. No special consideration is given either to distinguishing fish eggs and larvae from juveniles or adults, or distinguishing individual fish mortality from population impacts, or considering the role of confounding causal factors such as water temperature changes or overfishing on population health. AEI includes virtually any level of fish mortality caused by a CWIS. A version of this strategy has been proposed by the Riverkeeper.

- *Biocriteria or other proxy-based.* Here, a key tool is the use of quasi-scientific metrics, defined nationally, for concepts such as the littoral zone, biologically sensitive areas, aquatic zones of CWIS influence, or biocriteria-style metrics; the latter typically integrate a suite of ecological measures or professional judgments.[32] In contrast to some human health risks, a challenge here is to make such policy constructs work for ecological impacts varying widely in their nature, causes, amplitude, and significance. In August 2000, the EPA proposed a version of this strategy that includes standards dependent on water body type (river, estuary, marine, lakes) for new CWISs. This can be seen as a hybrid proposal combining a general value judgment on water body importance and a lower-level ecological risk assessment process using proxies.

- *Site-specific and risk-based.* Here, AEI is characterized on a site-specific basis with fish population (rather than acute) impacts being primary. AEI is evaluated in the context of both fisheries management and overall comparisons of stakeholder benefits and costs (Fig. 1). AEI defined in this way is close to a general definition of aquatic and stakeholder ecological risk, the primacy of population metrics being perhaps the only significant restriction. UWAG has proposed an approach like as an AEI definition, but its important features are that the characterization of AEI is site-specific and is limited to population impacts.[33]

While these are simplified positions, they capture key features of the approaches proposed by three major 316(b) stakeholders. The first option is the simplest pure technology standard and the last is the broadest site-specific risk-based approach. The first is forced to characterize AEI as any fish mortality because it is based on minimizing BTA, independent of cost or net benefits. In contrast, the last option provides a general framework for defining AEI on a case-by-case basis, and considers fish mortality adverse only when it has significant population consequences. In the second approach, a methodological compromise is attempted through general criteria tailored to water body types. The second approach is less clear about AEI than the other two, although AEI is implicit in the proposed proxy measures and selection of water body types. Thus, each policy approach implies something about AEI.

But in each case, the role of AEI either is not especially informative, or is just one part of a broader approach to 316(b) policy. To compare these 316(b) policy options then requires criteria not defined solely in terms of AEI. Implications of the three approaches can be briefly assessed as follows, using the organizing ideas above: CWA 316(b) policy vision, regulatory implementation, and knowledge base.

"Strict" 316(b) interpretation. First, the policy is relatively easy to implement; for example, the Riverkeeper has argued against the introduction of what they consider to be a needlessly complex assessment or stakeholder process. They have also effectively designated dry cooling, which minimizes evaporative water loss, as a generic minimizing BTA. An advantage of this particular technology rule is that it assures a high degree of fish protection and appears minimally subject to gaming or delay tactics in implementation (the latter being a Riverkeeper criticism of a site-specific or stakeholder approach).

In itself, a technology-based rule is neither good nor bad. Nearly all regulation is necessarily based on simplifications or aggregate rules needed to assure desired outcomes through implementable regulation. A regulated society has to accept some error tolerance between, on the one hand, an idealized best practice by cases, and, on the other, the implementation of uniform regulatory policy. The wastewater provisions of the CWA based on technology criteria, for example, show that such an approach can be done reasonably well. Indeed, as a matter of CWA interpretation, it seems fair to at least raise the issue of a possible analogy or disanalogy with this feature of CWA philosophy (though congressional financial support to repair the nation's once-dilapidated wastewater system is also relevant). It is also somewhat counterintuitive that the CWA should be used to improve water quality and protect aquatic life, but then be interpreted to allow large acute impacts to fish life through entrainment and impingement. So a "strict" 316(b) interpretation and its technology rule should not be dismissed immediately on methodological grounds, the problems of the 316(b) text notwithstanding.

But what is important is whether the *particular* uniform technology rule is a good one. Is the certainty it provides balanced or outweighed by other consequences, i.e., those representative of all stakeholder interests in Fig. 1? To what extent does the implied CWA policy vision represent a sound environmental choice? And how does the fish protection philosophy promoted as part of 316(b) cohere with what is known about how fish survive and reproduce, the different consequences of impingement and entrainment, and other resource management practices? A "strict" 316(b) approach is not necessarily in contradiction with a scientific perspective on ecosystems, but such a policy should account for these basic issues. A variant might be a "strict" 316(b) interpretation for CWISs or water bodies judged in need of the strongest protection. In that way, the CWA philosophy is preserved but targeted only to those conditions where it is clearly judged worthwhile. If there are *no* situations for which a uniform 316(b) technology standard is justified, then that would be useful to know as well. As another variant, UWAG has proposed a two-track option for new CWISs through which a conservative BTA choice could be implemented speedily, or a site-specific risk assessment could be carried out to better understand local impacts. This too partly preserves a CWA technology-forcing philosophy, but with greater flexibility.

Biocriteria or proxy-based. Biocriteria or other quasi-scientific metrics have several advantages: (1) a coarse technology standard is improved by measures more closely related to the outcomes of interest; (2) an approach to regulatory implementation is provided that is hopefully not as complex or practically overwhelming as more open-ended risk assessment processes; (3) a certain amount of predictability and fairness in implementation is provided; and (4) the rationale for various biocriteria or proxy guidelines might possibly be more easily justified on scientific grounds.

Biocriteria and regulatory proxies generally are motivated by the idea that the right criteria can become adequate regulatory tools. Constructed measures are often created on a site-specific basis for many environmental problems, so why not just do that at a general regulatory level? Of course, whether the right outcomes can be captured by the constructed measures is the great challenge. It is also predictable that if more cases and variability (multiple fish species, competing causal factors, water body management approaches, ease of modeling, etc.) needs to be captured, the less likely it is that the criteria will be successful. Given that such criteria are intended to have some kind of scientific base, they are open to criticism for being motivated by expediency rather than explanatory principles. The approach of creating regulatory proxies also has a long history in environmental regulation, and sometimes creates complex policy choices when hard value judgments are replaced by quasi-technical solutions.[34]

So the use of biocriteria or proxies has clear implementation advantages, and is an improvement over some pure technology rules for representing policy. But a difficulty is that important policy consequences may have to be reverse-engineered from the proxy measures because desired policy outcomes are only stated indirectly through the proxy measures. Therefore the intended policy vision may have to be backed out by seeing just what the criteria imply. The motivation to create an implementable policy is important, as is making pragmatic use of science. But ambiguity and lack of transparency in policy vision leaves the approach open to objections of confounding scientific tools and policy choices. A single point of agreement between the Riverkeeper and UWAG in their responses to the EPA's August 2000 proposed new facilities rule was their shared rejection of the EPA's proposed proxy measures (which were indeed discarded in the final November 2001 rule).

There is a great deal of experience in data collection, modeling, and interpretation from 316(b) experience, some of it overly complex or of limited decision-making value. For some situations, proxy measures or biocriteria may be useful, such as for managed reservoirs. But in general, it may be wiser to provide regulatory guidance for using recognized scientific approaches rather than designing new regulatory concepts whose scientific merits and operational value are yet to be established. One may ask regulators to exercise judgment in using the best knowledge available, rather than preprocessing that knowledge through a filter whose rigor has not been openly established. The EPA has organized their use of proxy measures for new facilities by water body type as a high-level means of prioritizing fish protection. That same overall approach could just as well be designed with water body guidance for assessment, modeling, and protection, instead of, or perhaps along with, proxy measures. Implementation is an important and often neglected policy objective, but must balance regulatory usefulness and scientific validity.

Site-specific and risk-based. This approach makes primary the value judgment that fish mortality is significant only when it has an effect on fish (or ecosystem) population health or viability, and not acute impingement and entrainment mortality by themselves. Attempts to characterize AEI by fixed population percentage losses or absolute numbers (e.g., biomass loss) appear largely to have failed, and technology rules or proxy measures are judged, from the site-specific perspective, to be conservative or inaccurate. Therefore, case-by-case risk assessment is proposed as the best means for more accurately and fairly defining AEI, and this perspective does reflect the local character of many ecological science studies.[35] The approach presumably includes local judgments about ecological, cost, and other tradeoffs, thus emphasizing roles for ecological (and social) value and technology effectiveness poorly addressed by technology or proxy-based approaches. The site-specific approach therefore has clear policy vision and is intended to make good use of the knowledge base. This policy vision however comes from contemporary risk assessment methods and is largely a substitute for the technology-based BTA perspective of the 316(b) text. Proponents will note the traditional acceptance of probabilistic risk assessment methods for fisheries management of populations and not individual fish, and that for decades, dozens of 316(b) decisions have effectively been made on a site-specific basis. Other stakeholders, such as those favoring a "strict" 316(b) interpretation, will differ on whether those decisions were made correctly, or whether the fisheries perspective is appropriate.

Instead of proposing a substantive definition of AEI (i.e., one that does more than identify population effects), the site-specific approach instead proposes a process. Only broad advice is then also provided on how it is to be implemented in practice. For example, UWAG's AEI definition amounts to a compact definition of multiple values risk, in that each 316(b) decision is looked at as involving local tradeoffs among fish protection, 316(b) technology and energy production costs, and potential environmental side effects from alternative technologies. EPRI has recommended use of the EPA's own Ecological Risk Assessment guidelines as one approach (and it is odd that others should have to recommend the Agency's recommendations to itself). An important question about whatever process might be used is the nature of value judgments. UWAG, for example, while proposing a process inclusive of all stakeholder benefits and costs, does not indicate whether broad stakeholder participation is appropriate. Along with site-specific modeling or impact assessment, such a process may also entail a greater resource commitment than a uniform technology rule or a proxy-based approach. At the same time, more open-ended stakeholder-cum-risk-assessment processes are increasingly common, and 316(b) could be implemented using similar approaches.[36] The certainty achieved by the "strict 316(b)" approach for fish protection outcomes, or the uniformity of biocriteria or proxy rules, is here traded for a more open process. The latter is also just what the "strict" 316(b) approach may want to exclude because of a potential risk of underprotecting fish through process manipulation or delay.[37]

Finally, relevant to site-specific assessment is the value judgment of deciding on an appropriate dollar value for mitigation projects. It is often difficult to retrofit an existing CWIS to reduce entrainment at costs less than several tens or even hundreds of million dollars. In such situations, several 316(b) decisions have led to enhancement or mitigation projects including fish stocking or ecological improvements.[38] A key decision variable is the dollar amount to invest in the enhancement project, as there may be no obvious upper bound to how much should be spent. To create a 316(b) "nexus," some kind of ecological valuation may be used, but the uniqueness and, typically, nonmarketability of ecological resources makes ecological value itself highly site-specific and subject to stakeholder judgment. Lacking guidance or standards, the determination of a "just" enhancement budget can potentially become a negotiation exercise subject to a company's ability to pay, and therefore raises significant fairness questions for the CWIS owner.[39] Thus, site-specific AEI explicitly raises policy problems of ecological valuation, but ones which also have been implicitly conducted during all the years that 316(b) decisions have been made without explicit regulatory guidance.

The three simplified policy proposals and their evaluations are summarized in Table 1. In summary, none of the three simplified 316(b) approaches should be faulted on principle, and indeed each has something to recommend. But neither does any completely answer natural problems raised in the areas of CWA vision, regulatory implementation, the science base, and the allocation of stakeholder benefits and costs. In any case, the evaluation of 316(b) policy proposals can be carried out through a simple functional model of important policy dimensions. With no "silver bullet," where does this leave the problem of developing 316(b) policy? A site-specific, risk-based approach comes the closest to the consequentialist, outcomes-based approach of Fig. 1. To the extent that 316(b) policy should reflect regulatory implementation concerns, or aspects of CWA philosophy deemed important, features of the other policy approaches may have to be incorporated into a hybrid policy.

TABLE 1

Coarse Evaluation of Simplified 316(b) Policy Approaches

	CWA Policy Vision	Implementation	Knowledge Base
Strict 316(b) interpretation	Assure high protection by simplest technology rule with little regard for side effects or cost. Takes 316(b) at face value.	Straightforward (e.g., dry cooling or equivalent is BTA).	Emphasizes impingement and entrainment mortality. Population vs. individual distinction not too significant.
Proxy-based	Use specialized constructs to identify water bodies or populations for protection. Much 316(b) policy built into proxy constructs.	Straightforward, at least assuming constructs give meaningful results.	Compromise between policy definition and scientific concepts.
Site-specific and risk-based	Protect populations; aim for consistency with other fish management practices; otherwise assessments and tradeoffs are site-specific. Largely rewrites 316(b).	Open-ended, especially given ecological and modeling challenges. Effort-accuracy tradeoffs and stakeholder roles unclear.	In principle, makes best use of available science base, but process details unspecified.

That may be achieved by a more direct or inductive approach to setting 316(b) policy, meaning to more concretely identify undesirable environmental outcomes, as opposed to AEI generally, and to identify just what should happen through 316(b) regulation. For example, a 316(b) policy organized by water body types using proxy or technology rules helps ensure certain minimum levels of fish protection. If such an approach overshoots its goals for costs or benefits, it can, for example, be adjusted to allow for site-specific risk assessment under identified conditions, planning in a more integrated way for implementing policy for new vs. existing facilities, or ranking 316(b) problems as, for example, "eliminate," "do not repeat," "not significant," "possible concern," "monitor," "up to local judgment," etc. (All these may be addressed differently for new vs. existing facilities as well.) Regulation needs rules, but for 316(b) they need to either be sharply focused, so that they achieve desired outcomes without large error, or flexible, so that they can be adjusted in practice.

The challenge is to do both with sufficient certainty of achieving environmental progress while not consuming excessive regulatory or other social resources. That means being both focused and flexible, or combining and tailoring elements from different 316(b) proposals so that their various advantages are exploited without the shortcomings of each taken separately.

THE U.S. EPA'S PROPOSED AND FINAL NEW FACILITIES RULES

This paper was written following publication of a Clean Water Act Section 316(b) *proposed* new CWIS facilities rule by U.S. EPA in August 2000, and was completed just as the *final* new facilities rule was published in the *Federal Register* in November 2001.[40] The August 2000 proposed rule motivated some of the paper's discussion, and the November 2001 final rule differs from the proposed rule in important respects. Publication of the final new facilities rule provided an opportunity to apply the framework proposed above by seeing how the final and proposed rules differ.

In the final new facilities rule, proxy definitions such as the controversial littoral zone concept and biocriteria are given up in favor of what is essentially a technology standard. The final rule applies to all water body types uniformly, unlike the proposed rule, which included different criteria for rivers, lakes, oceans, and estuaries. The main feature of the final new facilities rule is that CWIS capacity and flow (above certain minimums) should be no greater than that for a closed-cycle, but "wet" or evaporative (tower or pond), cooling system. Thus, BTA is not dictated, but CWIS protection should be equivalent to closed-cycle technology. The final new facilities rule also allows a CWIS permit applicant to carry out a site-specific study, very much like a "two-track" approach proposed by UWAG in response to the EPA's proposed new facilities rule. However, the level of fish protection, primarily impingement and entrainment reduction, still has to be equivalent to that provided by closed-cycle cooling. Thus, a site-specific risk assessment can only conclude that an alternative CWIS technology provides a level of protection greater or less than closed-cycle cooling, not that impingement or entrainment does or does not create AEI.

The following points summarize changes between the proposed and final rules using the comparative framework discussed above. A principal change was from a proxy-based to a technology-based approach, as just mentioned.

- AEI is again not formally defined, but almost. The EPA enumerates various adverse outcomes, including population changes, but the list is headed by impingement and entrainment. Thus the rule mainly provides protection against acute impingement and entrainment. The basic ecological value judgment appears to be that fish should be protected from large anthropogenic causes of mortality different from fishing. Impingement and entrainment are also considered important contributors to the cumulative risks faced by aquatic populations created by overfishing, nonpoint source runoff, and other hazards. The possibility of further characterizing AEI through the NPDES 316(b) process is largely foreclosed by limiting the relevant outcomes of site-specific risk assessment in a "two-track" process (see above) to those equivalent to closed-cycle cooling protection.
- The weakness of proxies, mainly the important littoral zone concept of the proposed rule, is acknowledged. Also missing from the final rule is differential

protection based on water body types. Instead, an equivalent technology rule is considered appropriate for 316(b) along the lines of CWA philosophy generally. Thus the precise nature of the fish protection intended is somewhat less clear.

- The stakeholder value hierarchy of Fig. 1 is reflected in the EPA's consideration of cost and other environmental outcomes. Based on their economic analysis, the EPA found that, because only a small number of future plants were expected to now *not* use once-through cooling, and because closed-cycle technology is already widely used, economic impacts on new facilities were justified by the fish protection provided. Additional emissions associated with closed-cycle cooling and the aesthetic impacts of cooling towers or ponds were similarly judged acceptable tradeoffs. Economic analysis also led the EPA to favor wet vs. dry cooling, the latter having been supported by the Riverkeeper as BTA. It appears that while the EPA uses very approximate fish protection estimates, the cost tradeoffs are sufficiently small on a national level for the limited accuracy to be unproblematic. Setting the cost boundary below that for dry cooling may have cost implications for the forthcoming existing facilities rules. It could of course also be argued that the EPA's estimates of benefits and costs do not justify the tradeoffs implicit in the new rule.
- As indicated already, BTA (for intakes above certain threshold flows) is defined by comparison with wet closed-cycle cooling. Since AEI is not defined, that implies that the choice of minimizing BTA is a key piece of the 316(b) nexus. An important aspect of this definition is that the EPA does not identify particular technologies: "EPA emphasizes that it is not requiring wet cooling, but that it is establishing performance-based technology requirements on the dynamic flow of the cooling water intake structure that reduce impingement and entrainment at a level that is achieved by using closed-cycle cooling"[4, p. 263]. The use of a technology equivalent may also have implications for forthcoming existing facilities rules. Again, EPA appears to have made traditional CWA (wastewater) philosophy central to 316(b).
- Regulatory implementation is identified multiple times by the EPA as a concern for site-specific approaches that require the use of mathematical models. The complexities of fish population assessment and modeling, including the need for multiyear studies, and the problems in characterizing CWIS causality appear to have been major reasons that the EPA rejected risk assessment approaches and instead designed a technology standard.[41] Evidence provided by the EPA justifying the new facilities rule included studies on intake-flow velocity and (possibly dated) impingement and entrainment counts from various 316(b) studies. The EPA also states that forthcoming 316(b) rules for existing CWISs may be entirely different, with no precedent set by the new facilities rule.
- Fisheries management approaches, including the concept of maximum sustainable yield (MSY), appear to be rejected by the EPA as a paradigm relevant to 316(b). That is, they do not consider impingement and entrainment a resource harvest to be managed like fishing. Further, the problem of collapsed fish stocks is cited as reflecting the need for fish protection, with CWISs identified as a contributing factor along with nonpoint source runoff and other pollution. The

EPA appears to be upholding a perspective that fish may be managed in one way as a fisheries resource, and in another way as in need of protection. The recognized differences between adult fish mortality and that for eggs and larvae plays little or no role in the EPA's rule, but likely because there is no useful technology standard which differentiates the two, not because it is scientifically unimportant.

These comparisons show that the comparative approach developed in this paper is broadly useful. The issues differentiating the proposed and final rules were anticipated by the framework, though of course it could not be predicted which way policy options might go. For example, the EPA might also have incorporated elements of a site-specific and risk-based approach, rather than relying on a pure technology standard. The EPA can be argued to have made an improvement over its previous proxy-based proposed rule without substituting some weaknesses of the "strict" 316(b) interpretation used above. The absence of any proposed implementation philosophy for site-specific risk assessment, including the need to simplify and interpret modeling studies, appeared a strong reason to revert to a technology standard; that still implies that EPA choose here not to follow their own ecological risk assessment guidelines and, more broadly, approaches to risk assessment and environmental policy emphasizing tradeoffs among social values . The comparative process may provide further insight into policy options as debate evolves over rules for existing facilities.

REFERENCES

1. USEPA (2000) Economic and Engineering Analyses of the Proposed Section 316(b) New Facilities Rule. EPA-821-R-00-019. U.S. Environmental Protection Agency, Washington, D.C.
2. Cronin, J. and Kennedy, Jr., R. (1997) *The Riverkeepers*. Simon and Schuster, New York.
3. USEPA (2000). National Pollutant Discharge Elimination System. Regulations Addressing Cooling Water Intake Structures for New Facilities. www.epa.gov/owm/316b.htm and Federal Register Vol. 65, No. 155, August 10, 2000, Section VII.F. U.S. Environmental Protection Agency, Washington, D.C.
4. USEPA (2001) National Pollutant Discharge Elimination System. Regulations Addressing Cooling Water Intake Structures for New Facilities, Final Rule, 40 CFR Parts 9, 122, 123, 124, and 125, unofficial pre-publication version. U.S. Environmental Protection Agency, Washington, D.C.
5. Anderson, W. and Gotting, E. (2001) Taken in over intake structures? Section 316(b) of the Clean Water Act. *Columbia J. Environ. Law* **26**, 1–79.
6. May, J.R. and van Rossum, M.K. (1995) The quick and the dead: fish entrainment, entrapment, and the implementation and application of Section 316(b) of the Clean Water Act. *Vt. Law Rev.* **20(2)**, 375–493.
7. Dixon, D., Bailey, D., Jordan, C., Wisniewski, J., Wright Jr., J., and Zammit, K., Eds. (2000) Power Plants and Aquatic Resources: Issues and Assessment. *Environ. Sci. Policy* **3(Suppl. 1).**
8. Utility Water Act Group (UWAG). (2000) Comments of the Utility Water Act Group on EPA'S Proposed § 316(b) Rule for New Facilities and ICR No. 1973.01 65 Fed. Reg. 49,060 (Aug. 10, 2000). Available from Hunton & Williams, 951 E. Byrd St., Richmond, VA 23219.
9. Riverkeeper, Inc. et al. (2000) Comments on EPA'S Proposed Regulation for Cooling Water Intake Structures at New Facilities Under Section 316(b) of the Clean Water Act. Available from Natural Resources Defense Council, 40 West 20th Street, New York, NY 10011.

10. McEvoy, A. (1986) *The Fisherman's Problem: Ecology and Law in the California Fisheries.* Cambridge University Press, New York.
11. Dunning, D., Ross, Q., and Merkhofer, M. (2000) Multiattribute utility analysis for addressing Section 316(b) of the Clean Water Act. *Environ. Sci. Policy* **3(Suppl. 1)**
12. Gregory, R., Lichtenstein, S., and Slovic, P. (1993) Valuing environmental resources: a constructive approach. *J. Risk Uncertainty* **7**, 177–193.
13. Keeney, R. (1992) *Value-Focused Thinking: A Path to Creative Decisionmaking.* Harvard University Press, Cambridge, MA.
14. McDaniels, T., Gregory, R., and Fields, D. (1999) Democratizing risk management: successful public involvement in local water management decisions. *Risk Anal.* **19**, 497–510.
15. Bell, D., Raiffa, H., and Tversky, A., Eds. (1988) *Decision Making: Descriptive, Normative and Prescriptive Interactions.* Cambridge University Press, New York.
16. National Research Council (1986) Ecological Knowledge and Environmental Problem-Solving. National Academy Press, Washington, D.C.
17. California Energy Commission (2000) Biological Resources. Testimony of Richard Anderson and Michael Foster (www.energy.ca.gov/sitingcases/mosslanding/).
18. Bailey, D., Loos, J., Perry, E., and Wood, R. (2000) A retrospective evaluation of 316(b) mitigation options using a decision analysis framework. *Environ. Sci.Policy* **3(Suppl. 1)**, S25–S36.
19. American Fisheries Society (1992) Investigation and Valuation of Fish Kills. AFS Special Publication 24, Bethesda, MD.
20. American Fisheries Society (1993) Sourcebook for Investigation and Valuation of Fish Kills. Supplement to AFS Special Publication 24, Bethesda, MD.
21. Ambrose, R., Schmitt, R., and Osenberg, C. (1996) Predicted and observed environmental impacts: can we foretell ecological change? In *Detecting Ecological Impacts: Concepts and Applications in Coastal Habitats.* Schmitt, R.J. and Osenberg, C.W., Eds. Academic Press, New York. pp. 345–369.
22. Rose, K.A. (2000) Why are quantitative relationships between environmental quality and fish populations so elusive? *Ecol. Appl.* **10**, 367–385.
23. Hilborn, R. and Mangel, M. (1997) *The Ecological Detective.* Princeton University Press, Princeton, NJ.
24. Lohner, T.W., Seegert, G., Vondruska, J., and Perry, E. (2000) Assessment of 316(b) impact on Ohio River fish populations. *Environ. Sci. Policy* **3(Suppl. 1)**, S249–S259.
25. Lorda, E., Danila, D.J., and Miller, J.D. (2000) Application of a population dynamics model to the probabilistic assessment of cooling water intake effects of Millstone Nuclear Power Station (Waterford, CT) on a nearby winter flounder spawning stock. *Environ. Sci.Policy* **3(Suppl. 1)**, S471–S482.
26. Jasanoff, S. (1990) *The Fifth Branch: Science Advisers as Policymakers.* Harvard University Press, Cambridge, MA.
27. Van Winkle, W. and Kadvany, J. (2002) Modeling fish entrainment and impingement impacts: bridging science and policy. In *Using Models for Environmental Management*, Dale, V.H., Ed. Springer-Verlag, New York, in press.
28. Irwin, J., Slovic, P., Lichtenstein, S., and McClelland, G. (1993). Preference reversals and the measurement of environmental values. *J. Risk Uncertainty* **7**, 5–18.
29. Fischhoff, B. (1991) Value elicitation: is there anything in there? *Am. Psychol.* **45**, 835–847.
30. National Research Council (1996). Understanding Risk: Informing Decisions in a Democratic Society. National Academy Press, Washington, D.C.
31. USEPA (1992) Framework for Ecological Risk Assessment. EPA/630/R-92/001. U.S. Environmental Protection Agency, Washington, D.C.
32. USEPA (1998) Guidelines for Ecological Risk Assessment. EPA/630/R-95/002F. U.S. Environmental Protection Agency, Washington, D.C.
33. Adler, R., Landman, J., Cameron, D. (1993) The Clean Water Act: 20 Years Later. Island Press, Washington D.C.
34. Golding D. and Krimsky, S. (1993) *Social Theories of Risk.* Praeger, New York.
35. Slovic, P. (1987) Perception of risk. *Science* **236**, 280–285.
36. National Research Council (1989) Improving Risk Communication. National Academy Press, Washington, D.C.
37. Singer, P. (1990) *Animal Liberation.* 2nd ed. Jonathan Cape, London.

38. National Research Council (1987). Regulating Pesticides in Food: The Delaney Paradox. National Academy Press, Washington, D.C.
39. Payne, J., Bettman, J., and Johnson, E. (1993) *The Adaptive Decision-Maker*. Cambridge University Press, New York.
40. Lackey, R.T. (1994) Ecological risk assessment. *Fisheries* **19(9),** 4–18.
41. Lackey, R.T. (1998) Fisheries management: integrating societal preference, decision analysis, and ecological risk assessment. *Environ. Sci. Policy* **1**, 329–335.
42. Shrader-Frechette, K.S., and McCoy, E. (1993) *Method in Ecology*. Cambridge University Press, New York.
43. Landy, M., Roberts, R., and Thomas, S. (1990) *The Environmental Protection Agency: Asking the Wrong Questions*. Oxford University Press, New York.
44. Zajac, E. (1995) *Political Economy of Fairness*. MIT Press, Cambridge, MA.
45. Foster, K. and Huber, P. (1997) *Judging Science: Scientific Knowledge and the Federal Courts*. MIT Press, Cambridge, MA.

APPENDIX: SUMMARY OF RESPONSES TO U.S. EPA'S AUGUST 2000 PROPOSED NEW FACILITY RULE

Simple Comparisons Between the Riverkeeper's and Utility Water Act Group's Responses to EPA's August 2000 Proposed New Facilities Rule[1,3]

Issue	UWAG[8]	Riverkeeper[9]
Motivation of CWA	Emphasis on legalistic aspects of CWA interpretation; BTA a confusion; cost is relevant as expressed elsewhere in CWA.	Fairness in protections; aim for best technology; straightforward interpretation of CWA philosophy and 316(b).
AEI as focus	Correct approach; needed to make sense of what is "minimized" in 316(b) and identification of BTA; should be defined for a site; is the part of 316(b) that relates to actual outcomes; lack of AEI definition is central focus of critique.	Any impact counts. Do not need to finesse.
Site-specific study	Says is scientific approach. AEI defined by site-specific study.	Would be co-opted; overly complex; impossible to be fair.
Population impacts vs. direct counts	Population is the only relevant impact, except for endangered species.	Is included in value judgment that individual fish impact defines harm.
Impingement vs. entrainment	Might separate value judgments for adult fish and eggs/larvae.	No separation of roles of larvae and adult fish.
BTA as focus	Error to take technology perspective, as opposed to outcomes. CWA fundamentally ill-conceived here.	Correct approach, consistent with CWA philosophy for waste and stormwater.
Fisheries comparison	Relevant. Need consistent resource management policies for fisheries and other consumptive uses.	No comment made.
Compensation or density-dependence	Relevant in that populations can be healthy with significant acute mortality.	Bogus use to explain away impacts.
Cost	Should use cost/benefit criteria. Implicit elsewhere in CWA.	Not relevant; not a concern as is low for new plants; not part of 316(b).
Role for new vs. existing plants	Unclear.	Unclear.
History of 316(b) decisions	No serious ecological problems.	Several bad decisions made.
EPA use of water body type, littoral zone, etc.	Opposed to EPA definitions, but categories are ones that would be used for site-specific analysis.	Also opposed; seen as a diversion from technology standard.

Volumetric analysis	No basis for use as a proxy.	Key goal: leads to dry cooling; is seen as obvious influencing factor.
Flow limits	Against.	Not strong enough.
What is a CWIS?	Excludes other parts of system such as cooling towers.	Includes system all the way to cooling towers.
Technology basis	Opposes.	Supports and sees as consistent with CWA.
Risk-based approach	Desired as in fisheries management.	No comment made.
Stakeholder role	Implied, not explicit in proposal.	Too complex to implement.
Population modeling	Required.	Too complex.
Other environmental outcomes	Relevant, and should be included in cost/benefit criteria.	Not significant impacts, e.g., noting existing dry cooling practices.
Mitigation	Relevant.	Only for past harm.
Similarity to thermal	Similar in that effects differ from chemical pollution risks.	Not applicable.

NOTES

1 See [1, chapter 11] for a summary of impingement and entrainment data across various water body types, fish species, and CWIS facilities. About 1,000 steam-electric power plants are affected by 316(b), and approximately 70 trillion gallons of water per year are withdrawn by power plants and other facilities for cooling purposes; nearly all of that water is a nonconsumptive use, i.e., it is returned, warmed, to the source body. The principal events in the history of Section 316(b) include:

- 1972: The Federal Water Pollution Control Act Amendments become law including Section 316(b).
- 1973: EPA publishes a "Development Document" addressing 316(b).
- 1976: EPA publishes final 316(b) rules.
- 1977: 316(b) rules are remanded back to the EPA as the result of an electric utility industry lawsuit.
- 1979: EPA withdraws its 316(b) regulations (after this time, the 1973 Development Document is used widely for guidance in making 316(b) decisions).
- 1993: Several environmental groups, led by the Hudson Riverkeeper organization, bring suit against EPA in *Cronin v. Browner* to force EPA to issue regulations (John Cronin is a founder of the original Hudson Riverkeeper[2]).
- 1995: A consent decree in *Cronin v. Browner* provides a schedule for issuing regulations.
- August 2000: EPA publishes *proposed* new CWIS facility rules[1,3].
- November 2001: EPA publishes *final* new CWIS facility rules[4]; rules for existing CWIS facilities are scheduled for 2002–2003, and may depend on the volume of water processed by the CWIS.

See Anderson and Gotting[5] and May and van Rossum[6] for 316(b) history and Dixon et al.[7] for perspectives on 316(b) policy and science. An important observation, made often by the electric power industry and noted in the text, is that 316(b) decisions have been made *de facto* on a case-by-case, site-specific basis for decades; critics claim that many CWISs currently impinge or entrain more fish than is ecologically acceptable, while the industry claims zero or de minimus ecological impacts over decades of once-through CWIS usage.

2 See the Utility Water Act Group[8] and Riverkeeper[9] for responses to the EPA's August 2000 proposed new facilities rule[3].

3 "Adverse environmental impact is a reduction in one or more representative indicator species that (1) creates an unacceptable risk to the population's ability to sustain itself, to support reasonably anticipated commercial or recreational harvests, or to perform its normal ecological function, and (2) is attributable to the operation of the cooling water intake structure. Because this definition is drawn from scientific principles fundamental to natural resource management, it can be interpreted using the same concepts and analytical techniques used by fisheries scientists and resource managers .. The proposed definition turns on 'unacceptable risk.' What risk is 'unacceptable' is a function of a number of biological and social factors, which must be managed through a scientific risk assessment and risk management process"[8]. This definition leaves open the role for a stakeholder process to characterize acceptable risk and its tradeoffs, and the modeling or data collection simplifications that may be needed for a practical regulatory process. These implementation issues noticed by the EPA and the Riverkeeper are made part of the evaluative framework in below.

4 Terrestrial impacts are included by Anderson and Gotting[5], but not UWAG[8], but this seems an insignificant issue. For a rare fisheries example of an aquatic-terrestrial impact, see McEvoy[10].

5 General comparisons of benefits and costs have been suggested[5,6,8]. These authors do not refer to well-known methods of multiple values (or multiple stakeholder) decision analysis as in Dunning et al.[11], Gregory et al.[12], and Keeney[13]. The value hierarchy of Fig. 1 provides the foundation for a multiple values and stakeholder approach to 316(b) decisions; McDaniels et al.[14] use this type of high-level analysis to structure an environmental decision without complete quantification of measures, which could also be useful for many 316(b) decisions. The EPA's final new facilities rule[4] discusses the tradeoffs implied by Fig. 1, but does not endorse a stakeholder approach. The comparative policy framework put forward below identifies some 316(b) issues that complicate a values-focused methodology[13], and is "prescriptive," meaning it combines descriptive and normative views of decision making[15].

6 UWAG's proposal[8] is to make this case-by-case process more explicit by following general risk assessment guidelines. Thus their point that some AEI role is needed is valid. UWAG's contention that ecological impacts are inherently site-specific is supported by the National Research Council[16].

7 Defining BTA as dry cooling is a major part of the Riverkeeper proposal[9].

8 That is, tradeoffs may be more or less formal, ranging from heuristic judgments of an expert panel[17] to the use of codified dollar-equivalents based on local fisheries rules[18,19,20]. These judgments may become quite relevant as rules for existing facilities are debated. As mentioned in the text below, 316(b) ecological valuation may be hard to disentangle from stakeholder negotiations involving the determination of a mitigation or enhancement settlement amount.

9 This is similar to elements of the UWAG proposal and shows how close it is to the multiple values approach of Fig. 1; see also note 5 above.

10 For an example of adverse population decline, see Ambrose et al.[21]: "Besides entraining large numbers of larval and juvenile fish, SONGS [San Onofre Nuclear Generating Station] kills 21 to 56 tons of larger fish each year when they are impinged on the screens in the intakes of Units 2 and 3. The impingement loss itself was not considered by the MRC [Marine Review Committee] to be a substantial effect. However, this loss contributes to the significant decline in local midwater fish abundances detected by the MRC, although other factors (such as the local increase in turbidity, which might cause fish to leave the area) also appear to be involved. The MRC considered the reductions in local fish populations to be a substantial, but local, impact." For an example involving ecosystem health at lower trophic levels, see the California Energy Commission[17]. At the same time, see Rose[22]: "Examples of overfishing causing population declines are numerous. However, despite extensive efforts, definitive quantitative demonstrations of dramatic fish population declines (especially for coastal species) caused by anthropogenic changes in EQ [abiotic aspects of environmental quality, including entrainment

and impingement] are embarrassingly lacking.... As the stakes continue to rise, either we must show the importance of anthropogenic changes in EQ to fish population dynamics, or we must conclude that these changes are insignificant relative to other sources of variation."

11 See Dixon et al.[7] for several approaches to biocriteria and perspectives on their application.

12 Several of the EPA's proposals for new (vs. existing) CWISs fall under this category[3].

13 For example, at the Chesapeake Bay Chalk Point plant[18], several proposed hypotheses and tests about tidal flows and larval transport in the Patuxent River were needed to ascertain that the key consequence of anchovy entrainment was forage fish reduction. The complexity and cost of determining this outcome, and its relative insignificance, may be compared to the much more direct task of reducing impingement from several millions of fish annually to less than one-half million through successful installation of a barrier net. Historical studies such as these may become valuable input for developing guidelines for existing facilities rules. The importance of hypothesis formation and "confrontations" with data in ecological analysis is discussed in Hilborn and Mangel[23].

14 A good example of population modeling for the Ohio River using a basic Leslie matrix approach is in Lohner et al.[24], in which a largely uniform modeling and data collection approach is used to model multiple species at several independent river pools, as opposed to modeling each pool using a different approach. For a contrasting example of intensive, site-specific modeling, see Lorda et al.[25]. The latter is also an example in which stochastic modeling is used to argue for a relatively small role for entrainment, compared to regional overfishing, as a cause of local winter flounder population decline.

15 For an example, see the California Energy Commission[17], in which larval and egg depletions were equated to wetland acreage, which in turn was cost out as a restoration program, and then used to set a final dollar valuation. However, the final mitigation/enhancement choice involved land easements and control of nonpoint source runoff, but not wetland development; the former were judged the higher priorities for improving ecological health at Elkhorn Slough. By putting 316(b) in an overall context of water body management, the CEC process in this way illustrates a desirable feature of site-specific valuation and decision making.

16 For an example for which declines in a single local winter flounder species (Niantic River, Connecticut, stock) becomes a special focus, see Lorda et al.[25]

17 Lohner et al.[24] is an example in which the aggregation from stochastic process to outputs more useful to decision making is carried out in some detail. On the important role played by modelers in making science useful for policy see Jasanoff[26] and van Winkle and Kadvany[27].

18 For an example in which fisheries valuation approaches were utilized, see Bailey et al.[18] For a site-specific approach involving lower trophic levels, see California Energy Commission[17]. For additional cost examples, see May and van Rossum[6]. Nonmarket ecological valuation is not nearly as difficult as many seem to believe, but does depend on recognizing the role to be played by a stakeholder process and the constructive, or stakeholder-defined, nature of valuation judgments[11,12]. For empirical studies relevant to environmental measurement and value definitions see Irwin et al.[28] and Fischhoff[29].

19 On EPA's ecological risk assessment process see [31, 32]. UWAG comes closest to addressing process issues through an AEI definition entailing comparisons of all relevant stakeholder benefits and costs, though it is unclear whether UWAG includes that too as part of a scientific risk assessment process, which it is not. On the need for science-stakeholder interactions, see the National Research Council[30].

20 This issue appears not to be addressed in any 316(b) proposal, but see also Rose[22], p. 381: "Without modeling approaches, isolating EQ [abiotic environmental quality] effects on long-lived species, such as fish, requires years of monitoring that span a range of natural conditions. Often, sufficient data to detect effects are not obtained until it is too late and the population has dramatically declined or easy recovery is hindered. This is not to imply that data collection is not needed. The accuracy and precision of model predictions depends on the quality and quantity of the empirical data used to develop and corroborate the model. But maintaining funding for sufficient long-term monitoring and waiting decades for definitive signals in the data is not possible in many situations.... While population-level analyses tend to require fewer data that are more readily available, they also imply several simplifying assumptions that are often briefly stated and then ignored...[and] the effects of community-level interactions are too often dismissed early in analyses without consideration of how they would affect predicted population responses."

21 In the terminology of the EPA's Ecological Risk Assessment process[31,32], these measurement criteria define risk assessment endpoints.

22 See Adler et al.[33]. The terms "entrainment" and "impingement" do not appear in the book's index.

23 One meaning is that fish can be managed as a resource harvest for CWISs just as they are for fisheries. That view can certainly be defended, but two issues have to be addressed. First is that one can have a division of regulatory labor in which one agency protects and another sets fishing levels, assuming there are not absurd inconsistencies between the two. Depending on the water bodies involved, the health of the fisheries, and other environmental protection goals, it is not immediately obvious that a uniform "resource harvest" approach is necessarily correct. Second, compared to a straightforward "counting" of impacts, as a society we have biases about how we account for all kinds of risks, with a great lack of mathematical consistency. While such unreflective views can be inconsistent, they are not necessarily wrong, as they may be ways of expressing complex, multidimensional attitudes toward risk[34,35], combined with legal and other institutional constraints. Such views can certainly be criticized and changed, and attitudes toward fish protection in different contexts may need that. But then the resource harvest view might be combined with a kind of risk communication or broader policy campaign, and should not be seen as obviously right or wrong[36].

24 See Singer[37], who places fish just inside the phylogenetic boundary of animals able to suffer pain, and argues therefore against their use as a human resource. The argument would not, however, obviously extend to entrainment of eggs and larvae. Singer, however, is also a complete consequentialist who would not distinguish fish killed for food production from those killed to produce electric power.

25 Fish population models are often probabilistic, but the scientific basis for which stochastic model is correct and how to estimate parameters or design a model (fecundity and survival rates, habitat resource constraints, environmental factors, whether compensation is appropriate, etc.) can be subject still to considerable uncertainty. On challenges for satisfactorily establishing population declines in the fisheries industry, see McEvoy[10].

26 The comparison may be a bit strong, given that literal enforcement of the 1958 Delaney Clause could actually *increase* cancer risks[38]. Other differences are that 316(b) deals only with ecological outcomes, and "minimizes" BTA with respect to an undefined AEI, while the Delaney clause implicitly defines adverse as within the limits of detection.

27 The effort-accuracy tradeoff[39] is an important concept in behavioral decision theory that is useful for understanding the practical requirements of policy analysis. As an illustration, and as indicated below, the EPA made regulatory implementation (specifically with respect to the NPDES permitting process) an important factor in their final rule-making. This behavioral consideration is a main difference between a pure values-focused approach[13] and the evaluative framework proposed in the text. Keeney[13] argues against "alternatives-focused" decision making, which would include technology-based rules defined independently of their consequences. However, in creating regulations, due to the quality of scientific knowledge available and the resources and priorities for regulatory implementation, rule-makers may justifiably resort to alternative-based strategies.

28 Anderson and Gotting[5] correctly criticize the open-endedness of the current 316(b) case-by-case process, but they do not appreciate that their site-specific approach does little to limit the range of normal variability known in empirical and analytical studies of multiple values decision making; see the decision analysis references in note 5 above.

29 For a range of mitigation values see May and van Rossum[6].

30 Anderson and Gotting[5] argue that it is a scientific judgment that the population level is correct. But this view is mistaken: this is a value judgement, held by some scientists, perhaps not by others. It is not a consequence of any scientific theory, in contrast to whether one can easily influence a population by managing early life stages, or the relevance of high levels of natural mortality for eggs and larvae. Arguments that all ecological measures are value-laden by some implicit or explicit choices for ecological health or well-being are in Lackey[40,41] and Shrader-Frechette and McCoy[42].

31 An unfortunate simplification in this paper is not better distinguishing existing and new facilities. Indeed, the whole 316(b) debate is complicated by the sequential promulgation of rules (first new, then existing), making for considerable second-guessing and possible miscommunication

about the consequences of earlier rules for later ones. For example, existing facility rules are to be further broken down by water volumes, with that measure likely a key proxy for the largest, often nuclear, plants. In their final new facilities rule[4], the EPA also states that a completely different approach is possible for existing facilities. Given the large cost of retrofitting existing CWIS to reduce entrainment, their approach to existing facilities will likely be quite different than for new ones.

32 See Dixon et al.[7] for papers on biocriteria approaches, strengths, and weaknesses. The construction of such quasi-scientific metrics occurs elsewhere in environmental policy, such as for pesticides or contaminated site regulatory ranking. A key problem with such metrics is that the final "units," while often providing useful diagnostic information, have no clearly defined dimensions for use as significance measures.

33 See note 3 above for the UWAG definition.

34 See Landy et al.[43] on substituting technical problems for value choices. Both the Riverkeeper and UWAG have criticized some of the EPA's biocriteria proposals, perhaps the single point on which they agree. But these groups may not recognize the need for such simplifications in regulatory policy, and why such measures appear for many environmental decisions such as those involving contaminated waste sites and pesticide safety (see note 27 above). Proxy measures are one approach to making use of science in a regulatory setting, but another approach is to allow for more flexible judgmental process of scientific outputs, and not to try to "customize" science to policy using quasi-scientific constructs. The EPA's final new facilities rule[4] rejects the biocriteria and proxies approach.

35 See National Research Council[16] on the ecological method.

36 See note 5 above.

37 This concern is raised by Riverkeeper[9] and may reappear in debates about forthcoming existing facilities rules.

38 See May and van Rossum[6], California Energy Commission[17], and Bailey[18] for various mitigation or enhancement projects. Some mitigation examples may indicate that the large project budgets involved, and modest benefits achieved, leave room for improvement[6].

39 For relevant issues, see Zajac[44].

40 The EPA's *proposed* new facilities rule is USEPA 2000[3] and the *final* new facilities rule is USEPA[4].

41 See Foster and Huber[45] for the scientific, judgmental and legal difficulties in identifying causality; for a 316(b) example involving over-fishing on Long Island Sound, see Lorda et al.[24].

Using Attainment of the Designated Aquatic Life Use to Determine Adverse Environmental Impact

Greg Seegert
EA Engineering, Science, and Technology, 444 Lake Cook Rd., Suite 18, Deerfield, IL 60015; 847-945-8010

E-mail: gls@eaest.com

Received November 19, 2001; Revised March 18, 2002; Accepted April 8, 2002; Published February, 2003

Section 316(b) of the Clean Water Act requires that cooling-water intake structures (CWIS) use Best Technology Available (BTA) to minimize adverse environmental impacts (AEI). The U.S. EPA has not defined AEI, and there is no clear consensus regarding its definition. Nonetheless, operational definitions are necessary to evaluate design alternatives and to measure the success of mitigative measures. Rather than having to develop measures of aquatic health that are highly site-specific, controversial, and often unlikely to elicit agreement from all sides of the environmental "fence", " it may be more productive to use existing ecological assessment tools. Aquatic Life Uses (ALU) already provide a regulatory framework to assess the quality (health) of the aquatic community in various habitats (e.g., warmwater habitat, exceptional warmwater habitat). Attainment of the ALU indicates that further point source controls are unnecessary, whereas nonattainment indicates that those pollutants or stressors causing the nonattainment must be reduced. A similar approach for existing water intakes is recommended. That is, attainment of the designated ALU will be taken as an indication that there is no AEI. Although attainment of the ALU may not be a foolproof indicator of a lack of AEI, this approach seems more reasonable that using scarce monetary resources to fix problems that likely do not exist, or having both regulators and the regulated community expend their resources debating whether various observed biological responses do or do not constitute AEI.

KEY WORDS: use attainment, aquatic life use, use designations, adverse environmental impact, Clean Water Act, 316(b), cooling water intake structure

DOMAINS: freshwater systems, ecosystems and communities, water science and technology, environmental management and policy, environmental technology, environmental monitoring

---

INTRODUCTION

In 1972 the U.S. Congress passed Public Law (PL) 92-500, the Federal Water Pollution Control Act Amendments of 1972 (the "Act" or "CWA"). Section 316(b) of the Act addressed cooling-water intake structures (CWIS). This section required that "... the location, design, construction, and capacity of cooling water intake structures reflect the best technology available for minimizing adverse environmental impact (AEI)." Despite the passage of nearly 30 years, scientists, regulators, and resource managers have yet to reach a consensus on what constitutes AEI[1]. This lack of consensus is reflected in the title of this Symposium, "Defining and Assessing Adverse Environmental Impact." Based on the titles of their papers, more than half the authors at this symposium attempted to define what level of organization (individual, population, or community) was appropriate to assess AEI or how AEI should be measured at the chosen organizational level. Thus, in their titles there are words or phrases like "defining and assessing AEI"[2], "indicators of AEI"[3], "the challenge of defining endpoints"[4], and "defining AEI"[5].

As part of a 1995 Consent Agreement, the U.S. EPA is revising Section 316(b) of the CWA. In the final rule for New Sources intakes, signed 9 November 2001, the EPA did not define AEI. Because no definition has been established, the debate over AEI continues, in part because the context for such a definition involves societal as well as scientific considerations[6]. Indeed, one of the more contentious debates during public meetings held in conjunction with promulgating the new 316(b) rules has been how to determine what level of organization to assess. One faction believes that in a societal context, the loss of even one individual (the "one dead fish" position) is unacceptable, while many scientists believe that assessment must be at a higher level of organization (population or community). Given the chasm that exists between these two positions, the EPA will have to display Solomon-like wisdom to appease all parties. Interpretation of AEI is particularly important to the electric utility industry because of the volume of cooling water they use and, at least, the potential to affect fish populations adversely.

Historically, the EPA consistently made decisions regarding 316(b) compliance on a site-specific basis, relying on information provided as part of 316(b) "Demonstrations." However, for new sources, the EPA has recently established generic guidelines that apply regardless of the waterbody type (e.g., rivers and streams vs. lakes and reservoirs) on which the CWIS is located. This generic approach makes it even more difficult to define AEI since populations and communities vary from site to site. A final complication is that roughly half of the U.S.'s electric generation capacity is located on large rivers (unpublished EPRI data), but the EPA has no guidance document describing how to assess biological populations or communities in large rivers. Because of this lack of guidance, the EPA has decided against using biological criteria as part of the 316(b) assessment process. Thus, even if agreement is reached on what constitutes AEI, there are currently no standard protocols to measure and assess endpoints on the waterbody type that contains the highest percentage of the nation's generating capacity.

AQUATIC LIFE USES: A PROPOSED ALTERNATIVE

Rather than continuing the 30-year old debate over what constitutes AEI, or how, when, or where to measure it, I recommend an alternative approach, which is determining whether or not the Aquatic Life Use (ALU) for the waterbody in question has or has not been attained. Water-quality criteria consist of two parts, use designations and criteria to protect those uses. Examples of use designations include aquatic life, public water supply, agricultural water supply, and several others. However, ALU is the only use relevant to assessing AEI.

The ALU designation can be subdivided according to the type (e.g., warmwater, coldwater, seasonal salmonid, etc.) or the quality (general, limited, or modified) of the fishery. Some states have a limited number of ALU categories. For example, Illinois classifies almost all of its waterbodies as General Use, except for a few classified as secondary contact. However, other states have a more elaborate system that considers both the type and quality of the fishery. For example, Ohio classifies streams as warmwater or coldwater. The warmwater streams are further subdivided depending on the quality of the biota and various physical (habitat) constraints. Thus, the ALU in Ohio streams can range from limited-resource water (the poorest quality streams), to modified warmwater habitat, to warmwater habitat, and ultimately to exceptional warmwater habitat, the highest quality streams in the state. If habitat quality imposes certain long-term constraints on the quality of the biota due to factors such as channelization, impoundments, etc., then the stream may be given a "modified" classification, as Ohio has done.

Ohio relies exclusively on biological endpoints to determine whether or not each waterbody is in attainment with regard to its designated ALU. Section 3745-1-07(A)(5) of the Ohio Water Quality Standards indicates that biological criteria established by the state provide "direct measure of the attainment of the ... aquatic life uses." This same section goes on to indicate that "demonstrated nonattainment of the applicable biological criteria ... will cause the director to seek and establish, if possible, the cause of the nonattainment of the designated use." Lastly, it indicates that (given nonattainment), "the director shall, whenever possible and reasonable, implement regulatory controls or make other recommendations regarding water resource management to restore the designated use." This evaluation is used by the Ohio EPA during each NPDES permit renewal cycle to determine whether or not more stringent effluent limits are needed. If the area is found to be in full attainment of the applicable ALU, then additional point source controls are not recommended. However, if biological scores are low and the area is not in attainment, then further controls over point source discharges are typically required (assuming that the nonattainment is likely due to a water quality problem).

I recommend a similar approach be taken with regard to determination of AEI. There are two basic scenarios:

1. The waterbody containing the CWIS is in attainment of its ALU.
2. The waterbody is not in attainment.

If the waterbody is in attainment of the ALU (Scenario 1), then it should be assumed that there is no AEI, because if there was an AEI, attainment would be precluded. If the waterbody is not in attainment, then an AEI is possible, though not confirmed (Scenario 2). For this approach to work, mechanisms need to be in place for measuring attainment. As discussed later in this paper, these mechanisms are well established in some states but poorly defined in others. A lack of attainment provides empirical evidence that some kind of problem exists but does not establish causation[7]. The lack of attainment could be the result of AEI from a CWIS, but it also could be the result of poor water quality, habitat limitations, or site-specific considerations (e.g., drought conditions). Further studies would likely be needed to establish the cause(s) of the nonattainment. Several recent papers[8,9] address the problem of dealing with multiple stressors. If it can be established that the lack of attainment is due to the CWIS, then mitigative measures to reduce or eliminate the AEI would be required. In this regard, it is interesting to note that even though states are required to list the reason(s) for nonattainment in their biannual report to Congress, losses of aquatic organisms at CWIS have never been identified as a reason for nonattainment.

There are several advantages to the proposed approach. First, as part of Section 305(b) of the CWA, states are already required to determine which of their waterbodies are impaired (i.e., not in attainment). Second, it allows the EPA and the states to focus on those waterbodies that need help (i.e., are not in attainment) and not expend scarce resources on waterbodies that are already meeting their CWA goals (i.e., they are in attainment). Third, owners of CWIS can focus their resources on sites that really need it. If a site containing a CWIS is found to be in nonattainment, the owner of that CWIS may want to initiate a 316(b) demonstration to determine to what extent, if any, the CWIS is contributing to the nonattainment. Simultaneously or alternatively, the owner might choose to investigate other reasons for the nonattainment. If it is found that the nonattainment was entirely or partially caused by the CWIS, then it would be reasonable to expect the owner of the CWIS to take actions to mitigate the AEI (i.e., restore the designated use) as required by the 316(b) rule. Thus, the process follows a logical progression; (a) determine if there is a problem, (b) determine if the CWIS causes or contributes to the problem, and (c) if the CWIS causes or contributes to the problem, then the AEI would be established and mitigation of that AEI would be required. A major benefit of this approach is that is avoids the contentious debate over what constitutes AEI.

Another broad benefit of the approach recommended herein is that it provides flexibility. A state or tribe can establish a variety of ALUs, each with a unique series of indicators and benchmarks. Of necessity, these measures will be biological in nature, but the state could choose those that they believe are most appropriate (fish, benthos, shellfish, or some combination of these) for their particular waterbodies. They would also be free to choose the organizational level they wish to measure. For example, they could look at the individual level (e.g., external anomalies, condition, etc.), at the population level (e.g., the abundance of target species), or at the community level (e.g., multimetric indices like the Index of Biotic Integrity[10,11] or the Invertebrate Community Index[12]). Further flexibility can be gained by varying the benchmark(s) over some spatial scale (e.g., by ecoregion, river basin, or even waterbody-specific).

A CASE HISTORY EXAMPLE

To illustrate how this approach would work, biological data from two power plant sites on the Muskingum River in Ohio are discussed below. The Ohio EPA[12] uses three biological indices to determine attainment of ALU, the Index of Biotic Integrity (IBI)[10,11], the Index of Well-Being (IWBmod) (modified from Gammon[13]), and their own Invertebrate Community Index[12]. Recent benthic data are not available for the Muskingum River sites so this example will focus on the two fish indices. For this region of Ohio, the Ohio EPA has established biocriterion of 40 for the IBI and 8.6 for the IWBmod for collections made by boat electrofishing.

In 1999, from late June through September, boat electrofishing collections were made near two power plants on the Muskingum River, the Muskingum River Plant (MRP), and the Conesville Generating Station (CGS). Five stations were established near each plant, two upstream of the plant and three downstream of it. All collections were made following standard Ohio EPA protocols[12]. At MRP, the mean IBI score both upstream and downstream of the plant was 40. The mean IWBmod score upstream of the plant was 8.9, nearly identical to the score (8.8) downstream of it. Because mean index scores all met the respective biocriterion for this river, the logical conclusion is that the CWIS at the MRP does not cause AEI. This is not to say that the CWIS had absolutely no effect as it surely entrained and impinged a certain number of fish. However, the fact that the area continues to meet objectively established biocriteria and thus is in attainment of the designated use indicates that these effects are biologically insignificant, and thus do not constitute AEI.

Although this example was for an operating CWIS, this process could also be used for new sources. Baseline studies could be conducted before the CWIS was operational. If monitoring after the CWIS began operating demonstrated that the area was no longer in attainment, a logical conclusion would be that the subsequent lack of attainment was due to the CWIS (assuming other factors remained the same during the pre- and post-operational periods).

The situation at the CGS is somewhat more complex. The Ohio EPA has studied this reach of the river several times, most recently in the mid-90s[14]. Their studies suggested that several stressors were, or at least had been, present.

Thermal stress was identified as a problem in 1988 during extremely low flows. However, thermal load management by the CGS appears to have addressed this problem. Organic loadings from an upstream paper mill appear to be the most significant current problem[14]. During the 1999 studies, mean IWBmod scores upstream and downstream of CGS were similar (8.8 and 8.5, respectively) and both met the biocriterion of 8.6 (the Ohio EPA allows a deviation below the criterion of 0.5 IWB units to account for normal variability). IBI scores both upstream and downstream of the plant (44 and 37, respectively) met the biocriterion of 40 (for the IBI, the Ohio EPA allows a deviation of 4 units). However, the difference in mean scores between the two areas was statistically significant. The reasons for the upstream/downstream difference are unclear but did not appear to be related to the CWIS. This difference might have been attributable to the thermal component of the discharge, to a DO sag associated with the upstream paper mill, or other factors. Given the facts that (1) both areas meet the biocriterion, and (2) this spatial difference was likely not attributable to the CWIS, it is reasonable to conclude that there is no intake-related AEI at CGS.

Thus, studies near two large CWIS on the Muskingum River demonstrate that measuring attainment of the designed ALU is appropriate and workable, as well as a reasonable surrogate to establish AEI.

LIMITATIONS OF ALU APPROACH

According to Section 305(b) of the CWA, states are required to determine which waterbodies within their jurisdiction are in attainment and which are not (i.e., which are impaired). Theoretically, each state should already have in place objective quantifiable mechanisms for determining attainment/nonattainment. Some states (e.g., Ohio) do have such procedures in place and in the case of Ohio have even codified the process. However, in other states, the process is less defined and more subjective, thereby making the assessment process more difficult. In these situations, more refined ALUs and/or more detailed guidance regarding the process of how attainment is determined may be necessary. Although it is the responsibility of the states, tribes, and EPA to develop such guidance, the burden of these development efforts may fall to 316(b) applicants. Because, however, the states are already required to address the issue of attainment/nonattainment, the process recommended herein should add little to the regulatory burden of each state. It takes advantage of something the states already have to do. Furthermore, following this approach will side step the contentious issue of what constitutes AEI, and it will allow states to focus on those waterbodies where there are demonstrated problems.

ACKNOWLEDGMENTS

The author would like to thank Dr. Tim Lohner and Mr. Chris Yoder for helpful comments they provided on a draft of this paper.

REFERENCES

1. Mayhew, D., Muessing, P., and Jensen, L. (2002) Adverse environmental impact: 30-year search for a definition. *TheScientificWorldJOURNAL* **2(S1),** 21–29.
2. Hickman, G. (2002) Proposed methods and endpoints for defining and assessing adverse environmental impact (AEI) in Tennessee River reservoirs. *TheScientificWorldJOURNAL* **2(S1),** in press.
3. Barnthouse, L., Heimbuch, D., Anthony, V., Hilborn, R., and Meyers, R. (2002) Indicators of AEI applied to the Delaware Estuary. *TheScientificWorldJOURNAL* **2(S1),** in press.
4. Van Winkle, W. and Coutant, C. (2002) The challenge of defining endpoints and risk criteria for 316(b) assessments. *TheScientificWorldJOURNAL* **2(S1),** submitted.
5. Bailey, D., Bulliet, K., and Christman, J. (2002) Defining adverse environmental impact: a fisheries approach. *TheScientificWorldJOURNAL* **2(S1),** in press.
6. Strange, E., Snyder, B., Nagle, D., Morgan, J., Jr., Tudor, L. (2001) Scientific and societal considerations in selecting assessment endpoints for environmental decision-making. *TheScientificWorldJOURNAL* **2(S1),** 12–20.
7. Seegert, G. (2000) Considerations regarding development of Index of Biotic Integrity metrics for large rivers. *Environ. Sci. Policy* **3**, 599–606.
8. United States Environmental Protection Agency (2000) Stressor Identification Guidance Document. EPA/822/B-00/025. Office of Water, Washington, D.C.
9. Yoder, C.O. and Rankin, E.T. (1995) Biological response signatures and the area of degradation value: new tools for interpreting multimetric data. In *Biological Assessment and Criteria: Tools for Water Resource Planning and Decision Making*. Davis, W. and Simon, T., Eds. Lewis Publishers, Boca Raton, FL. pp. 263–286.
10. Karr, J.R. (1981) Assessment of biotic integrity using fish communities. *Fisheries* **6**, 21–27.
11. Karr, J.R., Fausch, K.D., Angermeier, P.L., Yant, P.R., and Schlosser, I.J. (1986) Assessing biological integrity in running waters: a method and its rationale. *Ill. Nat. Hist. Surv. Spec. Publ.* **5**, 28.
12. Ohio Environmental Protection Agency (1987) Biological Criteria for the Protection of Aquatic Life: Users Manual for Biological Field Assessment of Ohio Surface Waters. Vol. 2. Division of Water Quality Monitoring and Assessment, Surface Water Section. Columbus.
13. Gammon, J. (1976) The Fish Populations of the Middle 340 km of the Middle Wabash River. Tech. Report No. 32. Water Resources Center, Purdue University, Lafayette, IN, 73 p.
14. Ohio Environmental Protection Agency (1995) 1995 Biological and Water Quality Study of the Upper Muskingum River Basin. Ohio EPA, Division of Surface Water, Columbus.

Defining "Adverse Environmental Impact" and Making § 316(b) Decisions: A Fisheries Management Approach

David E. Bailey[1,*] and Kristy A.N. Bulleit[2]
[1]*Mirant Corporation, 8711 Westphalia Road, Upper Marlboro, MD 20772*
[2]*Hunton & Williams, 1900 K Street, N.W., Washington, D.C. 20006-1109*

Received November 8, 2001; Revised February 21, 2001; Accepted March 1, 2002; Published February, 2003

The electric utility industry has developed an approach for decisionmaking that includes a definition of Adverse Environmental Impact (AEI) and an implementation process. The definition of AEI is based on lessons from fishery management science and analysis of the statutory term "adverse environmental impact" and is consistent with current natural resource management policy. The industry has proposed a definition focusing on "unacceptable risk to the population's ability to sustain itself, to support reasonably anticipated commercial or recreational harvests, or to perform its normal ecological function." This definition focuses not on counting individual fish or eggs cropped by the various uses of a water body, but on preserving populations of aquatic organisms and their functions in the aquatic community. The definition recognizes that assessment of AEI should be site-specific and requires both a biological decision and a balancing of diverse societal values. The industry believes that the definition of AEI should be implemented in a process that will maximize the overall societal benefit of the § 316(b) decision by considering the facility's physical location, design, and operation, as well as the local biology. The approach considers effects on affected fish and shellfish populations and the benefits of any necessary best technology available (BTA) alternatives. This is accomplished through consideration of population impacts, which conversely allows consideration of the benefits of any necessary BTA modifications. This in turn allows selection of BTAs that will protect potentially affected populations in a cost-effective manner. The process also employs risk assessment with stakeholder participation, in accordance with EPA's Guidelines for Ecological Risk Assessment. The information and tools are now available to make informed decisions about site-specific impacts that will ensure protection of aquatic ecosystems and best serve the public interest.

KEY WORDS: entrainment, impingement, 316(b), adverse environmental impact, fishery, survival, intake technology, costs and benefits, maximum net benefit, cooling water, intake structure

* Corresponding author. Email: david.e.bailey@mirant.com; kbulleit@hunton.com

DOMAINS: freshwater systems, marine systems, ecosystems and communities, organisms, water science and technology, environmental technology, environmental management and policy, computational biology, environmental modeling, environmental monitoring, information management

INTRODUCTION

Generating electric power requires cooling water to condense steam after it is used in steam-powered turbines. Withdrawing cooling water from surface waters for this purpose can impinge fish on screens and entrain fish and shellfish, eggs, and larvae. Impingement is the entrapment of fish or shellfish on screens that are used to prevent condenser blockage. Entrainment is the passing of organisms through the cooling water system, which may cause mortality from exposure to heat, physical stress, or chemicals.

In § 316 of the Clean Water Act, Congress included a subsection (a) to allow variances from thermal standards, if it is demonstrated that there will be "protection and propagation of a balanced, indigenous population of shellfish, fish and wildlife in and on the waterbody." Immediately following is § 316(b), which states that any standard applicable to a point source under § 301 or § 306 of the Act "shall require that the location, design, construction, and capacity of cooling water intake structures reflect the best technology available for minimizing adverse environmental impact." The U.S. Environmental Protection Agency (EPA), driven by a lawsuit in federal district court in New York State, is conducting a rulemaking to implement § 316(b)[1].

The purpose of this paper is to contribute to the development of § 316(b) regulations that will both protect living aquatic resources and reflect sound social policy. It addresses the following topics:

- The history of § 316(b) and EPA's current approach to the rulemaking
- The need for a definition of "adverse environmental impact"
- The need for a rule based on the tools and principles of fisheries management science
- The need for a rule that maximizes net social benefit
- A suggested approach that meets these needs.

A BRIEF HISTORY OF § 316(B) AND EPA'S 316(B) RULEMAKING

Congress enacted § 316(b) of the Clean Water Act in 1972. The language of § 316(b) first appeared in the Conference Report on the 1972 Federal Water Pollution Control Act Amendments in a section called "Thermal Discharges." There was no comparable language in earlier House or Senate bills and little testimony or debate in the record explaining its sudden appearance. It appears, in fact, to have been an afterthought[2].

In December 1973, little more than a year after the statute was enacted, EPA proposed a rule to implement § 316(b). The rule was finalized in 1976. Both the proposed and final versions referenced EPA Development Documents, which described factors and design alternatives to consider when making a § 316(b) determination. A preamble to the 1976 final rule said that "decisions relating to the best technology available are to be made on a case-by-case basis." The rule was short-lived, for the Fourth Circuit Court of Appeals set it aside on procedural grounds.[1] In 1977, EPA published a draft guidance document, but this was never finalized[2,3]. For over 20 years, § 316(b) has been widely implemented on a site-specific basis, guided by the 1977 draft guidance rather than by regulations.

In 1993, several environmental groups filed suit against EPA in a U.S. district court in New York, seeking to compel EPA to issue regulations to implement § 316(b).[2] EPA and the environmental plaintiffs settled the case and agreed to a rulemaking schedule in a consent agreement entered by the court.

EPA's final rule for *new* facilities was published in the Federal Register, December 18, 2001, and a new proposed rule for *existing* facilities was published April 9, 2002. Although new and existing facilities do deserve different treatment under § 316(b), many issues raised by the proposed new facilities rule will be the same as or similar to the issues for existing sources.

THE NEED FOR A DEFINITION OF "ADVERSE ENVIRONMENTAL IMPACT"

In the "Phase I" rulemaking for new facilities, EPA reports that it has received numerous comments addressing how "adverse environmental impact" (AEI) should be defined[4]. A definition is important because it establishes the basis for resource protection and provides a standard for selecting best technology available (BTA), in cases where BTA is required.

While a number of possible definitions of AEI have been offered, the following definition, proposed by the Utility Water Act Group (UWAG), is both scientifically sound and socially relevant for § 316(b) decisionmaking: "Adverse environmental impact is a reduction in one or more representative indicator species that (1) creates an unacceptable risk to the population's ability to sustain itself, to support reasonably anticipated commercial or recreational harvests, or to perform its normal ecological function and (2) is attributable to the operation of the cooling water intake structure"[5].

This definition focuses on protection at the population level. As stated in AFS Policy Statement #1, a goal of fisheries management is "to ensure self-sustaining populations that would support commercial and recreational fishing both now and

---

[1] *Appalachian Power Co. v. Train*, 566 F.2d 451, 459 (4th Cir. 1977).
[2] *Cronin v. Browner*, 898 F. Supp. 1052 (S.D.N.Y. 1995).

in the future"[6]. As Suter and Barnthouse concluded, "(t)he reproducing population is the smallest ecological unit that is persistent on the human time scale, and hence the lowest level that we can meaningfully protect"[7].

Despite this emphasis on population-level effects, it is recognized that for species whose populations are at critically low levels, the population can become endangered, in which case the protection of individual organisms through the Endangered Species Act[3] is appropriate. In addition to the federal statute, many states have enacted similar endangered species legislation.[4] These statutes, already in place, should and will be applied no matter what § 316(b) regulatory process EPA ultimately adopts.

The proposed AEI definition set out above also acknowledges that ecosystem integrity, structure, and function must be protected and, from a fisheries management perspective, that reasonably expected harvests should not be impaired. Finally, the recommended definition of AEI incorporates the idea of risk and therefore invokes risk management as part of the AEI decisionmaking process.

THE NEED FOR A RULE BASED ON THE TOOLS AND PRINCIPLES OF FISHERIES MANAGEMENT SCIENCE

The effect of cooling water intake structures (CWIS) on fisheries is fundamentally similar to the effects of recreational and commercial harvesting of fish and associated effects of bycatch and bait collection. One primary difference is which species are affected. Fishery harvesting, of course, targets species that are desirable for human or animal food consumption and sport interest, while CWIS losses are a function of the interaction of fishery populations with the CWIS. CWIS vulnerability tends to be highly variable, depending on the CWIS location, design, and species' life history and behavior. Nevertheless, the similarities between losses from fishing and CWIS losses are such that CWIS effects on the fishery can be evaluated using the same basic approaches used by state and federal fishery managers to manage their commercial and recreational fisheries. The species and sizes of fish and shellfish impinged and entrained can be quantified and evaluated in the context of fishery management tools, including long-term populating monitoring, annual harvest levels, models, and natural resource protection regulations. As part of their management efforts, fisheries managers have learned to manage complex trade-offs. For example, increasingly they are being asked to weigh trade-offs between game, nongame, native, and nonnative species management[8].

The fisheries management approach views the fishery as a renewable resource that can be managed. It recognizes that the federal government need not protect

3 Endangered Species Act of 1973, as amended, 16 U.S.C. §§ 1531-44.

4 See, *e.g.*, South Carolina Nongame and Endangered Species Conservation Act, S.C. Code Ann. § § 50-15-10 to -90; New Hampshire Endangered Species Conservation Act, N.H. Rev. Stat., Title XVIII, chap. 212-A; California Endangered Species Act, CA Fish & GD 3, chap. 1.5, §§ 2050 - 2116; Massachusetts Endangered Species Act, M.G.L.A. 131A; Illinois Endangered Species Protection Act, 520 Ill. Comp. Stat. (ILCS) 10/1 - 10/11.

every fish (leaving aside endangered species, which require special treatment), let alone every egg, but should instead preserve the fishery resource itself. Fisheries managers know that a certain level of cropping of fish stocks can occur without destroying a population's ability to sustain itself.

How low is too low? While the fishery science literature does not provide a definitive answer to this question, NMFS believes that a prudent rule can be established as follows: Two of the best known models in the fishery science literature find that, on average, the stock size at MSY (maximum sustainable yield) is approximately 40% of the stock size that would be obtained if fishing mortality were zero (the pristine level). . . . Also, the fishery science literature contains several suggestions to the effect that any stock size below about 20% of the pristine level should be cause for serious concern. *In other words, a stock's capacity to produce MSY on a continuing basis may be jeopardized if it falls below a threshold of about one-fifth the pristine level* (emphasis added)[9].

Commonly Used Fishery Reference Points

Due to similarities of CWIS impacts and commercial and recreational fishing impacts, fishery management tools have been commonly applied to evaluate these impacts[57]. Regulations issued by NMFS and the Fish and Wildlife Service (FWS) incorporate the concept of "optimum yield" of a fishery, based in turn on the concept of "maximum sustainable yield" (MSY) (50 C.F.R. 600.310(c)(1)(i) (1999)). MSY is defined as "the largest long-term average catch or yield that can be taken from a stock or stock complex under prevailing ecological and environmental conditions" (*id.*). Currently, tools such as Biomass per Recruit (BPR) and spawning stock measures are more in favor than MSY.

NMFS recognizes that maximum productivity from a stock can be achieved by reducing the stock size by as much as 60% and that the population will be able to sustain or replace itself until the stock size is reduced by about 80%. Fishery managers consider removal of 70 to 80% of an unfished stock's biomass (Spawning Stock Biomass or SSB) and 65 to 80% of a stock's reproductive potential (Spawning Stock Biomass per Recruit or SSBPR) to be safe, given the compensatory reserve inherent in most fish stocks[10,11]. "Spawning Stock Biomass per Recruit" (SSBPR) is the total weight of a mature spawning stock that would be generated over the lifetime of an individual recruit[12].

When reliable estimates of the compensatory capacity of a population exist, spawner-recruit models can be used to develop more realistic and less conservative biological reference points[13]. As with the SSBPR approach, spawner-recruit analyses show that mortality due to entrainment and impingement is likely to have negligible effects on the abundance or yield of a fish population unless that population is already being fished at a level that greatly exceeds F_{msy}.

Biological reference points and quantitative assessment tools used in fisheries management can also be used to evaluate the likelihood that entrainment and impingement mortality will reduce the reproductive capacity of a fish population to a level that warrants management concern. Fisheries management concepts, therefore, provide scientifically sound principles for determining whether cooling-water withdrawals can cause "adverse environmental impact" to vulnerable fish populations.[5]

Risk Assessment

No matter how sound the definition of AEI and the available assessment tools, a decisionmaking process that must decide "how much is too much" cannot escape uncertainty[15]. Assessing AEI inevitably calls for an assessment of risk to affected populations (or, for new facilities, *potentially* affected populations), to the aquatic community, and to the fishery. EPA's Ecological Risk Assessment Guidelines[16] provide a three-phase process of problem formulation, analysis, and risk characterization useful for AEI decisionmaking. The final product is a risk description that includes an interpretation of ecological adversity and descriptions of uncertainty and lines of evidence.

In short, the effect of cooling water intake structures on fisheries has many similarities to the effects of commercial and recreational fishing and associated effects (bycatch and removal of bait fish). Thus, the same general field and analytical methods developed for use in fishery management can be and have been applied to assess the effects of a CWIS on fish and shellfish in waterbodies from which cooling water is withdrawn.

THE NEED FOR A RULE THAT MAXIMIZES NET SOCIAL BENEFIT

Balancing Fishery Protection and Other Uses

The CWA establishes the protection of fisheries as a national goal [Clean Water Act § 101(a)(2), 33 U.S.C. § 1251(a)(2)]. Many states have likewise adopted this goal.[6] However, society has many goals for management and use of water resources, such as flood control, public water supply, agriculture, industrial water supply, and commercial and recreational fishing. Each of these uses results in impacts to fisheries, and it would be irrational to manage or regulate water resources solely for a single use such as maximizing fish production.

While any of these uses could be eliminated, to do so would result in a significant social cost. To take just one example, hydroelectric power is one of the most

5 In addition to the standard fisheries management assessment tools, § 316(b) studies and other research have led to a wide range of analytical tools for assessing population-level effects. The Electric Power Research Institute recently published a catalog of analytical methods and models useful for § 316(b) decisionmaking[14].

6 See, *e.g.*, Cal. Fish & G. Code §§ 2851, 8230 (2001); Rev. Code Wash. (ARCW) §§ 77.04.012, 77.70.160 (2001); R.I. Const. Art. I, § 17 (2001); La. Rev. Stat. 56:579.1 (2000).

significant in terms of volume withdrawn from a waterbody, but it also provides significant benefits such as (1) flood protection, (2) preservation of water during high-flow periods for use during low-flow periods, (3) recreational benefits, (4) increased fish habitat, (5) power production, and (6) economic development. To be sure, hydropower has deleterious effects, such as habitat fragmentation, blocking of the passage of fish, and effects on dissolved oxygen. But massive efforts are underway to mitigate these effects through impact assessments under the National Environmental Policy Act and relicensing proceedings by the Federal Energy Regulatory Commission.

Perhaps the most significant impact on fish – particularly in estuarine and marine waterbodies – is fishery exploitation[17]. In addition to the direct harvest of fish, fishery impacts occur through bycatch and bait fish removal. Another manner in which fisheries can be affected is by the deliberate introduction of nonnative species into waterbodies to promote recreational fisheries – e.g., introduction of Pacific salmon into the Great Lakes to create a recreational trout fishery and introduction of gizzard shad into reservoirs as a food source to increase sport fish populations.

In addition to water withdrawals and fishery harvests, human activities can alter fish populations in other ways. For example, land development or agricultural activities can cause sedimentation, habitat loss, and nutrient enrichment and affect dissolved oxygen levels and/or water temperature and clarity[18] and ultimately impact fisheries. Water transportation can also impact fisheries as a result of construction of navigation channels and shipping (*e.g.*, the Welland Canal, which introduced the sea lamprey into the Great Lakes, affecting the lake trout fishery) and the associated navigational use of the waterways, which can introduce exotic species in ballast water.

It is in this broader context of multiple impacts on fisheries and competing societal costs and benefits that we should approach the task of protecting fisheries from entrainment and impingement, while still providing a reliable source of electric power. Fig. 1 illustrates the three key aspects of sound § 316(b) decisionmaking. These aspects are (1) evaluation of biological conditions in the vicinity of the CWIS and assessment of the impact or potential impact to the fishery; (2) analysis of the location of the CWIS (*i.e.*, waterbody type and local aquatic community where the facility is located); and (3) CWIS design considerations.

Biological Conditions and CWIS Impacts

Fishery management/assessment methods and tools that are available to assess fisheries and impacts from the interaction of the CWIS and the fishery were discussed earlier in this paper. Other authors – including EPA in the Economic and Engineering Analyses Report developed for the Phase I § 316(b) rule[19] – have documented that very large numbers of organisms may become entrained or impinged at a single facility. If this is so, why haven't CWIS impacts been a more prominent national issue? There are a number of reasons:

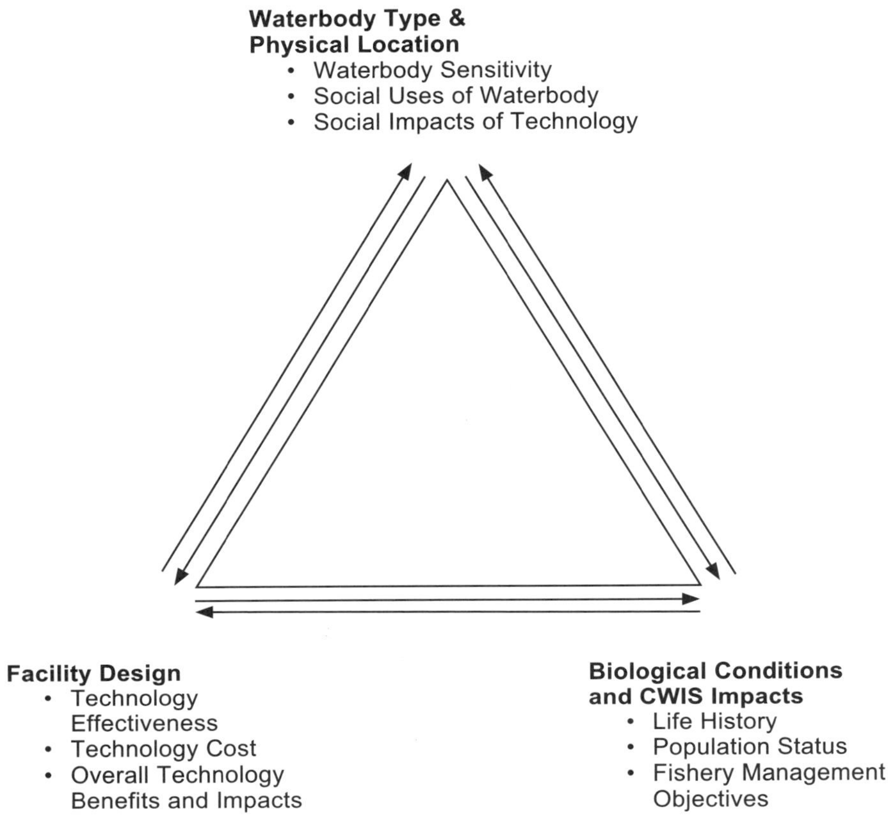

FIGURE 1. Key components for effective 316(b) decisionmaking.

- **§ 316(a) and (b) Studies** – Many states have already developed and implemented §§ 316(a) and (b) regulatory programs, including Maryland, Delaware, New York, New Jersey, Tennessee, South Carolina, California, Michigan, Ohio, Illinois, Alabama, Kentucky, Indiana, and others.[7] Studies conducted by companies located in these states (and, in some instances, independent studies conducted by the states themselves), including some long-term studies, provide a good baseline for understanding power plant fishery impacts[20]. Long-term data on one reservoir, Lake Wheeler, collected by the Tennessee Valley Authority[21], shed light on the relationship between long-term once-through cooling operation and the status of the fish community in the lake. Browns Ferry Nuclear (BFN) currently operates two units supported by

[7] For examples of state laws addressing impacts from cooling water intake structures, see RCSA § 22a, 430-4 (Connecticut); NJAC § 7:14A-11.6 (New Jersey); 6 NYCRR § 704.5 (New York); MRC § 26.08.03 (Maryland); 35 Ill. Admin. Code 306.201 (1998) (Illinois); 567 IAC 62.4 (455B) (Iowa); Cal. Wat. Code § 13142.5(b) (California).

six intake pumps with a rated total capacity of 2,312 MGD. BFN units were placed in operation between 1974 and 1977 (originally the plant supported three units). Reservoir-wide monitoring was discontinued in 1980, but cove rotenone samples were continued to provide a minimum data base on fish community in the vicinity of BFN, particularly in support of BFN's thermal variance monitoring program for the Alabama Department of Environmental Management. Cove rotenone samples have been collected annually during August and September at three sites since 1969. The data base, therefore, includes five years of pre-operational reservoir data (1969 to 1974) against which the long-term operational impacts of the plant can be compared. Details on sampling, species examined (19 species were examined, and, for each species, data were collected for three size classes: young-of-year, intermediate, and harvestable or adult), results, and analyses performed on the data are provided in TVA[21]. Although standing stock estimates for the reservoir exhibit extreme fluctuations, regression analysis revealed no significant increasing or decreasing trend for either total numbers (fish/hectare) or biomass (kg/ha) during the 30 years of monitoring.

- **Survival** – Early § 316(b) studies assumed 100% mortality to entrained organisms. Later studies, however, evaluated the survival rate of entrained organisms, many of them considering both immediate and latent mortality. EPRI recently completed a comprehensive review of entrainment mortality studies[22]. Fig. 2 presents a summary of findings demonstrating significant survival, in some cases exceeding 90%. Many of the recreationally important species had high survival rates, such as striped bass (mean survival rate 61%) and weakfish (mean survival rate 79%), while others, such as herrings and anchovies, had survival rates of approximately 25%[22,23]. Likewise, an entrainment mortality study for zooplankton at the Anclote power station in Florida demonstrated that the survival rate was quite high[24,26].
- **Stakeholder and Regulator Judgment** – Many biologists working for stakeholders, and regulatory and resource agencies as well, have judged that waterbodies where cooling water intakes operate are not impaired by entrainment and impingement. This view is reflected in the previous Administration's Clean Water Action Plan, which does not identify entrainment or impingement as a source of resource degradation[25].
- **Empirical Information** – Examples of successful fisheries in cooling ponds show that CWIS do not necessarily create adverse impact. Cooling ponds are constructed solely for the purpose of providing condenser cooling water, thereby eliminating the need for large withdrawals from a major source waterbody. Although a very high percentage of cooling pond water normally passes through the CWIS, many of these ponds support naturally reproducing fisheries[27,28,29]. While in some instances studies resulted in actions by facilities to modify their intake structures to reduce impingement or entrainment or both, or to implement offsite enhancements to avoid AEI, in most cases *no significant adverse environmental impact was identified.*

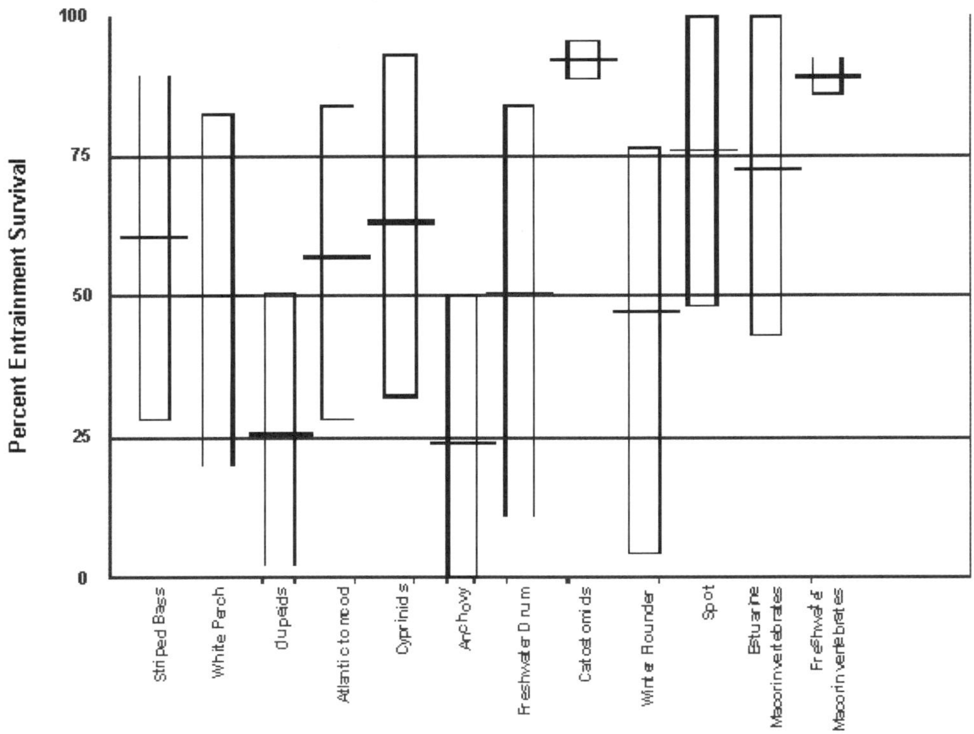

FIGURE 2. An illustration of the range in entrainment survival observed across various groups of fish (Source: EPRI 2000, Figure 3-1).

- **Behavioral or Life History Factors** – By virtue of their behavior or life history, many fish are able to avoid CWIS impacts[30,31,32]. For example, in freshwater many fish species lay eggs in nests or attached to substrate or vegetation, making them unavailable for entrainment. At Chalk Point, a power plant located on a tidal portion of the Patuxent River, it was initially assumed that up to 76% of each year's population in the river could be lost to entrainment. As a result of behavioral studies, however, the station determined that, due to regional movement, diurnal position in the water column, and the ability of larvae to avoid entrainment, the estimates of losses were reduced to 10 to 20%[33].
- **Compensation** – As noted by Myers[34], the concept of "compensation" is fundamental to understanding and managing biological resources. For any biological population to persist, reductions in population size caused by natural environmental fluctuations must result in increased survival, growth, or fecundity of the remaining individuals[35,36,37]. Mechanisms of compensation have been well studied in both terrestrial and aquatic systems. The compensatory response to reductions in population size is the key factor that permits fish populations to sustain themselves despite enormous natural mortality for early life stages and even intensive harvesting of adults.

Long-term research surveys have demonstrated compensation in a variety of marine, estuarine, and freshwater fish species. Field experiments in which fish population sizes are artificially manipulated have also been used to demonstrate compensation[34]. UWAG has identified approximately 50 recent scientific studies (many published in the last 10 years) demonstrating specific mechanisms responsible for compensation in a variety of fish species[38].

The National Research Council (NRC) has recognized the importance of compensation for modern fisheries management:

> Many species appear to have strongly compensatory (spawner-recruit) relationships; that is, per capita recruitment increases significantly as stock size decreases. Reference levels are now more commonly based on a % (SSBPR), but the percentage is often specified by analogy with other stocks or by using the results (of comparisons among other biological reference points). A knowledge of the compensatory capacity of the stock is necessary to define the most appropriate (biological reference points) for a stock. Even without such knowledge, however, a conservative % (SSBPR) still can be selected. (Citation omitted).[13, p. 44]

Spawner-recruit relationships of the type discussed by the NRC are used to manage two estuarine-dependent fish species, striped bass, and weakfish[39,40]. Methods discussed by the NRC can be used to incorporate the concept of compensation in management strategies for species for which spawner-recruit data are not available.

Fisheries scientists have demonstrated the importance of compensation for ensuring the continued persistence of fish populations, and fisheries managers routinely consider compensation when establishing harvesting regulations. While the precise quantification of compensation can be difficult, its occurrence cannot be disputed.

The above factors are presented not to suggest that CWIS impacts are always insignificant, but rather to put impingement and entrainment impacts in perspective. In the vast majority of cases, CWIS impacts have *not* been determined to be a substantial limiting factor for fisheries; thus, in most cases the elimination of these impacts would not be expected to substantially improve fisheries.

Facility Design

Where adverse environmental impacts are identified, a wide range of CWIS technologies designed to reduce impacts are available, as documented in a recent EPRI report[41] and summarized in Taft 2000[42]. The EPRI report identifies a wide array of technologies available for protecting impingeable organisms, including barrier nets, angled screens, and technologies designed to take advantage of fish behavior. For protecting *entrainable* organisms, wedgewire screens, fine mesh screens and, more recently, the Gunderboom[8] have been demonstrated to be effective in certain waterbody types and for certain species, although these technologies have limitations in some waterbody types or for protection of certain species or have not yet been evaluated in a full range of waterbody conditions[43,44,45,46,47,48,49].

In addition, while not part of the CWIS, wet closed-cycle and dry cooling systems significantly reduce or eliminate the need for condenser cooling water. While some have advocated that these systems be designated as "best technology available" (BTA) for § 316(b) purposes, they can have significant negative environmental effects that would preclude their universal application. Both types of system also have significant energy requirements that reduce the efficiency and increase the fuel consumption of the generating facility. This inefficiency results in increased fuel use and air pollutant emissions, which in turn can affect water quality and fisheries by deposition of nitrogen emissions.

Wet closed-cycle cooling, which can reduce cooling water requirements by up to 98%, causes consumptive water use with fishery consequences during low-flow conditions in freshwater. Wet closed-cycle cooling towers may also be unsuitable due to their noise and vapor plumes[50]. Additionally, both wet closed-cycle and dry cooling systems have significant space requirements and aesthetic impacts. The associated increase in impervious surface (especially from dry cooling systems) can impact water quality and fisheries. Wet closed-cycle cooling systems are frequently used as components of new generation construction projects, but due to their potential environmental disbenefits, they would be a poor choice for universal BTA from a net social benefit perspective[50,51].

Waterbody Type and Physical Location

Location considerations include characteristics such as waterbody type (marine, estuarine, riverine, or lake), the aquatic physical environment (*e.g.*, hydro- and thermodynamics, depth, and water quality conditions in the vicinity of the facility), and the local terrestrial setting (*e.g.*, urban, rural, or industrial; topography, space constraints, and proximity to facilities such as airports, historically important sites, *etc.*). Such factors directly affect the feasibility of certain CWIS technologies. In particular, use of wet closed-cycle or dry cooling systems – with their associated space requirements, noise, and aesthetic issues – can have significant effects on local communities.

It is therefore important that decisions balance tradeoffs in these factors to make sound decisions. The Dickerson Station on the freshwater free-flowing

[8] Fine-mesh screens have a mesh size of 0.5 to 1.0 mm. Wedgewire screens use wire with a vee- or wedge-shaped cross-section, welded to a frame to form a slotted screen. The screens are constructed in cylinders up to 7 feet in diameter, which can be attached to a common header. A Gunderboom is manufactured by mixing individual polyester fiber strands into a mat. The mat is then rolled to a specified density and pressed further by a process called needle punching, which mixes the fiber layers and improves fabric strength and durability. The permeable curtain that results can be floated and anchored in place around a cooling water intake structure.

Potomac and the Chalk Point Station on the tidal Patuxent, both facilities located in Maryland, provide a useful example of this importance. Maryland uses the AFS fishery replacement values to quantify the value of economic losses for BTA impingement decisionmaking. These values were $11,281/yr for Dickerson and $28,450/yr (after barrier net deployment) at Chalk Point. The Department of Natural Resources estimated that the economic value of entrainment losses was approximately $1000/yr (1981 dollars) at Dickerson and had a *net present value* of $1.3 million (*i.e.*, 1989 dollar loss projected over the life of the facility) at Chalk Point. The values are low in contrast to the cost of wet closed cycle cooling, estimated to be on the order of $100 million at Chalk Point and somewhat less at Dickerson, even without considering the environmental disadvantages of this technology.

MAKING § 316(B) DECISIONS: A PROPOSED PROCESS THAT MEETS THE NEEDS IDENTIFIED ABOVE

An approach to § 316(b) decisions that takes advantage of fishery management tools and balances multiple waterbody uses and social considerations must also be manageable and implementable from a regulatory perspective. The major components of an approach currently under development that incorporates these needs are described below. This approach establishes some distinctions between § 316(b) decisions for existing facilities and those for new facilities.

Decision Process for Existing Sources

The proposed approach is based on the definition of AEI presented earlier, which focuses on population- and community-level impacts and fishery use protection. It includes the elements listed below.

Use of Representative Indicator Species

Previous work has demonstrated that it is not necessary to study each and every species in a waterbody. Rather, species can be selected based on recreational or commercial importance, roles in the food chain, and/or vulnerability. Previous work has identified most of the species typically vulnerable to CWIS impacts, and site-specific screening studies can confirm the selection of species for further study as necessary.

Determination of Adverse Impact

Three alternative approaches are proposed for making § 316(b) adverse impact decisions. The first approach uses explicit criteria that are sufficiently stringent to support a decision that the facility presents no risk of adverse impact. The second approach uses a process based on the principles of EPA's risk assessment/risk management framework. The third allows decisions based on previously conducted site-specific § 316(b) studies, if it can be shown that they meet certain standards.

Use of Screening Criteria

It is important that the decisionmaking criteria be clear and explicit to facilitate easy implementation. The criteria are not performance standards. Instead, they are designed to be well below a level that could reasonably be expected to result in AEI. This approach addresses the issue of uncertainty by setting criteria at these low thresholds. Several specific criteria being evaluated include:

- **Location.** This criterion is based on determining if a CWIS is located in a waterbody or portion of a waterbody that cannot support aquatic life at any significant level due to poor water quality, such as anoxia, or lack of habitat. For example, if the CWIS withdraws its intake water from an anoxic zone which cannot support impingeable and entrainable organisms important to the fishery, it would be very unlikely to result in AEI.
- **Facility design.** If a facility employs a CWIS which is designed or has features to minimize impingement and/or entrainment, or makes use of technologies such as wet closed-cycle cooling, it would present no appreciable risk of AEI. In this situation, the technology must be demonstrated to be effective. If the technology is known to be effective only for impingement, for example, then the issue of entrainment will still need to be assessed.
- **Percentage of waterbody used.** Use of this criterion is suggested for entrainable organisms in smaller waterbodies such as freshwater rivers, lakes and reservoirs. A criterion value of 5% (or less) of the 90% exceedance flow of a river or of the volume of the biological zone of influence[9] in a lake or reservoir, measured when entrainable life stages of RIS are present, is proposed. This approach essentially is based on a 95% protection standard, which is believed to be adequately protective for freshwater locations.
- **Biological criteria.** The low-risk biological criteria being evaluated again are limited to use in freshwater rivers, reservoirs, and lakes other than the Great Lakes. Criteria of 5% (or less) loss of a non-harvested species and 1% (or less) loss of a harvested species are being considered as values that would generally

[9] The biological zone of influence is the zone within a waterbody that is occupied by an RIS. For freshwater rivers the biological zone of influence for RIS entrainable life stages is the portion of the cross-sectional flow of the river occupied by the RIS where the river flows past the CWIS. If an RIS is found primarily along the shoreline, the biological zone of influence is the sum of the flow along the shorelines on both sides of the river. For smaller freshwater lakes and reservoirs and controlled-flow rivers that have lake-like characteristics, the biological zone of influence is the volumetric area occupied by the RIS during the time when the RIS is vulnerable to entrainment. Reservoirs and large rivers with controlled flow may have either riverine or lake-like characteristics, several types of spawners, and perhaps disproportional distribution of habitat (upstream versus downstream of the plant). For such waters the more appropriate of the above two criteria must be determined and applied.

pose no risk of adverse impact. These values are low compared to generally allowable fishery harvest management levels. Other biological criteria, also being considered, would take advantage of well-designed long-term monitoring programs and measures such as the multi-metric criteria developed by Duke Power and TVA, in use at many southern reservoirs.

Use of Risk Assessment and Risk Management Principles

A second method of AEI decisionmaking involves use of EPA's ecological risk assessment/risk management guidelines. This approach includes active stakeholder participation, in which natural resource managers and interested members of the public identify populations of special interest for assessing the potential impact of the CWIS. These form the basis of the next step, which is identification of appropriate methods and analytical approaches. Finally, study endpoints are established to allow easy AEI decisionmaking after data collection. The process can address uncertainty by balancing comprehensiveness of study design, use of fishery information for species of concern, and modeling assumptions.

Use of Previously Conducted § 316(b) Studies

This approach makes use of the extensive § 316(b) studies already conducted at many facilities. Any studies that are not reasonably current would have to be evaluated to ensure that the studies are representative of the current facility design and biological conditions and that the data collection methods and analytical tools remain valid. In particular, sufficient information must be provided to show that the populations examined continue to be appropriate in terms of fishery management objectives. The objective of this approach is to take full advantage of the previous investment in data collection and evaluation conducted by regulators. Fishery managers and other stakeholders would be able to participate in the evaluation through the NPDES process.

The above three decisionmaking approaches could be used independently or in combination. For example, screening criteria could be used initially to provide focus to determine appropriate RIS, with the final decisionmaking done using the risk-based approach. Finally, the decision process for existing facilities would incorporate three additional features: using cost-benefit analysis to maximize net benefit, allowing "environmental enhancements" in appropriate circumstances, and reviewing BTA determinations if new information showed that circumstances had changed.

Maximum Net Benefit

If the decisionmaking process outlined above shows that an existing CWIS is creating, or will create, an appreciable risk of AEI, then the decisionmaker must decide what is "best technology available" or BTA to "minimize" AEI. UWAG's economic consultant advises that the most rational way to make this decision is to choose the technology that maximizes net benefits (that is, benefits minus costs). To use this approach, the permit applicant would have to identify all reasonably available intake structure technologies that would reduce the impact to the aquatic community and be feasible at the site. The applicant would also estimate the costs and benefits of each such technology, including the impacts of the CWIS on aquatic biota, in addition to the monetary costs of construction and operation, energy costs, and environmental costs such as air pollution, aesthetics, and land use. Summing the costs and benefits for each "available" technology, the permittee would choose as "best" the one that had the highest net benefit. Industry believes that cost-benefit analyses suitable for BTA selection can be developed based on existing tools and methods, such as by adopting some features of EPA's BEN model for evaluating the benefits of violating environmental laws or of the methods used to evaluate natural resource damages[52].

Environmental Enhancements

"Environmental enhancements" are actions taken by a facility determined to cause AEI (or a facility that wishes to settle a dispute over its permit) to compensate for the CWIS losses to the affected RIS species rather than install a CWIS technology. Environmental enhancements – such as wetlands creation or fish stocking – are one means of compensating for CWIS losses. In some cases, the *most* limiting factor for the aquatic populations is not the CWIS, but rather (1) low dissolved oxygen as a result of nutrient enrichment or (2) lack of habitat for spawning and nursery functions[53]. In such cases, by investing dollars addressing the most limiting environmental factors, the facility may spur a more significant recovery to the population than could be achieved through installation of a CWIS technology. Such actions, as long as they are directly related to the fishery, can result in a greater net social benefit than installation of BTA. Enhancements have been used effectively at a number of stations, including Salem, John Sevier, and Chalk Point. Florida Power Corporation's Crystal River fish hatchery is another successful enhancement program. Such environmental enhancements are not "intake technologies" (and therefore cannot be "BTA" nor be required by authority of § 316(b)), but the § 316(b) regulatory framework is flexible enough to allow them to be used, if offered by the permit applicant. They can be employed as a cost-effective means of addressing adverse environmental impact, potentially resulting in environmental benefits greater than use of BTA alone.

Periodic BTA Review

Once an existing facility has gone through the process of determining that the CWIS is BTA, the BTA status would need to be revisited at the time of permit renewal if the regulatory agency had information showing that the previous studies were no longer valid (for example, that biological conditions had changed). Factors that might result in a change of BTA status would include modifications to the CWIS design or operation or significant changes in the waterbody.

Decision Process for New Facilities

The process described above is for existing facilities. For new facilities, a "Two Track" approach has been proposed that would allow a company seeking to build a new facility that will require use of surface cooling water either (1) to commit to a highly protective (indeed often *over*-protective) technology at the outset or (2) to engage in a site-specific analysis to determine whether the intake would create an appreciable risk of AEI and, if so, what would be the BTA for the site.

Track 1: The "Fast Track"

Track 1, the "Fast Track," would allow the applicant to commit to one of the following highly protective technologies, in return for expedited permitting without the need for pre-operational or operational studies in the source waterbody by using one of two options:

- Option 1: Employ a technology that limits intake flow to the flow that would be required by wet closed-cycle cooling for a given amount of generation at that site and design the average approach velocity (measured in front of the intake screens or the opening to the cooling water intake structure) to be no more than 0.5 ft/s; or
- Option 2: Employ a technology that will achieve a level of protection from impingement and/or entrainment that is reasonably consistent with Option 1. This option is intended to permit facilities to use either standard technologies, or new ones, that have been demonstrated to be effective for the species, type of waterbody, and flow volume proposed for their use. Examples of candidate technologies include:
  a. Wedgewire screens, where there is constant flow, as in rivers;
  b. Traveling fine mesh screens with a fish return system designed to minimize entrainment and impingment mortality; and
  c. Gunderbooms, at sites where they would not be rendered ineffective by high flows or fouling.

"Reasonably consistent with" means that an acceptable alternative technology should provide a level of protection within the range expected under Option 1 achieved by flow reductions associated with wet closed-cycle cooling and a 0.5 ft/s approach velocity for the type of waterbody on which the facility is to be sited. Use of highly protective technologies should eliminate the need for periodic BTA review.

The effectiveness of wet closed-cycle cooling is well documented. The other technologies listed above promise a level of protection reasonably consistent with that of wet closed-cycle cooling. To prevent impingement, the Gunderboom is designed to have a low approach velocity (almost unmeasurable) and uses a very fine mesh to provide entrainment protection[42]. Wedgewire screens are designed to minimize entrainment and impingement through a combination of small slot width (0.5 to 2 mm) and an approach velocity of less than 0.5 ft/s[42,54].

For fine mesh screens, the survival of fish collected on the screens is species- and life- stage-specific[41,42]. Survival of many species can be very high, exceeding 90% even at velocities above 0.5 ft/s. As for entrainment, again the effect of fine mesh screens varies by species, but the data indicate that, if control mortality is taken into account, fine mesh screens can reduce entrainment mortality by 90% or more for some species. Other species, such as bay anchovy, have a high mortality both naturally and after encountering fine mesh screens. Nevertheless, given the present state of knowledge, it is reasonable to include fine mesh screens (with a properly designed fish return system) as a candidate technology for some sites that can reduce overall losses to a level (*i.e.*, 90% or better) reasonably consistent with wet closed-cycle cooling.

Option 2 of Track 1 encourages alternative or innovative intake structure technologies. A proponent of a new alternative technology would conduct a laboratory or site-specific study appropriate for the waterbody type and species of concern prior to employment of the technology. If the demonstration was successful, after the facility deployed the new technology, monitoring would be conducted as appropriate to validate performance.

At a few sites, there could potentially be unusual species-specific circumstances in which Fast Track technologies meeting the above criteria would not be sufficient to avoid AEI. While the number of such sites is likely to be very small, the evaluation process should give permit writers the authority to require additional protective measures *if* the permitting agency has information to support a finding that exceptional conditions exist such that the proposed facility could affect one or more populations in a way that would not be prevented by other federal or state requirements (such as the Endangered Species Act) and thus has the potential to cause AEI.

Track 2: A More Tailored Approach

Track 2 of the proposed Two Track approach is similar to the decisionmaking process for existing facilities summarized above. It differs in that Track 2 for new facilities can make use only of predictive fishery management tools, rather than retrospective ones. Track 2 would be for facilities that wished to pursue use of a less stringent BTA.

In these cases the applicant could evaluate AEI using the risk screening criteria or the risk assessment/risk management AEI evaluation methods for existing sources. For the population percent reduction criteria, source waterbody type, data availability and assessment, and analytical tool availability will determine the difficulty of predicting impingement rates in a sufficiently quantitative manner. Where this cannot be done, new facilities will need to plan for some kind of technology to protect fish from impingement.

The Two Track decision process, then, is both efficient and flexible, and it has one very important advantage: it avoids worsening the already-present "energy crisis" now affecting California and possibly soon other states[55,56,57]. Track 1, the Fast Track, is available for speeding new generating facilities online in parts of the country where they are needed most, in return for a commitment to highly (often overly) protective intake structure technology, and also encourages innovative technologies. Track Two allows a close look at the features of any proposed site and avoids arbitrary, less efficient, restrictions.

CONCLUSION

The Clean Water Act requires that cooling water intake structures minimize, where it exists, "adverse environment impact." In order to be able to determine whether AEI exists or is threatened, and if it exists to decide how to minimize it, one must first have a definition. The definition needs to ensure the protection of living resources. And the process for "minimizing" AEI needs to strike a balance among competing social needs.

Tools are available today to accomplish both these goals. The science of fisheries management provides concepts (like maximum sustainable yield), tools (like biological modeling), and knowledge (such as knowledge of how fish populations compensate for losses) that will allow cooling water users and regulatory agencies to make sound § 316(b) decisions that will protect the living fishery resources. Cost-benefit analysis, drawing on experience of calculating the benefits of environmental violations and natural resource damages, provides a tool for choosing an intake technology that maximizes the net benefits to society.

Given a workable definition of AEI and the tools to assess and minimize it, one needs, finally, a decisionmaking process that allows the tools to be used appropriately. The electric utility industry has proposed such a process, one that provides both the opportunity to bring new generating plants online quickly, in return for installing highly (often overly) protective intake technology, and the flexibility to look closely at site characteristics when assessing the risk of AEI, and taking advantage of site characteristics as appropriate to concurrently protect the environment and produce energy efficiency.

REFERENCES

1. U.S. Environmental Protection Agency (2000) National pollutant discharge elimination system--regulations addressing cooling water intake structures for new facilities; proposed rule. 65 Fed. Reg. 49,060 (August 10, 2000).
2. Anderson II, W.A. and Gotting, E.P. (2001) Taken in over intake structures? Section 316(b) of the Clean Water Act. 26 Colum. *J. Envtl. L.* **1,** 1–79
3. U.S. Environmental Protection Agency (1977) Guidance for evaluating the adverse impact of cooling water intake structures on the aquatic environment: Section 316(b) P. L. 92-500 (Draft).
4. U.S. Environmental Protection Agency (2001) Notice of data availability; national pollutant discharge elimination system–regulations addressing cooling water intake structures for new facilities. 66 Fed. Reg. 28,853 (May 25, 2001).
5. Utility Water Act Group (UWAG) (2000) Comments of the Utility Water Act Group on EPA's proposed § 316(b) rule for new facilities and ICR No. 1973. 01.
6. American Fisheries Society (2000) AFS Policy Statement #1: North American Fisheries Policy (Revised). http://www.fisheries.org/Public_Affairs/Policy_Statements/ps_1.shtml

7. Barnthouse, L. (1993) Ecological Risk Assessment 26. (in the section called "Assessment Concepts") Suter, G., Ed. cited in Anderson, II, W.A. and Gotting, E.P. (2001) Taken in over intake structures? Section 316(b) of the Clean Water Act. 26 Colum. *J. Envtl. L.* **1,** 3, 19–21.
8. Beamesderfer, R. (2000) Deciding when intervention is effective and appropriate. *Fisheries* **25(6),** 18–23.
9. National Marine Fisheries Service (1998) Magnuson-Stevens Act Provisions; National Standard Guidelines; Final Rule. 63 Fed. Reg. 24,212, 24,219 (May 1, 1998).
10. Mace, P. and Sissenwine, M. (1993) How much spawning per recruit is enough? In Smith, S.J., Hunt, J.J., and Rivard, D., Eds. Risk Evaluation and Biological Reference Points for Fisheries Management. *Can. Spec. Publ. Fish. Aquat. Sci.* **120,** 101–118.
11. Smith, S.J., Hunt, J.J., and Rivard, D. (Eds.). (1993) Risk Evaluation and Biological Reference Points for Fisheries Management. *Can. Spec. Publ. Fish. Aquat. Sci.* **120,**
12. Goodyear, C.P. (1993) Spawning stock biomass per recruit in fisheries management: foundation and current use. In Smith, S.J., Hunt, J.J., and Rivard, D., Eds. Risk Evaluation and Biological Reference Points for Fisheries Management. *Can. Spec. Publ. Fish. Aquat. Sci.* **120,** 67–81.
13. National Research Council (1998) *Improving Fish Stock Assessments.* National Academy Press, Washington, D.C. 177 pp.
14. Electric Power Research Institute (1999) Catalog of Assessment Methods for Evaluating the Effects of Power Plant Operations on Aquatic Communities. Report No. TR-112013.
15. Weeks, H., and Berkeley, S. (2000) Uncertainty and precautionary management of marine fisheries: Can the old methods fit the new mandates? *Fisheries* **25(12),** 6–14,15.
16. U.S. Environmental Protection Agency. (1998) Guidelines for Ecological Risk Assessment. 63 Fed. Reg. 26,845–26,924 (May 14, 1998).
17. Auster, P.J. (2001) Defining thresholds for precautionary habitat management actions in a fisheries context. *North Am. J. Fisheries Mgmt.* **21,** 1–9.
18. Virginia Secretary of Natural Resources (2000) Annual Report on Status of Tributary Strategies, Chesapeake Bay Agreement and Water Quality for Virginia's Chesapeake Bay and Tributaries, p. 5.
19. U.S. Environmental Protection Agency (2000) Economic and Engineering Analyses of the Proposed § 316(b) New Facility Rule. EPA-821-R-00-019.
20. Richkus, W.A. and McLean, R. (2000) Historical overview of the efficacy of two decades of two decades of power plant fisheries impact assessment activities in Chesapeake Bay. In Wisniewski, J., Ed. Power Plants & Aquatic Resources: Issues and Assessments. *Env. Sci.Policy* **3,** S283–S293.
21. Tennessee Valley Authority (1998) Browns Ferry Nuclear Plant Thermal Variance Monitoring Program Final Report. TVA Water Management Environmental Compliance. Norris, TN, 54 pp; including supplemental statistical analyses, 10 pp.
22. Electric Power Research Institute (2000) Review of Entrainment Survival Studies: 1970-2000. Report No. 1000757.
23. Cannon, T.C., Jinks, S.M., King, L.R., and Lauer, G.J. (1978) Survival of entrained ichthyoplankton and macroinvertebrates at Hudson River power plants. In Jensen, L.D., Ed. Proceedings of the Fourth National Workshop on Entrainment and Impingement. EA Communications, Melville, NY. pp. 71–89.
24. Melton, B.R. and Serviss, G.M. (2000) Florida Power Corporation – Anclote Power Plant entrainment survival of zooplankton. In Wisniewski, J., Ed. Power Plants & Aquatic Resources: Issues and Assessments. *Env. Sci. Policy* **3,** S233–S248.
25. U.S. Environmental Protection Agency and U.S. Department of Agriculture. (1998) Clean Water Action Plan: Restoring and Protecting America's Waters, pp. 7–9. http://cleanwater.gov/action/toc.html
26. Ronafalvy, J., Cheesman, R.R., and Matejek, W.M. (2000) Circulating water traveling screen modifications to improve impinged fish survival and debris handling at Salem Generating Station. In Wisniewski, J., Ed. Power Plants & Aquatic Resources: Issues and Assessments. *Env. Sci. Policy* **3,** S377–S382.
27. Electric Power Research Institute (1979) Synthesis and Analysis of Ecological Information from Cooling Impoundments. Report No. EA-1054, Vol. 1.

28. Electric Power Research Institute (1980) Evaluation of a Cooling-Lake Fishery. Volume 1: Introduction, Water Quality & Summary. Prepared by Illinois Natural History Survey. Report No. EA-1148.
29. Electric Power Research Institute (1986) Sport Fishery Potential of Power Plant Cooling Ponds. Prepared by Southern Illinois University at Carbondale. Report No. EA-4838.
30. Taber, C.I. (1969) The Distribution and Identification of Larval Fishes in the Buncombe Creek Arm of Lake Texoma with Observations of Spawning Habits and Relative Abundance. [PhD dissertation] University of Oklahoma. 120 pp.
31. Balon, E.K. (1975) Reproductive guilds of fishes: A proposal and definition. *J. Fish. Res. Bd. Can.* **32,** 821–864.
32. Jones, P.W., Martin, F.D., and Hardy, Jr., J.D. (1978) Development of fishes of the Mid-Atlantic Bight. Prepared by Chesapeake Biological Laboratory of the University of Maryland for U.S. Fish and Wildlife Service. FWS/OBS-78/12.
33. Bailey, D.E., Loos, J.J., and Perry, E.S. (1998) Studies of cooling water intake structure effects at Potomac Electric Power Company Generating Stations. EPRI Clean Water Act Section 316(b) Technical Workshop. Coolfont Conference Center, Berkeley Springs, WV.
34. Myers, R. (2000) Appendix B to Comments of the Utility Water Act Group on EPA's Proposed § 316(b) Rule for New Facilities and ICR No. 1973.01.
35. Rose, K.A., Cowan, Jr., J.H., Winemiller, K.O., Myers, R.A., and Hilborn, R. (2001) Compensatory density-dependence in fish populations: Importance, controversy, understanding, and prognosis. In press.
36. Van Winkle, W. (2000) A perspective on power generation impacts and compensation in fish populations. In Wisniewski, J., Ed. Power Plants & Aquatic Resources: Issues and Assessments. *Env. Sci. Policy* **3,** S425–S431.
37. Boreman, J. (2000) Surplus production, compensation, and impact assessments of power plants. In Wisniewski, J., Ed. Power Plants & Aquatic Resources: Issues and Assessments. *Env. Sci. Policy* **3,** S445–S449.
38. Meronek, T.G., et al. (1996) A review of fish control projects. *North Am. J. Fisheries Mgmt.* **16,** 63–74.
39. National Marine Fisheries Service (1998) Report of the 26th Northeast Regional Stock Assessment Workshop. Stock Assessment Review Committee Consensus Summary of Assessments. Northeast Fisheries Science Center Reference Document 98-03. Woods Hole, Massachusetts.
40. National Marine Fisheries Service (2000) Report of the 30th Northeast Regional Stock Assessment Workshop. Stock Assessment Review Committee Consensus Summary of Assessments. Northeast Fisheries Science Center Reference Document 00-03. Woods Hole, Massachusetts.
41. Electric Power Research Institute (1999) Fish Protection at Cooling Water Intakes: Status Report. Report No. TR-114013.
42. Taft, E.P. (2000) Fish protection technologies: A status report. In Wisniewski, J., Ed. Power Plants & Aquatic Resources: Issues and Assessments. *Env. Sci. Policy* **3,** S349–S359.
43. Taft, E.P., Horst, T.J., and Downing, J.K. (1981) Biological evaluation of a fine-mesh traveling screen for protecting organisms. Presented at the Workshop on Advanced Intake Technology, San Diego, CA, April 22–24, 1981.
44. Brueggemeyer, B., Cowdrick, D., and Durrell, K. (1988) Full-scale operational demonstration of fine mesh screens at power plants. In Proceedings of the Conference on Fish Protection at Steam and Hydro Plants, San Francisco, CA, October 28–30, 1987. Electric Power Research Institute CS/EA/AP-5663-SR.
45. Veneziale, E.J. (1991) Fish protection with wedge wire screens at Eddystone Station. In Proceedings of the American Power Conference.
46. Hanson, B.N., Bason, W.H., Beitz, B.E., and Charles, K.E. (1978) A practical intake screen which substantially reduces entrainment. In Fourth National Workshop on Entrainment and Impingement, Chicago, IL, December 5, 1977.
47. Lifton, W. (1979) Biological aspects of screen testing on the St. Johns River, Palatka, Florida. Prepared for Passive Intake Screen Workshop, Chicago, IL, December, 1979.
48. Lawler, Matusky & Skelly Engineers (1997) Lovett Generating Station Gunderboom Evaluation Program 1996. Prepared for Orange and Rockland Utilities, Inc.
49. Lawler, Matusky & Skelly Engineers (1998) Lovett Generating Station Gunderboom Evaluation Program 1998. Prepared for Orange and Rockland Utilities, Inc.

50. U.S. Nuclear Regulatory Commission (1996) Generic Environmental Impact Statement for License Renewal of Nuclear Plants. Main Report. Final Report. NUREG-1437, Vol. 1. pp. 163–183.
51. Argonne National Laboratory (1992) Impact on the Steam Electric Power Industry of Deletion of § 316(a) of the Clean Water Act: Phase 2, Energy and Environmental Impacts.
52. U.S. Environmental Protection Agency (1999) BEN Users Manual. Office of Enforcement and Compliance Assurance.
53. U.S. Environmental Protection Agency. (2000) National Water Quality Inventory: 1998 Report to Congress. EPA 841-R-00-001.
54. Ehrler, C. and Raifsnider, C. (2000) Evaluation of the effectiveness of intake wedgewire screens. In Wisniewski, J., Ed. Power Plants & Aquatic Resources: Issues and Assessments. *Env. Sci. Policy* **3,** S361–S368.
55. Smith, R. and Emshwiller, J.R. (2001) Why California isn't the only place bracing for electrical shocks. *Wall St. J.* Apr. 26, 2001. Page A1 col. 6.
56. National Energy Policy Development Group (2001) National Energy Policy. U.S. Government Printing Office, Washington, D.C. ISBN 0-16-050814-2.
57. North American Electric Reliability Council (2001) 2001 Summer Assessment: Reliability of the Bulk Electricity Supply in North America. 64 pp.

Indicators of AEI Applied to the Delaware Estuary

Lawrence W. Barnthouse[1,*], Douglas G. Heimbuch[2], Vaughn C. Anthony[3],Ray W. Hilborn[4], and Ransom A. Myers[5]
[1]LWB Environmental Services, Inc.,105 Wesley Lane, Oak Ridge, TN 37830; [2]PBSJ, 12101 Indian Creek Court, Beltsville, MD 20705; [3]P.O. Box 459, Gaecklin Rd., Boothbay, ME 04537; [4]University of Washington, School of Fisheries, Box 355020, Seattle, WA 98195; [5]Dalhousie University, Department of Biology, Halifax Nova Scotia, Canada B3H 4J1

Received November 15, 2001; Revised April 8, 2002; Accepted April 17, 2002; Published February, 2003

We evaluated the impacts of entrainment and impingement at the Salem Generating Station on fish populations and communities in the Delaware Estuary. In the absence of an agreed-upon regulatory definition of "adverse environmental impact" (AEI), we developed three independent benchmarks of AEI based on observed or predicted changes that could threaten the sustainability of a population or the integrity of a community.

Our benchmarks of AEI included: (1) disruption of the balanced indigenous community of fish in the vicinity of Salem (the "BIC" analysis); (2) a continued downward trend in the abundance of one or more susceptible fish species (the "Trends" analysis); and (3) occurrence of entrainment/impingement mortality sufficient, in combination with fishing mortality, to jeopardize the future sustainability of one or more populations (the "Stock Jeopardy" analysis).

The BIC analysis utilized nearly 30 years of species presence/absence data collected in the immediate vicinity of Salem. The Trends analysis examined three independent data sets that document trends in the abundance of juvenile fish throughout the estuary over the past 20 years. The Stock Jeopardy analysis used two different assessment models to quantify potential long-term impacts of entrainment and impingement on susceptible fish populations. For one of these models, the compensatory capacities of the modeled species were quantified through meta-analysis of spawner-recruit data available for several hundred fish stocks.

All three analyses indicated that the fish populations and communities of the Delaware Estuary are healthy and show no evidence of an adverse impact due to Salem. Although the specific models and analyses used at Salem are not applicable to every facility, we believe that a weight of evidence approach that evaluates

* Corresponding author. E-mails: lwb.env@worldnet.att.net; dgheimbuch@pbsj.com; Anthony@wiscasset.net; Rhilborn@hotmail.com; Ransom.Myers@Dal.ca

multiple benchmarks of AEI using both retrospective and predictive methods is the best approach for assessing entrainment and impingement impacts at existing facilities.

KEY WORDS: 316(b), adverse environmental impact, AEI, biological indicators, fish populations

DOMAINS: ecosystems and communities, environmental management, environmental modeling, marine systems

INTRODUCTION

Section 316(b) of the Clean Water Act requires that "the location, design, construction, and capacity of cooling water intake structures reflect the best technology available for minimizing adverse environmental impact." However, neither the act itself nor any applicable regulatory guidance provides an explicit definition of the term "adverse environmental impact" (AEI) or explicit criteria for determining when an AEI has occurred or might potentially occur.

The draft Section 316(b) guidelines[1], which were never formally approved, contain language suggesting that the focus of AEI determinations should be on the impairment of populations and communities:

"Adverse environmental impacts occur when the ecological function of the organism(s) of concern is impaired or reduced to the level which precludes maintenance of existing populations; a reduction in optimum sustained yield to sport and/or commercial fisheries results; threatened or endangered species of aquatic life are directly or indirectly involved; and/or the magnitude of the existing or proposed damage constitutes an unmitigatable loss to the aquatic system."

Because all organisms have finite life spans, the reproducing population is the smallest ecological unit that is persistent in time. Assessments of the impacts of entrainment (defined as the transport of fish eggs or larvae and other small aquatic organisms through a cooling-water system) and impingement (defined as the trapping of fish on the screens that prevent large debris from being drawn into a cooling-water systems) at Hudson River power plants, which were the focus of intensive study during the 1960s and 1970s, focused on potential reductions in the abundance and yield of fish populations[2,3]. In this and other studies, assessments relied on concepts and methods that are ultimately grounded in resource management, especially fisheries science.

We designed an impact assessment approach for the Salem Generating Station, located in Lower Alloways Creek, NJ, on the Delaware Estuary (Fig.1), based on these precedents and on more recent guidance from the U.S. Environmental Protection Agency (EPA) on the use of multiple lines of evidence in ecological risk assessments. Our approach was intended to address impacts of Salem on the balance of the fish community present in the Delaware Estuary and on the sustainabil-

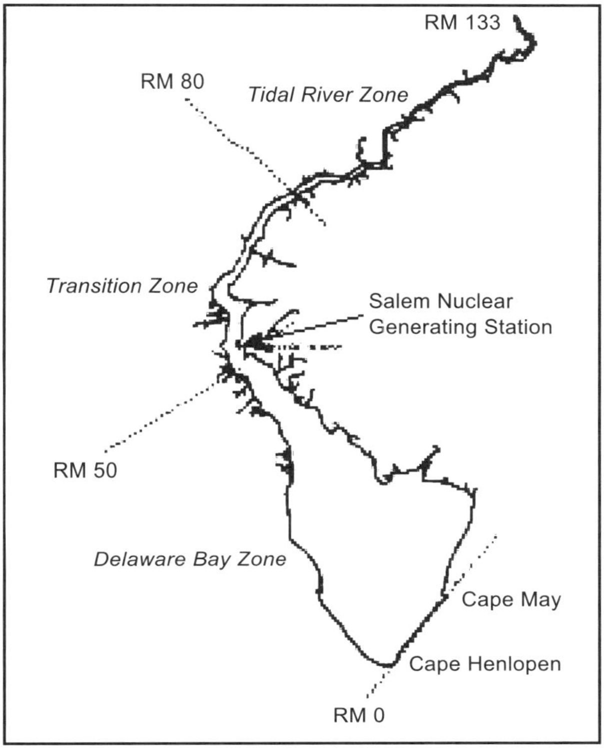

FIGURE 1. Longitudinal zones of the Delaware Estuary.

ity of specific representative fish populations that utilize the estuary. Salem began commercial operation in 1977 and, except for outages for maintenance, refueling, and system upgrades, has operated continuously since that time. Because of its size, location, and cooling-water withdrawal rate, Salem has long been identified as a source of potential intake structure-related impacts, and was the subject of an extensive Section 316(b) Demonstration in 1984, which was updated in 1991 and 1993. All of these documents relied heavily on assessment approaches, especially the use of population-level assessment models, which were previously used in the Hudson River assessment studies. The assessment described in this paper was prepared to support a New Jersey Pollutant Discharge Elimination System Permit Renewal Application submitted by Public Service Electric and Gas (PSEG) in 1999.

The Delaware Estuary extends 133 mi (214 km) from the head of tide at Trenton, NJ, to the mouth of Delaware Bay. It is one of the largest estuaries on the Atlantic Coast, with an open-water area of approximately 725 mi^2, not including tributaries and fringing marsh-plain areas. The estuary is divided into three longitudinal zones (Fig. 1) based on salinity, turbidity, and biological productivity: the Tidal River Zone, a zone of low salinity, low turbidity, and moderate biological productivity extending from the head-of-tide at Trenton, NJ (RM 133), to Marcus Hook, PA (RM 80); the Transition Zone, a zone of high turbidity, variable salinity, and variable biological productivity extending from Marcus Hook (RM 80) to Artificial Island (RM 50); and the Delaware Bay Zone, a zone of high salinity, low turbidity, and high biological productivity extending from Artificial Island to the mouth of the bay.

Salem has been operating for more than 20 years. During this period an enormous quantity of data has been collected concerning environmental conditions in the Delaware Estuary and concerning the status and biological characteristics of fish populations that utilize the estuary. We used all available and relevant data for the assessment, including in-plant sampling data, ichthyoplankton and finfish surveys, and coastwide stock assessment data. These sources of data are summarized in Table 1. The finfish species evaluated as representative important species (RIS) included bay anchovy, weakfish, striped bass, white perch, American shad, alewife, blueback herring, spot, and Atlantic croaker.

The availability of an extensive time series of in-plant, riverwide, and coastwide data allowed us to utilize empirical, retrospective approaches rather than relying only on the predictive, model-based approaches used in previous studies. Moreover, in performing modeling studies to supplement the retrospective approaches, we were able to use advanced methods and data sources that were unavailable for earlier Section 316(b) assessments at Salem.

DEFINITION OF BENCHMARK OF AEI

We used three benchmarks of AEI – two relating to past operations, and one relating to current and future operations – to evaluate whether Salem operations may have caused or could cause AEI. For the first benchmark termed the "Balanced Indigenous Community" (BIC) benchmark, we evaluated whether the operation of Salem had upset or modified the balance of fish species present in the Delaware Estuary, as reflected in species presence-absence data. For the second benchmark, termed "Continuing Decline in Abundance of Aquatic Species" (Trends), we evaluated trends in the abundance of age 0 fish belonging to the nine RIS, using one or more of the three available long-term trends data sets available for the estuary. For the third benchmark, termed the "Stock Jeopardy" benchmark, we used widely recognized models and fishery management reference points to evaluate potential current and future effects of Salem on the sustainability of the nine populations evaluated.

We evaluated all three benchmarks using a weight of evidence approach to determine the presence or absence of an AEI. It would be reasonable to conclude that Salem was not causing an adverse impact on the estuary if the fish community appeared to be in balance, if no species were exhibiting a long-term decline that

TABLE 1

Sampling Programs Used to Assess Impacts of Salem on Delaware Estuary Fish Populations and Communities

Program	Spatiotemporal Coverage	Gear (s)	Use in Impact Assessment
PSEG in-plant sampling and soecial studies	Salem cooling water system 1977-1998	Various gears and collection methods	Estimation of entrainment and impingement loss rates; CMR estimates for stock jeopardy analysis
PSEG Nearfield Bottom Trawl Survey	RM 40 - RM 61 1970-1982; 1986-1998	16 ft. otter trawl	BIC analysis; trends analysis
PSEG Baywide Survey	RM 0 - RM 73 1979-1982; 1996-1998	16 ft. otter trawl 4 ft. x 6 ft. fixed frame pelagic trawl 1.6 ft. plankton net	CMR estimates for stock jeopardy analysis
DNREC Juvenile Trawl Survey	RM 6 - RM 59 (Delaware side only) 1980-1998	16 ft. otter trawl	Trends analysis
NJDEP Beach Seine Survey	RM 59 - RM 140 1980-1998	100 ft. beach seine	Trends analysis
PSEG White Perch Mark-Recapture Program	RM 50-119 1980-83, 1996-98	16 ft. otter trawl	CMR estimates for stock jeopardy analysis
New Jersey – Delaware American Shad Monitoring Program	Upper river 1975-1983, 1986, 1989, 1992, 1995, 1996	Haul seine (marking), angler returns (recapture), hydroacoustic survey (1995-96)	CMR estimates for stock jeopardy analysis
ASMFC and NOAA stock assessments and coastal curveys	Atlantic coast and major estuaries		Cumulative impact assessment

could be attributable to Salem's operation, if stock assessment models indicated no jeopardy due to Salem's operations, and if the results from the three benchmarks were consistent with other data concerning the species in question.

THE BALANCED INDIGENOUS COMMUNITY ANALYSIS

The purpose of the BIC analysis was to determine whether the fish community of the Delaware Estuary has changed during the period of operation of Salem in a way that might indicate the presence of an AEI. Ecologists use the term "community" to denote the entire assemblage of species present in a given location or habitat. Data on the number and relative abundance of species present in different communities are used to draw inferences concerning their evolutionary history, successional status, temporal stability, or degree of disturbance. Communities are said to be "diverse" if many species are present. Some ecologists[4,5] have argued that disturbances – including physical disturbances, pollution, and overharvesting – generally cause reductions in the diversity of communities. Although the general validity of this proposition has been questioned[6], empirical observations have demonstrated that the diversity of many types of biological communities is, in fact, reduced by a wide variety of environmental stresses[5]. For this reason, measures of species diversity are still used as indicators of the influence of environmental stress on biological communities.

Many indices of fish species composition have been proposed and used by ecologists[6,7]. As noted by Gotelli and Graves[6], most of these indices are highly correlated with one another. Moreover, many indices lack valid statistical tests and biologically meaningful interpretations. Hurlburt's[8] measures of species richness, defined as "...the number of species present, without any particular regard for the exact area or number of individuals examined," do not suffer from these defects. Hurlburt[8] distinguished two types of species richness measures: *numerical richness,* meaning the number of species present in a collection containing a specified number of individuals, and *areal richness*, meaning the number of species present in a given area or volume of the environment. Areal richness is also termed "species density." Our analysis used both numerical richness and species density as measures of fish community structure in the Delaware Estuary.

PSEG's Nearfield Bottom Trawl Program (Table 1) provides the only data set for which both prestartup and poststartup data are available; therefore, this data set was the only data set used for the BIC analysis. The bottom trawl samples the benthic stratum of the estuary, and it could be argued that this gear samples only part of the fish community present at any given location. However, due in part to the turbulence and high turbidity of the estuary in the vicinity of Salem, both pelagic and benthic fish species are represented in the bottom trawl collections. Moreover, since the BIC analysis is based on counts of the number of species present in a sample, irrespective of their relative abundance, the results should be relatively insensitive to gear-related sampling biases.

Unit 1 began commercial operation in 1978. Although preoperational testing was conducted prior to 1978, all of the years 1970 through 1977 are considered to be preoperational years for the purpose of the BIC analysis. If station operations were adversely affecting the fish community of the Delaware Estuary, it is unlikely that all possible effects would occur immediately. Moreover, for long-lived, slow-maturing species, several years would be required before mortality imposed on early life stages could result in reduced abundance of older fish. The years 1978 through 1985 are considered to be a "transitional period" for the purpose of the BIC analysis. The years 1986 through the present are considered to be the "operational period" for the purpose of the BIC analysis. The impact of station operations on the fish community of the Delaware Estuary was evaluated by comparing species richness and species density, as determined from the near-field bottom trawl data, between the 1970–1977 preoperational period and the post-1985 operational period.

A standard collection size of 650 was selected for the numerical richness calculations. Because the finfish species composition of the estuary changes seasonally, separate analyses were performed for the spring (April–May), summer (June–August), and fall (September–October) seasons. Fig. 2a shows the results for the summer season. The results show no apparent trend in species richness over time. Statistical tests performed using Heck's[9] method for calculating the variance in species richness estimates confirm that there is no statistically significant difference in species richness between the preoperational and the operational periods. Preoperational vs. operational differences in richness for the spring and fall seasons (not shown) are also statistically insignificant.

Figure 2b shows trends in species density for the summer season in preoperational, transitional, and operational periods. Fig. 2b shows an apparent increase in species density over time, from a mean of 3.98 species per sample in the preoperational period to a mean of approximately 4.82 in the operational period. Results for the spring and fall seasons are similar. For all three seasons, a two-sided t-test shows that the mean number of species per sample has been significantly higher in the operational period than in the preoperational period.

THE TRENDS ANALYSIS

We reviewed all available fish monitoring data sets for the Delaware Estuary to determine their appropriateness for analyzing long-term abundance trends. The available data sets included the New Jersey Department of Environmental Protection (NJDEP) Beach Seine Survey, the Delaware Department of Natural Resources and Environmental Conservation (DNREC) Juvenile Trawl Survey, the DNREC Large Trawl Survey, the PSEG Nearfield Bottom Trawl Survey, and the PSEG Beach Seine Survey. The PSEG Beach Seine Survey was excluded because this survey began only in 1995. The DNREC Large Trawl Survey was excluded because of the numerous changes in sampling gear, locations of sampling, and months of sampling that have occurred since the program began in 1966. The PSEG Bottom Trawl Survey has been conducted using the same gear, and in the same locations and months since 1970; however, because of changes in gear deployment methods that affect sampling efficiency (use of a fixed-length towline prior to 1979; change in tow direction beginning in 1996), only data for the years 1979 through 1995 were considered sufficiently comparable to be used in the trends analysis. The DNREC Juvenile Trawl Survey has maintained consistent sampling methods since 1980; therefore, this data set was selected for analysis. The NJDEP Beach Seine Survey has been conducted since 1980; however, prior to 1986, the months and locations of sampling varied among years. Since 1986 sampling has been consistent from July through October, from RM 60 (RKM 96) to RM 140 (RKM 224); therefore, only the data from 1986 onward were used in the trends analysis. The estuary was divided into six sampling regions, and the data for each of the selected surveys were sorted by region and month. For the DNREC and PSEG surveys, species-specific age-length keys were used to select age 0 fish for analysis. The NJDEP data do not include lengths for species other than striped bass; therefore, for all other species, trends results for this data set reflect all ages present in each collection.

We constructed trends indices for each RIS by (1) determining the regions and months within which young-of-the-year fish of that species are found, (2) calculating the average catch per haul (CPH) for each of the selected regions and months, and then (3) calculating an average CPH over all regions and months. Index values were calculated for a given data set only if data were available for all of the selected months and regions. We evaluated each data set separately, so that up to three independent trends analyses could be performed for each species.

We used two types of statistical analyses for detecting trends from the relative abundance indices, depending on the nature of the data from the sampling program. If sampling had been discontinuous (i.e., a break of one or more years in the time series for a given data set), then a test for differences in average CPH before and after the break was used. If sampling had been continuous over the entire data set, then a linear regression was used to test for a slope significantly different from 0. In addition to the statistical tests, we summarized the changes in abundance as percent change in the population per year.

Table 2 presents the percent change in abundance per year of age 0 fish (for the NJDEP survey, all fish collected) in the Delaware Estuary by species and program. Changes that are statistically significant are shown in boldface. Table 2 shows that most of the RIS have increased in abundance over the period covered in the trends analysis. For alewife, Atlantic croaker, striped bass, weakfish, and white perch, all three programs indicate an increase. Only the NJDEP Beach Seine Program samples American shad in adequate numbers; the NJDEP American shad index

(a)

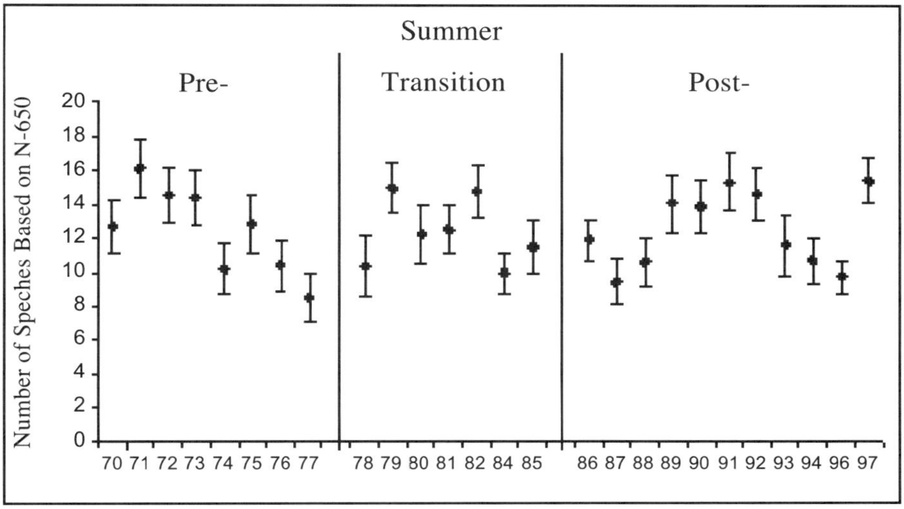

Post- Post- Post-

(b)

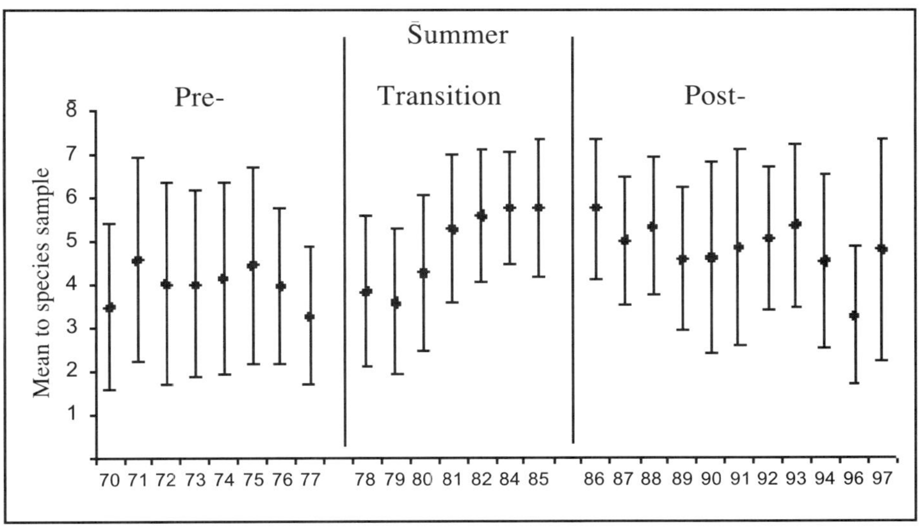

FIGURE 2. Species richness (a) and species density (b) in the vicinity of Salem during preoperational, transitional, and operational years.

shows a statistically significant upward trend. Two of the three indices for bay anchovy indicate an increase in abundance; only the PSEG bottom trawl survey indicates a decrease.

Decreased abundance is indicated only for blueback herring and spot. The decline in abundance of blueback herring in the Delaware Estuary parallels a coastwide decline in the abundance of this species that began prior to the startup of Salem in 1977[10]. New Jersey and Delaware are at the northern end of the range of spot and Atlantic croaker; the abundances of both species in New Jersey and Delaware fluctuate with changing oceanic conditions. The abundances of these two species as demonstrated in both the Delaware Estuary surveys and the New Jersey and Delaware commercial landings appear to be inversely related (high abundance of one species is generally associated with low abundance of the other); the reason for this pattern is unknown.

THE STOCK JEOPARDY ANALYSIS

We used population-level assessment models that extend the approaches used in previous assessments, drawing on recent advances in fisheries science and management practice. Previous assessments used the conditional mortality rate (CMR) as a measure of impacts of power plants on fish populations[11,12,13]. The CMR is a rate of "fishing" mortality that is closely related to the instantaneous fishing mortality rate (F) used to quantify effects of fishing:

$$CMR = 1 - e^{-F_P} \qquad (1)$$

where F_P = instantaneous rate of mortality due to entrainment or impingement at a power plant.

Boreman and Goodyear[12] and Barnthouse and Van Winkle[13] discuss procedures for calculating CMRs for entrainment and impingement. The CMR does not account for compensatory mechanisms that can offset entrainment or impingement mortality, and the CMR cannot be used to project long-term effects of entrainment and impingement on fish populations. The term "compensatory mechanisms" refers to biological processes (e.g., competition for limited food resources or habitat) that reduce the growth rates of large populations and increase the growth rates of small populations.

Because of this limitation, we chose methods that use the CMR as an input to more advanced assessment approaches: the spawning stock biomass per recruit (SSBPR) approach and the spawner-recruit (S-R) approach. The SSBPR approach estimates the lifetime reproductive output of a recruit (usually defined as a 1-year-old fish), accounting for the expected reproduction of the fish at each future age and the probability that the fish will survive to reach that age[14]. When a population is subjected to mortality caused by fishing, the reproductive output of a typical recruit is decreased, because the probability of each recruit surviving

TABLE 2

Percent Change in Abundance per Year of Age-0 Fish (DNREC and PSEG Programs) or of All Ages Collected (NJDEP Program) for Each RIS Finfish Species

Species	Program		
	DNREC Juvenile Trawl	NJDEP Beach Seine	PSEG Nearfield Bottom Trawl
Alewife	**55,4**[a]	2.1	38.7
American shad	NI[b]	7.3	NI
Atlantic croaker	+[c]	+	**3610.3**
Bay anchovy	1.3	**24.4**	**-4.8**
Blueback herring	**-5.5**	**-7.6**	NI
Spot	**-2.4**	**-8.1**	NI
Stripe bass	**40.4**	**5.3**	NS[d]
Weakfish	**18.7**	28.6	0.1
White perch	**91.4**	**12.6**	**41.7**

[a] Boldface indicates statistically significant change ($p < 0.05$).
[b] NI indicates that no index of abundance was calculated because of insufficient abundance in the regions sampled by a program.
[c] + indicates an increasing trend; percent change could not be calculated because the predicted initial population size from the regression model was zero.
[d] NS indicates no significant trend; percent change could not be calculated because the predicted initial population size from the regression model was zero.

to reproduce is decreased (in the case of fishing). For a population subjected to mortality due to cooling-water withdrawals, the probability of a spawned egg surviving to age 1 is decreased. This mortality is equivalent to removing the reproductive output that would have produced the lost recruits. To offset the losses and sustain the population, the survival or reproduction of the remaining fish must increase. Fisheries scientists have found that many fish populations can continue to sustain themselves when fishing has reduced SSBPR to as low as 20% of the level found in an unfished population[15]. Information needed to estimate a biological threshold below which SSBPR should not be reduced was unavailable for most of the species evaluated in our assessment. For this assessment, we assumed that an AEI could not occur as long as SSBPR remained above 30% of the unfished value. We then estimated, for all species for which data sufficient to calculate a CMR were available (all RIS species except striped bass and Atlantic croaker), the SSBPR in the presence of the combined effects of Salem and fishing and compared that SSBPR to a reference point of 30% of the unfished SSBPR.

Fisheries managers also use the total spawning stock biomass (SSB) of a fish population as a measure of population status. The SSBPR approach does not explicitly calculate the influence of reduced egg production on future recruitment or spawning stock biomass (SSB). Any such calculation requires an S-R model, which expresses the relationship between the size of the spawning stock in a given year – measured in terms of total biomass – and the number of recruits that will be produced by that stock. To estimate the effect of Salem on SSB for each of the modeled fish populations, we employed an S-R modeling approach termed the "equilibrium spawner-recruit analysis" (ES-RA). The ES-RA model extends the SSBPR approach by considering: (1) uncertainty concerning the values of critical life history parameters and (2) the relationship between SSB and recruitment. Since the ES-RA model includes more information than the SSBPR approach and should involve a lower degree of uncertainty, we used a less conservative reference point of 20% of the unfished value for the SSB analysis.

The ES-RA model requires the same data used in the SSBPR approach, i.e., age-specific estimates of natural mortality, fishing/power plant mortality, and weight or fecundity. In addition, the ES-RA requires an estimate of the slope of the spawner-recruit curve at the origin, i.e., the number of recruits produced by each spawner at very low population sizes. This slope is measured by the α parameter of the Ricker[16] or Beverton-Holt[17] spawner-recruit models.

Ideally, estimates of α would be obtained for each population of interest from the analysis of observations of spawner and recruit abundance for that population. However, such data are rarely available for species that are vulnerable to entrainment and impingement. Recently, Myers and Mertz[18] demonstrated that a statistical approach termed "meta-analysis" can be used to estimate α for a population of interest from estimates of α derived from long-term spawner recruit data for other populations.

Meta-analysis was used to derive estimates of α for use in the ES-RA analysis. Myers et al.[19] have compiled S-R data for more than 600 fish populations. Of the 600 data sets, 246 were considered sufficiently complete to provide reliable estimates of α. Of these 246 fish populations used, 109 were anadromous, 11 were freshwater, and 126 were marine or estuarine. The data set included 57 species belonging to 21 families and 8 orders.

Distributions of the parameter α were generated from the available S-R data using the methods described by Myers et al.[18,20]. Although either the Ricker or the Beverton-Holt models could have been used to estimate α from the S-R data sets, the Ricker model was chosen because it provided more conservative (precautionary) fits than the Beverton-Holt; i.e., on average it provided lower estimates of α for each species. Two approaches were employed in the choice of data sets to be used in the meta-analysis for each of the individual RIS. The first approach was to select closely related species, e.g., species within the same genus as the individual RIS. This approach was used for the alewife, American shad, and blueback herring. The second approach was to select species with similar life history characteristics and environmental tolerances, e.g., with the same type of reproduction (i.e., oviparous vs. ovoviviparous), habitat (i.e., anadromous, freshwater, or marine), natural mortality rates, longevity, fecundity, age at maturity, latitudinal distribution, and ambient temperature. This approach was used for all other species evaluated.

Given the above parameters, the ES-RA model was used to calculate equilibrium SSB for each species as a function of plant and fishing mortality. The Beverton-Holt model was used for this step in the analysis because it is more precautionary than the Ricker model. Specifically, for any given level of entrainment or impingement mortality, the Beverton-Holt model predicts a greater reduction in equilibrium spawner abundance than does the Ricker model. Assuming a Beverton-Holt type S-R relationship, the equilibrium biomass and yield equations derived by Lawson and Hilborn[21] were modified to include entrainment and impingement. The model assumes a life history in which power plant impacts are divided into two phases: a "precompensation" phase, and a "postcompensation" phase. Precompensatory mortality due to entrainment or impingement can be partly or largely offset by subsequent density-dependent mortality, resulting in a much smaller impact on age 1 abundance than would be predicted by a model that assumes no compensation (e.g., the CMR model). Postcompensatory mortality cannot be offset by compensation and translates directly into proportional reductions in age 1 abundance and subsequent recruitment to the adult population. Compensation in this case is delayed until the subsequent generation of fish, in which age 0 survival increases as a result of reduced adult abundance. Fishing mortality, according to this model, is always postcompensatory mortality.

The meta-analysis produced probability distributions of α for each fish species; these distributions were incorporated in the ES-RA calculations using Monte Carlo analysis. Uncertainty concerning the actual values of other key parameters, such as natural mortality rates, fishing mortality rates, and the fraction of the population present in the Delaware Estuary, was also addressed using Monte Carlo analysis. Probability distributions for these parameters were derived from the scientific literature or from best professional judgment concerning the possible range of values.

The results of the ES-RA calculations were expressed as ratios of spawning stock biomass, including the combined effects of fishing and power plants, to the spawning biomass of an unexploited population (SSB_0). CMRs were available for weakfish, bay anchovy, spot, white perch, alewife, American shad, and blueback herring; the ES-RA model was applied to each of these species. To demonstrate the approach, the discussion below focuses on weakfish.

Figure 3 presents results of the stock jeopardy analysis for weakfish. Life history parameters used in the SSBPR and ES-RA calculations are presented in Table 3. These parameters were taken from the most recent stock assessment available at the time this assessment was performed[22].

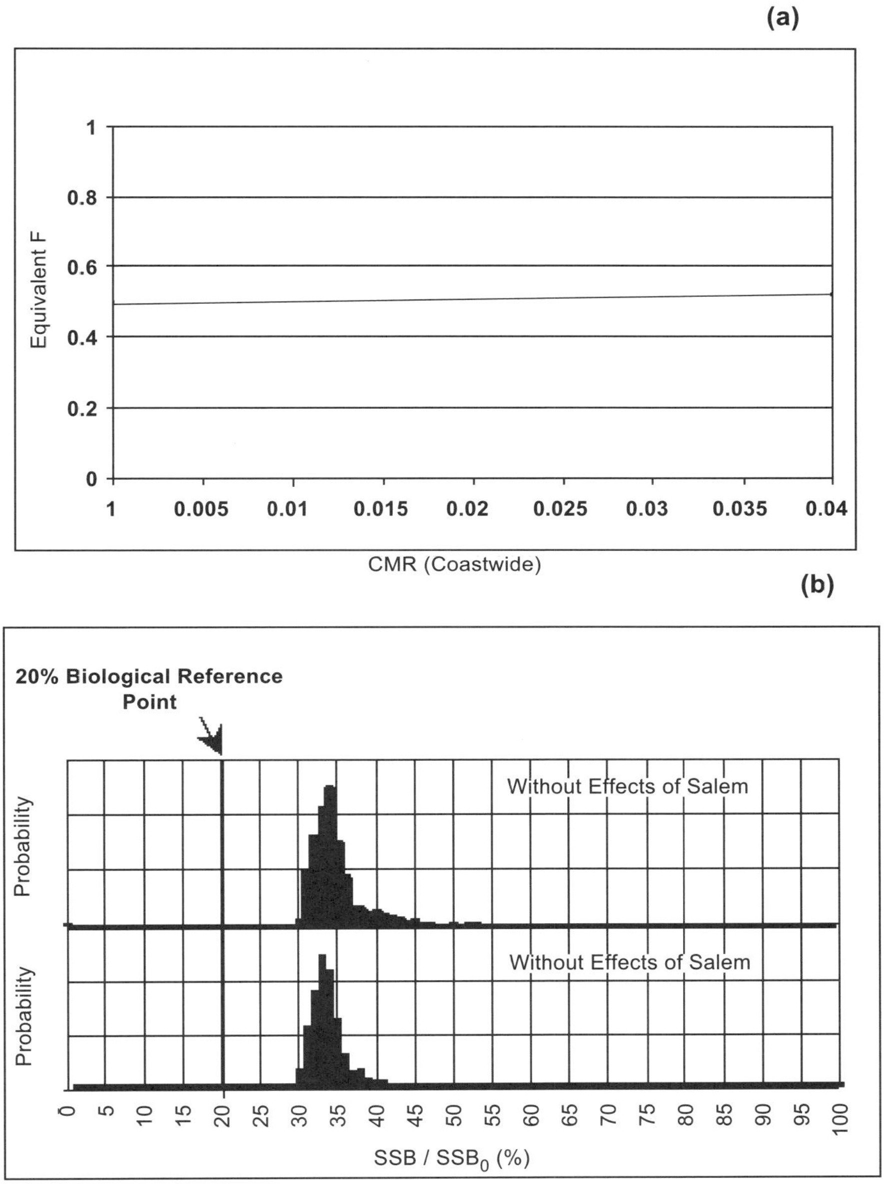

FIGURE 3. Results of stock jeopardy analysis for weakfish. Fig. 3(a) shows the influence of Salem on weakfish SSBPR, expressed as an incremental addition to the target rate of fishing for weakfish, for a range of assumptions concerning the coastwide CMR due to Salem. Fig. 3(b) shows the influence of Salem + fishing on weakfish SSB, expressed as a fraction of the equilibrium SSB for an unfished population.

TABLE 3

Life History Parameters for Weakfish[22]

Age	M	Vulnerability to Fishery	% Female	% Mature	Fecundity	Weight (lbs.)
1	0,25	10%	50%	30%	6824	0.26
2	0,25	50%	50%	85%	32973	0.68
3	0,25	100%	50%	90%	71387	1.12
4	0,25	100%	50%	100%	130848	1.79
5	0,25	100%	50%	100%	272716	2.91
6	0,25	100%	50%	100%	1041839	6.21
7	0,25	100%	50%	100%	1454325	7.14
8	0,25	100%	50%	100%	2147645	9.16
9	0,25	100%	50%	100%	2778770	10.83
10+	0,25	100%	50%	100%	3547138	12.50

Weakfish are managed as a mixed coastwide population ranging from Nova Scotia to Florida, although recent evidence[23] suggests that there is substantial fidelity to natal estuaries. We assumed that the weakfish population from North Carolina to Massachusetts constitutes a distinct breeding population. Based on analysis of state landings data, we assumed that the Delaware Estuary contributes 10–20% of the coastal stock. A reliable estimate of the rate of fishing mortality for weakfish was unavailable when this assessment was performed[22] However, a target fishing rate (F) of $F = 0.5$ had been established by the Atlantic States Marine Fisheries Commission (ASMFC). We used the SSBPR model to evaluate the potential impact of Salem in the context of this target fishing rate. Combining mortality due to Salem with fishing mortality is, in terms of impact on SSBPR, equivalent to incrementally increasing the rate of fishing on the adult stock. Fig. 3a shows the relationship between the coastwide CMR and the equivalent fishing rate for weakfish. Assuming that the Delaware Estuary contributes 20% of the coastwide weakfish population, the estimated coastwide CMR (0.034) is equivalent to raising the fishing rate for weakfish from the target rate of 0.5 to 0.517.

Fig. 3b shows the influence of fishing and Salem on total SSB for weakfish, including the stock-recruitment relationship estimated using the ES-RA. Because neither empirical estimates of F nor an approved S-R model for weakfish were available at the time this assessment was performed, we assumed that the weakfish population would in the future be fished at a rate close to the MSY level computed from the ES-RA model (approximately 35% of the unfished SSB). Fig. 3b shows that, even including uncertainty in key parameters, SSB is predicted to remain close to the MSY level and well above the overfishing reference point (20% SSB).

Results of the stock jeopardy analyses for other species showed even smaller potential impacts than were found for weakfish. In no cases did predicted impacts on SSBPR or SSB approach or exceed the overfishing reference points.

CONSISTENCY WITH OTHER INFORMATION CONCERNING POPULATION STATUS

The results of our analyses are consistent with other information concerning the status and trends of the Delaware Estuary's biological resources, and together they indicate that the operation of Salem over the past 20 years has not had an adverse impact on those resources.

Water quality in the Delaware River has improved dramatically since 1970. The Delaware Estuary's watershed contains one of the heaviest concentrations of industrial facilities in the world[24]. Until the 1950s, most of the wastewater generated in the watershed was discharged to the estuary without primary treatment, and secondary treatment only became prevalent in the 1980s[25]. Water quality was especially poor in the vicinity of Philadelphia, where dissolved oxygen concentrations frequently fell to zero[25]. Low DO blocked the passage of migratory fish such as American shad and other alosids, and destroyed or degraded much of the spawning and nursery habitat utilized by striped bass[26]. Since the mid-1980s, DO levels in the vicinity of Philadelphia have greatly improved. High densities of striped bass ichthyoplankton were observed in 1988, in a region of the river where ichthyoplankton had been absent in 1970[27]. Striped bass juvenile density, according to both the DNREC and NJDEP monitoring programs, was near zero during the early 1980s but increased rapidly thereafter. Weisberg et al.[28] attributed increases in the abundance of striped bass, white perch, and American shad in the Delaware Estuary to improvements in water quality.

Fisheries data are also consistent with the results discussed above. In response to restrictions placed on commercial and recreational fishing, spawning stock biomass of both striped bass and weakfish have risen to the highest levels observed over the past 20 years[29,30]. Coastal landings of American shad and river herring (alewife + blueback herring) have declined steadily throughout this century, in part due to overfishing but also due to declining water quality and blockage of tributaries by dams[10,31]. Within the Delaware River, however, both annual recruitment and spawning stock size are increasing[32]. No recent stock assessments are available for river herring. Stock assessments are also unavailable for spot and Atlantic croaker; however, landings data indicate that the abundance of both species fluctuates greatly in New Jersey and Delaware waters[33]. These fluctuations are probably related to fluctuations in coastal water temperatures that expand and contract the ranges occupied by these species. Even in peak years, catches in New Jersey and Delaware comprise no more than a few percent of total coastwide landings[33]. Available data for white perch in the Delaware Estuary show that commercial landings have increased since the mid-1980s, coincident with observed improvements in water quality.

It could be argued that improvements in water quality and reductions in fishing effort are simply confounding influences that mask adverse effects due to Salem, and therefore that data demonstrating that populations and communities are healthy are irrelevant to determining the presence or absence of adverse impacts. Perhaps the observed increases would have been even greater if not for entrainment and impingement at Salem. Arguments of this type cannot be logically disproved. They can, however, be evaluated qualitatively using the "risk hypothesis" approach discussed in EPA's Guidelines for Ecological Risk Assessment[34], as discussed below.

Of the finfish species we evaluated, the species that appear to be the most vulnerable to Salem include bay anchovy, weakfish, striped bass, and white perch. If entrainment and impingement were depleting vulnerable populations, then the abundance of one or more of these populations should decline. If the depleted populations were prey species such as bay anchovy – by far the most abundant prey species in the Delaware Estuary – then the abundance of predator species such as weakfish, striped bass, and white perch might also be expected to decline. If, on the other hand, the depleted populations were predators such as weakfish, striped bass, and white perch, then the abundance of prey species such as bay anchovy might be expected to increase. These types of changes have not occurred during the 20-year operational history of Salem. Instead, the abundances of all of the above species has increased.

The changes that have been observed over the last 20 years are inconsistent with the expected effects if Salem had been having an AEI on fish populations, but are consistent with the expected effects of water-quality improvements and reduced fishing effort. Species that utilize areas of the estuary that were formerly affected by poor water quality, including striped bass, white perch, American shad, and alewife, experienced substantial population growth when water quality improved in the 1980s. Species such as striped bass, weakfish, and American shad, for which harvests were restricted to promote recovery of depleted populations, increased in abundance following the enactment of those restrictions. Any impacts due to Salem clearly were too small to affect the responses of these populations to management changes intended to improve the Delaware Estuary and its biological resources.

CONCLUSIONS

Investigation of responses of fish populations to entrainment and impingement should involve evaluation of multiple lines of evidence, including both empirical observations and model-derived predictions. For the 1999 Salem Section 316(b) Demonstration, we developed three independent benchmarks of AEI and used multiple sources of data and modeling approaches to evaluate each benchmark. All three analyses indicated that the fish populations and communities of the Delaware Estuary are healthy and show no evidence of an adverse impact due to Salem, at least as defined in terms of changes in community balance or reduced sustainability of representative populations. Although conclusions derived regarding each individual benchmark are subject to multiple uncertainties, the concordance of all of our analyses appears sufficient to demonstrate that the influence of Salem is small compared to the influence of other major factors that affect the estuary.

The exact form of our analysis was constrained by the types of data available; the approach would have to be modified for application to other power plants and ecosystems. However, the general principal of evaluating multiple benchmarks, within a clearly defined analytical framework, should be applicable to all Section 316(b) determinations.

ACKNOWLEDGEMENTS

The authors acknowledge the support of the John Balletto, Maureen Vaskis, and other members of the PSEG staff for the support and encouragement they provided during the conduct of this assessment. We also gratefully acknowledge the assistance of Jennifer Field, John Posey, John Seibel, and Lorraine Read, without whom the assessment would never have been completed.

REFERENCES

1. U.S. Environmental Protection Agency. (1977) Guidance for Evaluating the Adverse Impact of Cooling Water Intake Structures on the Aquatic Environment: Section 316(b), P.L. 92-500. U.S. Environmental Protection Agency, Washington, D.C.
2. Barnthouse, L.W., Boreman, J.G., Christensen, S.W., Goodyear, C.P., Van Winkle, W., and Vaughan, D.S. (1984) Population biology in the courtroom: the Hudson River controversy. *BioScience* **34,** 14–19.
3. Barnthouse, L.W., Klauda, R.J., Vaughan, D.S., and. Kendall, R.L. (1988) Science, law, and Hudson River power plants: a case study in environmental impact assessment. *Am. Fish. Soc. Monogr.* **4**.
4. Margalef, R. (1968) *Perspectives in Ecological Theory*. University of Chicago Press, Chicago.
5. Rapport, D.H., Regier, H.A., and Hutchison, T.C. (1985) Ecosystem behavior under stress. *Am. Nat.* **125,** 617–638.
6. Gotelli, N.J. and Graves, G.R. (1996) *Null Models in Ecology*. Smithsonian Institution Press, Washington, D.C.
7. Peet, R.K. (1974) The measurement of species diversity. *Annu. Rev. Ecol. System.* **5,** 285–307.
8. Hurlburt, S.H. (1971) The non-concept of species diversity: a critique and alternative parameters. *Ecology* **52,** 577–586.
9. Heck, Jr., K.L., van Belle, G., and Simberloff D. (1975) Explicit calculation of the rarefaction diversity measurement and the determination of sufficient sample size. *Ecology* **56,** 1459–1461.
10. Atlantic States Marine Fisheries Commission. (1998) Amendment 1 to Interstate Fishery Management Plan for Shad and River Herring. ASMFC, Washington, D.C.
11. PSEG. (1984) Salem Generating Station 316(b) Demonstration. NPDES Permit No. NJ0005622. PSEG, Newark, NJ.

12. Boreman, J. and Goodyear, C.P. (1988) Estimates of entrainment mortality for striped bass and other fish species inhabiting the Hudson River estuary. *Am. Fish. Soc. Monogr.* **4,** 152–160.
13. Barnthouse, L.W. and Van Winkle, W. (1988) *Am. Fish. Soc. Monogr.* **4,** 182–190.
14. Goodyear, C.P. (1993) Spawning stock biomass per recruit in fisheries management: foundation and current use. *Can. Spec. Pub. Fish. Aquat. Sci.* **120**, 67–82.
15. Mace, P.M. and Sissenwine, M.P. (1993) How much spawning per recruit is enough? *Can. Spec. Pub. Fish. Aquat. Sci.* **120**, 101–118.
16. Ricker, W.E. (1954) Stock and recruitment. *J. Fish. Res. Board Can.* **11,** 559–623.
17. Beverton, R.J.H. and Holt, S. (1957) On the dynamics of exploited fish populations. Fishery Investigations, Series II, Marine Fisheries, Great Britain Ministry of Agriculture and Food 19.
18. Myers, R.A. and Mertz, G. (1998) Reducing uncertainty in the biological basis of fisheries management by meta-analysis of data from many populations: a synthesis. *Fish. Res.* **37,** 51–61.
19. Myers, R.A., Bridson, J., and Barrowman, N.J. (1995) Summary of worldwide stock and recruitment data. *Can. Tech. Rep. Fish. Aquat. Sci.* **2024**. Updates of the original database are available at http://www.mscs.dal.ca./~myers/welcome.html)
20. Myers, R.A., Bowen, K.G., and Barrowman N.J. (1999) Maximum reproductive rate of fish at low population sizes. *Can. J. Fish. Aquat. Sci.* **56,** 2404–2419.
21. Lawson, T.A. and Hilborn, R. (1985) Equilibrium yields and yield isopleths from a general age-structured model of harvested populations. *Can. J. Fish. Aquat. Sci.* **42,** 1766–1771.
22. National Marine Fisheries Service. (1998) 26th Northeast Regional Stock Assessment Workshop (26th SAW). Stock Assessment Review Committee (SARC) Consensus Summary of Assessments. Northeast Fisheries Science Center Reference Document 98-03.
23. Thorrold, S.R., Latkoczy, C., Swart, P.K., and Jones C.M. (2001) Natal homing in a marine fish metapopulation. *Science* **291,** 297–299.
24. Sutton, C., O'Herron, II, J.C., and Zappalorti, R.T. (1996) The Scientific Characterization of the Delaware Estuary. The Delaware Estuary Program, DRBC Project No. 321, HA File No. 93.21, 200 p. and appendices.
25. Albert, R.C. (1988) The historical context of water quality management for the Delaware Estuary. *Estuaries* **11,** 99–107.
26. Wang, J.C.S. and Kernohan, R.J. (1979) Fishes of the Delaware Estuaries: A Guide to the Early Life Histories. EA Communications, Inc., Towson, MD.
27. Weisberg, S.B. and Burton, W.H. (1993) Spring distribution and abundance of ichthyoplankton in the tidal Delaware River. *Fish. Bull.* **91,** 788–797.
28. Weisberg, S.B., Himchak, P., Baum, T., Wilson, H.T., and Allen, R. (1996) Temporal trends in abundance of fish in the tidal Delaware River. *Estuaries* **19(3),** 723–729.
29. National Marine Fisheries Service. (2000) 30th Northeast Regional Stock Assessment Workshop (30th SAW). Stock Assessment Review Committee (SARC) Consensus Summary of Assessments. Northeast Fisheries Science Center Reference Document 00-04.
30. Atlantic States Marine Fisheries Commission. (2000) 2000 Status of the Atlantic Striped Bass. ASMFC Striped Bass Technical Committee, Atlantic States Marine Fisheries Commission, Washington, D.C.
31. Atlantic States Marine Fisheries Commission. (1998) American Shad and Atlantic Sturgeon Stock Assessment Peer Review. ASMFC, Washington, D.C.
32. Santoro, E.D. (1998) Delaware Estuary Monitoring Report. Delaware Estuary Program, West Trenton, NJ.
33. National Marine Fisheries Service. (1998) Report to Congress: Status of Fisheries of the United States. National Marine Fisheries Service, 88 p.
34. U.S. Environmental Protection Agency. (1998) Guidelines for Ecological Risk Assessment. EPA/630-R-95/002F. U.S. Environmental Protection Agency, Washington, D.C.

Adverse Environmental Impact: A Consultant's Perspective

Alan W. Wells* and Thomas L. Englert
Lawler, Matusky & Skelly Engineers LLP, One Blue Hill Plaza, Pearl River, NY 10965

Received November 26, 2001; Revised April 16, 2002; Accepted April 19, 2002; Published February, 2003

Environmental consultants are in a unique position to address the practical aspects of a working definition of "adverse environmental impact" (AEI) within Section 316(b) of the Clean Water Act. In our work with the electric utility industry, attorneys, and regulatory agencies, we have encountered numerous and sometimes conflicting interpretations as to what constitutes AEI. In our over 30 years of experience, we have applied most of the approaches suggested for addressing this issue, including biostatistical methods, trend analysis, time series methods, conditional mortality rate models, stock-recruitment models, equivalent adult models, and ecosystem models. In our experience, the paradigm most helpful in bringing about agreement among stakeholders is to (1) create a model of operating scenarios, (2) use empirical data from on-site studies to parameterize the model, (3) convert losses by life stage to equivalent adult losses, (4) convert equivalent adult losses to economic value, and (5) compare scenarios on an economic basis.

KEY WORDS: adverse environmental impact, 316(b) demonstration, Clean Water Act, impact assessment methods, biological models, conditional mortality rate

DOMAINS: freshwater systems, marine systems, ecosystems and communities, organisms, water science and technology, environmental management and policy, environmental technology, methodological approach, environmental modeling, environmental monitoring

---

* Corresponding author. Emails: awells@lmseng.com, tenglert@lmseng.com

INTRODUCTION

Scientists, engineers, and economists working as environmental consultants are in a unique position to address the practical aspects of a working definition of "adverse environmental impact" (AEI). In our contacts with the electric utility industry, various regulatory agencies, and the legal community, we have encountered numerous and sometimes conflicting interpretations of what constitutes AEI. Over the last 30 or more years, Lawler, Matusky & Skelly Engineers LLP (LMS) has applied most of the suggested approaches with varying degrees of acceptance. In this paper, we share some of our experiences with the hope of providing insight into the development of a useful methodology for meeting the statutory requirements of §316(b) of the Clean Water Act.

To establish a common understanding, we must first ask, "Why do we care about a definition of AEI?" The answer to this requires that we examine the wording of the 316(b) regulation itself.

Section 316(b) of the Clean Water Act (33 USC §1251 et seq.) requires that: "...the location, design, construction, and capacity of cooling water intake structures reflect the best technology available for minimizing adverse environmental impact."

There are no explicit definitions of AEI in the regulations. Although the U.S. EPA drafted §316(b) regulations in 1976 (40 CFR Parts 401.14, 402.10-402.12 [1976]), those rules were later declared ineffective (Appalachian Power Company v. Train, 566 F.2d 451, 457 [4th Cir. 1977]) and were quickly withdrawn (44 Fed. Reg. 32,956 [June 7, 1979]). In the vacuum of Federal guidelines, two schools of interpretation have evolved around the §316(b) wording.

In our experience, regulatory agencies tend to interpret the wording as "the best technology available for *minimizing* adverse environmental impact" (emphasis added) while the regulated community, guided by their legal advisors, tends to interpret the wording as "the best technology available for minimizing *adverse* environmental impact." The former interpretation stresses that any impact, adverse or otherwise, must be minimized as long as the cost of the remedy is not wholly disproportionate to the benefit achieved (Decision of the General Counsel on Matters of Law Pursuant to 40 CFR Section 125.36(m), No. 63, July 29, 1977, at 23, 27). The latter interpretation effectively establishes a two-tiered hierarchical test (Fig. 1). Only if the impact is deemed "adverse" is the "cost-benefit" analysis necessary.

In practice, what really happens is *not* a hierarchical test. Without a basis for determining AEI, our clients generally elect to submit to the regulatory agencies a 316(b) demonstration with both elements. First, an assessment of the significance of the impact is made. In the absence of any established guidelines defining AEI, the outcome of the assessment is usually that "no adverse impact" has, or will, occur. But, by the same token, industries requesting the operating permit cannot be

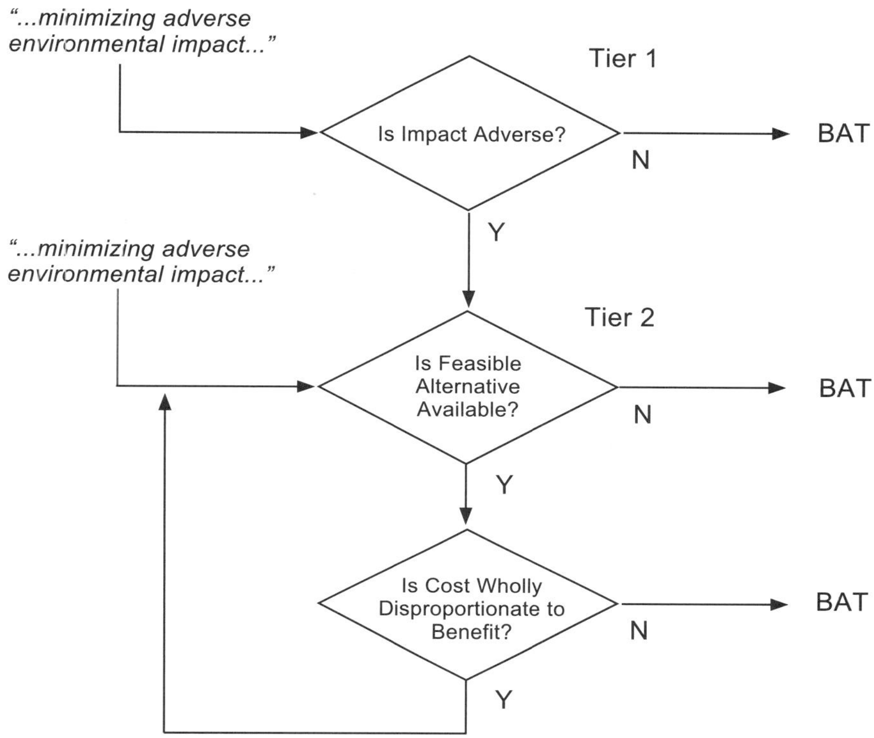

FIGURE 1. Interpretative pathways for the phrase "minimizing adverse environmental impact" in Section 316(b) of the Clean Water Act.

assured that their arguments will be accepted, and therefore, they feel compelled to complete the cost-benefit analysis as well.

In the following discussion, we focus on the various approaches LMS has used in the first tier of this process – i.e., the test for AEI – and how effective these approaches were in reaching a mutually agreeable operating permit. As will be seen, we digress periodically to consider aspects of the cost-benefit analysis.

AEI ASSESSMENT METHODS

Assessment methods fall into three broad categories. *Biostatistical Methods* generally apply classical statistical techniques to test null hypothesis models. These models are retrospective and cannot be used for *proposed* intakes or for modeling alternative operating scenarios. *Predictive Biological Methods* typically use intake flow characteristics and population parameters to estimate the near-term and long-term effects of the intake on a single population. Predictive methods can be used in both a retrospective and prospective manner. *Community Response Methods* attempt to examine the intake effects on the entire biological community (or some part thereof). While community response techniques can be used in a prospective mode, they are more commonly applied in a retrospective mode.

Biostatistical Methods

During the initial flurry of 316(b) activity in the mid 1970s through early 1980s, among the first methods to be applied in 316(b) assessments for the determination of AEI were biostatistical methods. These methods fell into two major groups: *classical hypothesis testing* using analysis of variance (ANOVA) style designs and *trend analysis*.

ANOVA-style analyses seem to be an outgrowth of the "no prior appreciable harm" model used in 316(a) demonstrations[1] and popularized by a book on environmental sampling by Green[2]. The ideal design was often referred to as a BACI design, meaning before/after/ control/impact. A statistically significant interaction term was taken as evidence of AEI. In cases where the full BACI model could not be used, less persuasive before/after or control/impact comparisons were made. Statistically significant differences between the before and after samples *or* between the control and impacted samples were taken as AEI.

By 1977, papers began to appear that described deficiencies in the ANOVA-style approach[3,4] and our experienced reinforced these warnings. Problems encountered included the following:

- Failure to define the effect size to be detected
- Insufficient statistical power to detect biologically meaningful impacts
- Failure of untransformed data to meet requirements of normality, additivity, independence, and heteroscedasticity
- Failure of transformed data to have real-world meaning
- Interpretation confounded by numerous statistically significant high-order interaction terms
- Inability to use the model in a predictive mode (a critical problem when trying to compare operational alternatives).

During the early stages (early 1980s) of working with the Technical Advisory Group (an advisory group representing numerous regulatory agencies involved in the licensing of the facility) for the Salem Generating Station, in New Jersey, it was very evident that the sampling design necessary for a biostatistical analysis was markedly different from that necessary for other AEI approaches. Given the finite dollar resources available to conduct studies, the TAG readily agreed with the refocusing of sampling priorities to a predictive biological modeling approach (see below) and the abandonment of the biostatistical analysis.

Trend analysis typically requires a relatively lengthy time series of catch-per-unit-effort (CPUE) data. The most commonly used procedure for trend analysis is linear regression. A statistically significant negative slope suggests the potential for AEI. Such an analysis, however, implies a linear population response – a response unlikely for any sustained period of time. A more appropriate but much less often used approach is intervention analysis[5]. Essentially, this analysis is simply the before/after analysis with a longer series of observations and adjustments made for autocorrelated data. We should note that we have rarely encountered data sets of sufficient length (both pre- and postintervention) and consistency to support intervention analysis.

In either the regression or intervention analysis approaches, it is difficult to separate the effects of an intake from other factors, such as overfishing, natural population cycles, or a series of bad year classes due to poor weather. This, in fact, is the major problem with either method. The analysis may describe the overall status of the stock, but it fails to discriminate among causative agents. A good example of the problem is evident in the Hudson River striped bass population. Consultants for the Hudson River Utilities have argued that the population is not significantly affected by the power plants, pointing to increasing stock size documented in CPUE data. Opponents have been quick to point out that the most likely reason for the phenomenal population growth is that the Hudson River striped bass fishery has been closed since 1976 due to PCB contamination (e.g., [6]).

Predictive Biological Methods

Predictive biological models fall into three general categories, sometimes referred to as: *Absolute Damage Models*, *Relative Damage Models*, and *Long-Term Damage Models*. As one progresses from absolute, to relative, to long term, the number of input parameters, sampling program cost, and relevance to AEI determination increase. With each new input parameter, however, overall relative model precision tends to decrease and the potential for acceptance by regulatory agencies decreases (Fig. 2).

Absolute damage models yield the estimated number of organisms killed by entrainment and impingement. These losses generally are presented as number by life stage, but, we have found that expressing losses as equivalent adults[7,8] is more helpful in understanding the biological significance of the losses and in assessing the economic value:

$$N_A = \sum_{i=1}^{n} N_i S_i$$

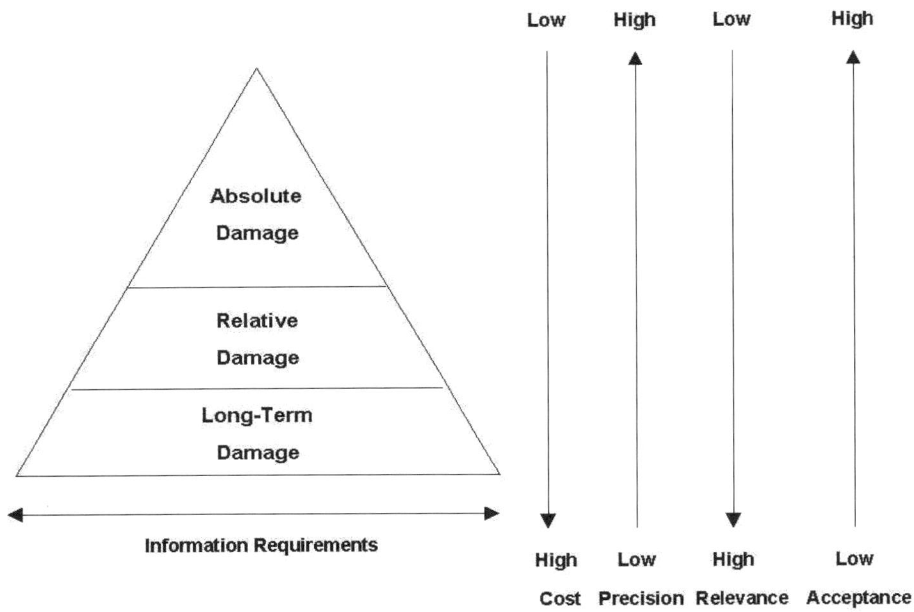

FIGURE 2. Relationships among various Section 316(b) damage assessment modes.

where n = number of life stages, N_i = number of life stage *i* killed, and S_i = survival probability from life stage *i* to adulthood.

Losses expressed in this manner are more readily compared to other known sources of mortality, particularly to a commercial or recreational fishery. It is generally a relatively easy matter to convert numerical losses expressed in this manner to standing stock biomass (usually pounds), production, or economic value. We have not had regulatory agencies object to this form of analysis, and in several instances, they have specifically requested this approach.

In general, the absolute damage model approach requires only routine on-site data collection. An additional advantage of this approach is that the derived concentrations of entrained or impinged organisms in the intake water can be used to compare different operating scenarios. One need only alter the intake flow, through-plant exposure temperature and exposure duration, or impingement mortality rate. Operating alternatives can then be compared directly on the basis of numeric loss, biomass, or dollars.

Clearly, basing an AEI assessment on absolute losses alone can be misleading. For example, large numeric losses are potentially more harmful to a small source population than to a large population. Relative damage models attempt to estimate the loss as a fraction of the source population. They are sometimes referred to as fractional loss models or conditional mortality rate (CMR) models. (The conditional mortality rate is the fraction of the year class lost in the absence of all other sources of mortality).

One of the first attempts at estimating "relative damage" was through the use of hydrodynamic models[9]. These models are based on the assumption that the early life stages of fish are transported by the hydrodynamics of the source waterbody. Hydrodynamic-based models were hotly contested during the early days of the Hudson River hearings[10,11], primarily because of the inability to adequately capture fish behavior. It was soon realized that even the very young larvae of many species do not behave strictly as passive particles, but instead have developed behavioral mechanisms to adapt to their particular environment. Hydrodynamic models soon gave way to more empirically based models such as the Empirical Transport Model[12,13] and the Empirical Impingement Model[14]. In our experience, the use of these empirically based models is generally well accepted by regulatory agencies and, in many cases, are explicitly required as part of the permitting process. Empirical models still have a number of limitations, however. Among them are:

- CMR values are relative to an arbitrarily defined study area.
- The models often unrealistically assume a closed system (modifications to relax this assumption are available).
- Confidence intervals are difficult to establish.
- Models such as the Empirical Impingement Model require estimates of population size and mortality rates, both of which are difficult to obtain with precision.
- Models do not include compensatory or depensatory processes.

The primary problem with "relative damage" models, however, is the translation of the CMR to AEI. Once the CMR is known, is it adverse? We have frequently run across a "rule of thumb" stating that a CMR value greater than 10% is adverse (15% for forage species) or has the potential for adverse impacts (e.g., [15]). On occasion we have also encountered agencies that consider CMRs of less than 1% adverse. There is, however, no firm biological foundation for either of these thresholds. The "adversity" of the value is highly dependent on the circumstance and no general guideline seems possible.

The only practical way to assess the consequences of a given CMR on a population is to incorporate the value into a "long-term damage" model. Among the first of these models to be used was the Leslie Matrix population projection model. We found this approach to be of little help in the assessment process and not well received by guidance committees. Without the incorporation of density-dependent mechanisms, even the slightest loss drives the modeled population to eventual extinction. Stock-recruitment models based on Ricker, Beverton-Holt, or similar functions[16,17] are somewhat more useful, but they were poorly suited to nonequilibrium populations or iteroparous spawners[18]. The stock-recruitment approach met with considerable resistance during the initial Hudson

River hearings in the 1970s, primarily because of the difficulty in estimating the model parameter[19,20]. In our recent experience, RAMAS[21] and approaches derived from Hilborn and Walters[22] seem to have gained some acceptance. These models combine population projection matrices with flexible density-dependent recruitment models. Bayesian maximum likelihood methods[23] coupled with meta-analysis[24] have made important gains in obtaining better estimates of the critical density-independent response parameters. Individual based models, such as Rose et al.[25], can also be useful in overcoming limitations of estimating density-dependent parameters.

"Long-term" damage models allow us to examine the effects of various CMR values on a population over time. Additionally, by incorporating Monte Carlo techniques, we can use the models to assess the probability of certain outcomes. But this increased usefulness comes at a price-increased data requirements. Information on age-specific growth, mortality and fecundity values, year (and sometimes age) specific fishing rates, carrying capacity (K) and the intrinsic rate of population increase (r) are needed. It is also being widely recognized that K and r can change in response to density-dependent and density-independent factors[26,27].

We have encountered widely varying responses to the inclusion of "long-term" damage models. In the Hudson River hearings during the 1970s, much of the nearly 10-year-long debate was spent arguing over these types of models – in particular, the degree of density-dependent compensation. In the long run, these models contributed little to the final outcome. We should point out that a consensus was reached on CMR values and that they did, in fact, play an important role in settling the case[28]. In at least one major case in which we were involved during the early 1980s after the Hudson River case, the technical advisory body explicitly cautioned us not to use "long-term" models. The advisors did not want to repeat the Hudson River turmoil, and they believed they could determine AEI without this information.

In several more recent cases, opposition to "long-term" models seemed to have waned[29,30]. In one instance, while we were not discouraged from using the approach, the reviewers ignored the results in reaching their final recommendations. In another ongoing case, the technical advisory body actively encouraged use of these models, but as the case progresses, it is becoming increasingly unclear how important this technique will ultimately be.

A "long-term" damage assessment is a complex undertaking and close cooperation with resource agencies seems to be required for the effort to be credible. Such a close relationship, however, can lead to several difficulties. One is a gradual change in scope and focus. In a recent application, we worked closely with the utility, resource agencies, and other consultants, all with varying degrees of understanding of the modeling process. Each model input parameter was thoroughly discussed and agreed upon before incorporation into the model. As understanding of the model grew, various members would suggest changes or additional output. While some changes were easily incorporated, others would have required fundamental changes to the model's underlying structure – changes that would have required considerable reprogramming and resulted in serious time delays. Without the changes, the proposing committee member(s) began to lose confidence in the unmodified model.

Despite the enhanced insight offered by "long-term" damage models, we are still saddled with the fundamental problem of interpretation of AEI. Is there a biological benchmark that can be considered adverse? One frequently suggested benchmark – stemming from the 1975 Draft Guidance Manual[31] – is a reduction in maximum sustainable yield (MYS) or optimum sustainable yield (OSY). Unfortunately, given any power plant CMR and realistic values of density-dependent response, "long-term" damage models will *always* show a reduction in MSY or OSY. Consequently, any loss, no matter how small, would be considered as adverse. Fisheries benchmarks – such as F_{MSY} F_{OSY}, or several other related measures – emphasize the fact that intake losses are clearly part of the larger picture of fisheries resource allocation. The loss of aquatic organisms can be allocated among all user groups, including commercial and recreational fishermen. This larger perspective, however, needs to be addressed by the resource agencies, not by the individual 316(b) applicant. It should also be noted that fisheries benchmarks are workable for species supporting commercial or recreational fisheries, but do not appear to be applicable to nonexploited species due to lack of data availability.

Community Response Methods

While most of the analyses we have conducted have been at the population level, it is also worthwhile to examine impacts from a community perspective. Ideally, populations selected for study under 316(b) represent a cross-section of the biological community that occupies the source waterbody. Population-level results can then be generalized to the community level. We have found that, in practice, "target species" disproportionately represent commercially or recreationally important species or reflect an advisory group's or group member's special interest in a particular species.

A number of techniques have been devised for looking at community-level effects. Among them are such simple methods as species richness, diversity, evenness, and percent similarity, and more complex forms such as indices of biotic integrity, loop analysis, and ecosystem models. We have found the simple forms to be useful in some circumstances. For example, in one recent application, richness, diversity, and evenness measures demonstrated an obvious change in fish community structure[30]. An examination of assemblages within the study community (e.g., benthic vs. pelagic, resident vs. migrant, and northern vs. southern origin) demonstrated nearly identical population trends among all groups. From this knowledge, we were able to narrow down the list of causative agents. We have had little call for the more complex community approaches; particularly loop analysis and ecosystem models. These methods are not readily understood by reviewers and generally require impractical amounts of data.

One of the primary difficulties with the application of a community-level approach is the inevitable blurring of the lines between 316(b) (with its AEI test for intake structures) and 316(a) (with its "protection of a balanced indigenous community" test for discharges). In the past, we have encountered several agencies that have required a clear distinction between 316(a) and 316(b). While intake and discharge effects certainly are interdependent, current statutes require that they be kept distinct. We generally conduct community analyses, such as described above, under 316(a).

DISCUSSION AND CONCLUSIONS

In our over 30 years of corporate experience, we have found that it is generally the simple approaches, in combination, that work best for addressing Section 316(b) issues. In particular, we have found:

- Absolute damage estimates are always required and are a necessary first step to any other analysis.
- Relative damage estimates are frequently requested by agencies, but they can be relatively less precise than absolute damage estimates and are seldom definitive in AEI assessment. There is no practical AEI benchmark because all situations are different.
- "Long-term" damage estimates may provide useful insights into the relative importance of various sources of impact and into the probability of outcomes of various actions. However, these methods, relative to the "absolute" and "relative" methods, require data on an increased number of parameters, some of which are difficult to estimate with precision. Uncertainty in the model input parameters leads to lack of confidence in the results. In the past, we have found "long-term" damage assessment models even less influential than "relative" damage assessments in 316(b) negotiations.
- Simple "community" level analyses can be important, but are best left to 316(a) demonstrations. This is not to say that community level analyses should not be conducted; certainly there are good ecological reasons for examining community responses. Under the current statutes, however, the analysis is more appropriately placed in context of 316(a) (i.e., maintenance of a balanced indigenous community). If both 316(a) and 316(b) Demonstrations are submitted simultaneously, as is often the case, the community analysis can be easily referenced from the 316(b) portion.

The most effective approach we have found can be broken down into the following steps:

1. Construct a simple model of plant operations that incorporates alternative operating modes.

2. Using the empirically observed concentrations of organisms, compute "absolute" losses by life stage (or age class) for the existing operational mode and for several alternative scenarios.

3. Use the equivalent adult model to translate losses by life-stage to losses of adults.

4. Convert the adult losses to dollars. (Conversion to economic value is not without problems and controversy. In some cases the American Fisheries Society monetary values for fishes have been used[32] or, more frequently, the dockside value for commercially important species. Maryland, under Natural Resources Article Section 8-1405(c) incorporates economic loss values directly into their statute. None of these methods capture the all of the indirect environmental and social values of the resource. Methods attempting to incorporate these measures are infrequently encountered.)

5. Compare alternative operating modes on a dollar value basis.
   The preceding approach requires little more than an understanding of the plant operation, routine on-site monitoring data, an economic value for organism losses, and a few basic life-history parameters. What the approach does *not* require is a definition of "adverse." Our paradigm shifts the focus from one of biology to one of economics. Ultimately, we must recognize that "adverse" has no true biological meaning. It is only from a human and societal perspective that can we say what is adverse. As scientists and engineers, we can contribute to the understanding of various alternatives by expressing the consequences of alternative actions in terms that stakeholders understand.

REFERENCES

1. U.S. Environmental Protection Agency (USEPA) (1977) Draft Interagency 316(a) Technical Guidance Manual and Guide for Effects Sections of Nuclear Facilities Environmental Impact Statements. Mimeo,77 p.
2. Green, R.H. (1979) Sampling Design and Statistical Methods for Environmental Biologists. John Wiley & Sons, New York, 257 p.
3. McCaughran, D.A. (1977) The quality of inferences concerning the effects of nuclear power plants on the environment. In *Proceedings of the Conference on Assessing the Effects of Power-Plant-Induced Mortality on Fish Populations*. Van Winkle, W., Ed. Pergamon Press, New York. pp. 229–242.
4. Thomas, J.M. (1977) Factors to consider in monitoring programs suggested by statistical analysis of available data. In *Proceedings of the Conference on Assessing the Effects of Power-Plant-Induced Mortality on Fish Populations*. Van Winkle, W., Ed. Pergamon Press, New York. pp. 243–255.
5. Box, G.E.P. and Tiao, G.C. (1975) Intervention analysis with application to economic and environmental problems. *J. Am. Stat. Assoc. (Theor. Methods Sect.)* **70,** 70–90.
6. Coastal Conservation Association New York (CCANY) (1999) Press Release: March 10, 1999, Coastal Conservation Association Opposes Reopening of Commercial Striped Bass. CCA Online, http://www.ccany.org/archives/pressrelease9.cfm.
7. Horst, T.J. (1975) The assessment of impact due to entrainment of ichthyoplankton. In *Fisheries and Energy Production: A Symposium*. Saila, S.B., Ed. Lexington Books, D.C. Heath and Company, Lexington, MA. pp. 107–118.
8. Goodyear, C.P. (1978) Entrainment impact estimates using the equivalent adult approach. U.S. Fish and Wildlife Service. FWS/OBS-78/65, 14 p.
9. Lawler, Matusky & Skelly Engineers (LMS) (1975) Report on development of a real-time, two-dimensional model of the Hudson River striped bass population. LMS Project No. 115-149, 71 p.
10. Swartzman, G., Deriso, R., and Cowan, C. (1977) Comparison of simulation models used in assessing the effects of power-plant-induced mortality on fish populations. In *Proceedings of the Conference on Assessing the Effects of Power-Plant-Induced Mortality on Fish Populations*. Van Winkle, W., Ed. Pergamon Press, New York. pp. 333–361.
11. Christensen, S.W. and Englert. T.L. (1988) Historical development of entrainment models for Hudson River striped bass. *Am. Fish. Soc. Monogr.* **4**, 133–142.
12. Boreman, J., Goodyear, C.P., and Christensen, S.W. (1978) An empirical transport model for evaluating entrainment of aquatic organisms by power plants. U.S. Fish and Wildlife Service, Biological Services Program, National Power Plant Team, FWS/OBS-78/90, 67 p.
13. Boreman, J., Goodyear, C.P., and Christensen, S.W. (1981) An empirical methodology for estimating entrainment losses at power plants sited on estuaries. *Trans. Am. Fish. Soc.* **110(2),** 253–260.
14. Barnthouse, L.W., DeAngelis, D.L., and Christensen, S.W. (1979) An empirical model of impingement impact. ORNL/NUREG/TM-290, and NUREG/CR-0639. Oak Ridge National Laboratory, Oak Ridge, TN, 28 p.
15. Versar, Inc. (1988) Technical Review and Evaluation of Thermal Effects Studies and Cooling Water Intake Structure Demonstration of Impact for the Oyster Creek Nuclear Generating Station. Final Report. Prepared for New Jersey Department of Environmental Protection, Division of Water Resources, Trenton, NJ.
16. Riker, W.E. (1975) Computation and interpretation of biological statistics of fish populations. *J. Fish. Res. Board Can.* **191**, 1–382.
17. Beverton, R.J.H. and Holt, S.J. (1957) On the dynamics of exploited fish populations. Fishery Investigations, Series II, Marine Fisheries, Great Britain Ministry of Agriculture Fisheries and Food 19, 533 p.
18. Lawler, J.P. (1988) Some considerations in applying stock-recruitment models to multiple-age spawning populations. *Am. Fish. Soc. Monogr.* 4, 204–215.
19. Christensen, S.W. and Goodyear, C.P (1988) Testing the validity of stock-recruitment curve fits. *Am. Fish. Soc. Monogr.* **4**, 219–231.

20. Fletcher, R.I. and Deriso, R.B. (1988) Fishing in dangerous waters: remarks on a controversial appeal to spawner-recruit theory for long-term impact assessment. *Am. Fish. Soc. Monogr.* 4, 232–243.
21. AkHakaya, H.R. (1998) RAMAS GIS: Linking Landscape Data with Population Viability Analysis (version 3.0). Applied Biomathematics, Setauket, NY.
22. Hilborn, R. and Walters, C.J. (1992) Quantitative Fisheries Stock Assessment: Choices, Dynamics and Uncertainty. Chapman and Hall, New York.
23. Hilborn, R. and Mangel, M. (1997) *The Ecological Detective, Confronting Models with Data.* Monographs in Population Biology 28. Princeton University Press, Princeton, NJ, 315 p.
24. Myers, R.A., Bowen, K.G., and Barrowman, N.J. (1999) Maximum reproductive rate of fish at low population sizes. *Can. J. Fish. Aquat. Sci.* **56**, 2404–2419.
25. Rose, K.A., Tyler, J.A., Chambers, R.C., Klein-MacPhee, G., and Danila, D.J. (1996) Simulating winter flounder population dynamics using coupled individual-based young-of-the-year and age-structured adult models. *Can. J. Fish. Aquat. Sci.* **53(5),** 1071–1091.
26. MacCall, A.D. (1990) Dynamic Geography of Marine Fish Populations. Washington Sea Grant Program, Seattle, Washington, 153 p.
27. Englert, T.L., Wells, A.W., and Norris, R.A. (2000) Incorporation of changes in habitat quantity and quality into density-dependent population models. *Environ. Sci. Policy* 3, S451–S458.
28. Englert, T.L., Boreman, J., and Chen, H.Y. (1988) Plant flow reductions and outages as mitigative measures. *Am. Fish. Soc. Monogr.* **4,** 274–279.
29. Public Service Electric and Gas Company (1999) Salem Generating Station, Permit Renewal Application NJPDES Permit N. NJ0005622. March 4, 1999.
30. U.S. Gen New England (USGNE) (2001) Brayton Point Station, Permit Renewal Application, NPDES Permit No. MA0003654.
31. U.S. Environmental Protection Agency (USEPA) (1975) Guidelines to Determine Best Available Technology for the Location, Design, Construction, and Capacity of Cooling Water Intake Structures for Minimizing Adverse Environmental Impact, Section 316(b) P.L. 92-500. December 5, 1975. Mimeo
32. American Fisheries Society (1991) A Handbook of Monetary Values of Fishes and Fish-Kill Counting Guidelines. American Fisheries Society Socioeconomics Section, AFS Southern Division Committee on Pollution, Special Publication Number 13. 73 pp.

Proposed Methods and Endpoints for Defining and Assessing Adverse Environmental Impact (AEI) on Fish Communities/Populations in Tennessee River Reservoirs

Gary D. Hickman* and Mary L. Brown
River System Operations & Environment, Tennessee Valley Authority, 17 Ridgeway Road, Norris, TN 37828

Received November 8, 2001; Revised March 14, 2002; Accepted March 15, 2002; Published February, 2003

Two multimetric indices have been developed to help address fish community (reservoir fish assemblage index [RFAI]) and individual population quality (sport fishing index [SFI]) in Tennessee River reservoirs. The RFAI, with characteristics similar to the index of biotic integrity (IBI) used in stream fish community determinations, was developed to monitor the existing condition of resident fish communities[1,2,3]. The index, which incorporates standardized electrofishing of littoral areas and experimental gill netting for limnetic bottom-dwelling species, has been used to determine residential fish community response to various anthropogenic impacts in southeastern reservoirs.

The SFI is a multimetric index designed to address the quality of the fishery for individual resident sport fish species in a particular lake or reservoir[4]. The SFI incorporates measures of fish population aspects and angler catch and pressure estimates. This paper proposes 70% of the maximum RFAI score and 10% above the average SFI score for individual species as "screening" endpoints for balanced indigenous populations (BIP) or adverse environmental impact (AEI). Endpoints for these indices indicate: (1) communities/populations are obviously balanced indigenous populations (BIP) indicating no adverse environmental impact (AEI), or are "screened out"; (2) communities/populations are considered to be potentially impacted; and (3) where the resident fish community/population should be considered adversely impacted. Suggestions are also made concerning how examination of individual metric scores can help determine the source or cause of the impact.

KEY WORDS: biocriteria, biological indices, fish community assessment, reservoir fish assemblage index (RFAI), sport fishing index (SFI)

* Corresponding author. Emails: gdhickman@tva.gov; mlbrown3@tva.gov.

DOMAINS: ecosystems and communities, environmental management, environmental monitoring, environmental technology, freshwater systems, structural biology, water science and technology

INTRODUCTION

Karr[5] suggested that multimetric indices are robust enough and are more representative of biological responses to anthropogenic influences than traditional water quality monitoring programs. The index of biotic integrity (IBI) originally developed by Karr was used by the Tennessee Valley Authority (TVA) as the basis for development of a fish community quality index in TVA reservoirs. This index was then applied to biomonitoring programs in other geographical regions and aquatic systems[1,2,6,7,8,9,10,11,12]. Jennings[1] first described the multimetric reservoir fish assemblage index (RFAI) as a cost-effective method to address quality of resident fish assemblages as a reflection of environmental quality. The index was further refined[2,3], reducing sampling variability and substituting some metrics that were more reflective of reservoir conditions. Fish community quality is defined as how close resident communities approach the community structure and function anticipated without anthropogenic influence (based on best observed conditions along with professional judgment of biologists familiar with biotic indices and the zoogeography of the Tennessee River). Additional testing of RFAI performance was completed[13] in four reservoirs of both the Catawba and Cumberland River systems to determine the applicability of the index outside the Tennessee River system. Additional minor modifications were made to index metrics. The resulting RFAI was able to distinguish differences between various fish communities in these systems, and results were repeatable. Differences were more difficult to detect within reservoir fish communities, indicating that the biological zone of influence may include large sections of an individual reservoir, including the entire reservoir on smaller impoundments (< 10,000 acres), or that this technique is not sufficiently robust for this application.

Colvin and Vasey[14] first introduced the concept of using multiple metrics in the determination of fishing quality for individual species within a water body. Hickman[4] proposed use of a series of commonly collected population and angler success measures to derive a sport fishing index (SFI) as a measure of recreationally important individual species population quality within a reservoir.

Adverse environmental impact (AEI) endpoints were not adequately developed in section 316(b) of the Clean Water Act of 1972. The definition of AEI required "use of best management practices (BMP) to minimize AEI," but never defined what constituted "AEI" with respect to cooling water intake structure losses. Under 316(a), the thermal effluent endpoint was described as "maintenance of balanced indigenous populations (BIP)," but again did not describe how to determine if BIP existed in the vicinity of a plant. The U.S. Environmental Protection Agency was sued in 1997, requiring a more precise definition of AEI.

The objective of this paper is to suggest potential endpoints for the RFAI and SFI indices. These endpoints suggest (1) where there is no appreciable risk (i.e., no reasonable or significant risk) that resident communities/populations are adversely impacted, (2) levels where adverse impacts are possibly occurring, or (3) communities/populations with obvious unacceptable levels of impact. Suggestions are also made concerning examination of individual metric scores to help determine the source or cause of the impact(s).

METHODS

Two recent reports[2,3], contain detailed explanations of methods used to arrive at RFAI scores. In general, 15 boat electrofishing samples (each 300 m in length) located proportional to existing shoreline habitat and ten overnight experimental gill net sets (five 6.1 m panels with bar mesh sizes of 2.5, 3.8, 5.1, 6.4, and 7.6 cm) were used to obtain standardized samples of the fish community. Sampling results are compared to reference conditions (i.e., those anticipated from a reservoir in the same physiographic region[15] and reservoir zone in the absence of human-induced impacts other than impoundment and operational characteristics such as winter drawdown). As mentioned previously, reference conditions against which individual samples are compared were derived from best observed conditions of numerous samples (5-year period at several sites in geographically and hydrologically similar reservoirs), with adjustments made by groups of knowledgeable biologists making the criteria more conservative. Scores for individual metrics are assigned using three levels (least degraded-5; intermediate-3; and most degraded-1)[1]. Individual metric scores are then summed to obtain the final RFAI score. RFAI scores from 1993–2000 from upstream and downstream areas in the general vicinity of TVA fossil plants were compared to demonstrate use of these endpoints. Examination of individual metrics was performed to determine potential for plant operation to be contributing to, or causing, adverse impacts.

Hickman[4] described in detail the development and composition of the SFI. The SFI includes information on population parameters and angler success/use routinely collected by many state fishery agencies (Fig. 1). Both population and angler statistic metrics include quality and quantity aspects. Population quantity measures are simply catch per unit effort by the most appropriate gear type (i.e., electrofishing, gill netting, or trap netting) for the species being addressed. Catch results from only one gear type are used for SFI determination. Population quality measures include size distribution parameters (proportional stock density [PSD] and relative stock density [RSD] of preferred, memorable, and trophy-size groups) and relative weight (Wr) values. Angler catch per hour of intended species addresses the quantity aspect of creel data, and angler use (hours fished for intended species) represents the quality aspect. When creel results were not available, population results were doubled. Though not ideal, this does provide an indication of population quality.

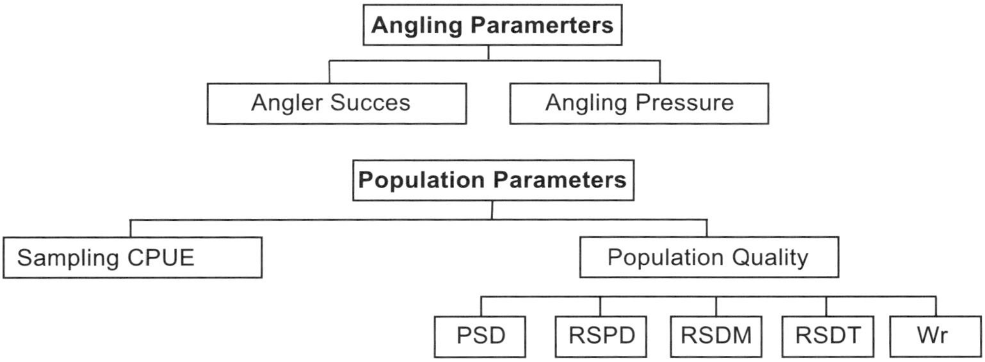

CPUE - Catch per Unit Effort
PSD - Proportional Stock Density
RSDP - Relative Stock Density of Preferred-Sized Individuals
RSDM - Relative Stock Density of Memorable-Sized Individuals
RSDT - Relative Stock Density of Memorable-Sized Individuals
Wr - Relative Weight

FIGURE 1. Parameters used to calculate the sport fishing index.

Population and angler results are scored against reference values. Reference values for population quality aspects are those suggested by Gablehouse[16] for maintenance of a balanced multispecies fishery. Reference conditions for both population and angler catch rates were obtained by trisecting historical observed values in Tennessee and Cumberland River reservoirs. As with RFAI, scores were assigned based on the scale of least degraded-5; intermediate-3; and most degraded-1. Metric scores are summed to obtain the SFI score for each important sport fish species.

RESULTS

Determination of a screening level endpoint (no additional sampling required to demonstrate community AEI or existence of BIP) requires a conservative "no-risk" approach. This was accomplished in three ways. First, RFAI metric scoring criteria were developed on a conservative basis. Reference conditions were based not only on maximum observed values over a large data base, but species expectations were elevated to include any that were historically within the geographic range and were determined to be able to thrive in a reservoir environment.

Second, RFAI scores are made even more conservative by removing the calculated sample variability (to prevent "false positives"). This was done by comparison of RFAI scores from 54 paired sample sets (repeat samples within one week) collected over the past seven years. Differences range from 0 to 18 points – the

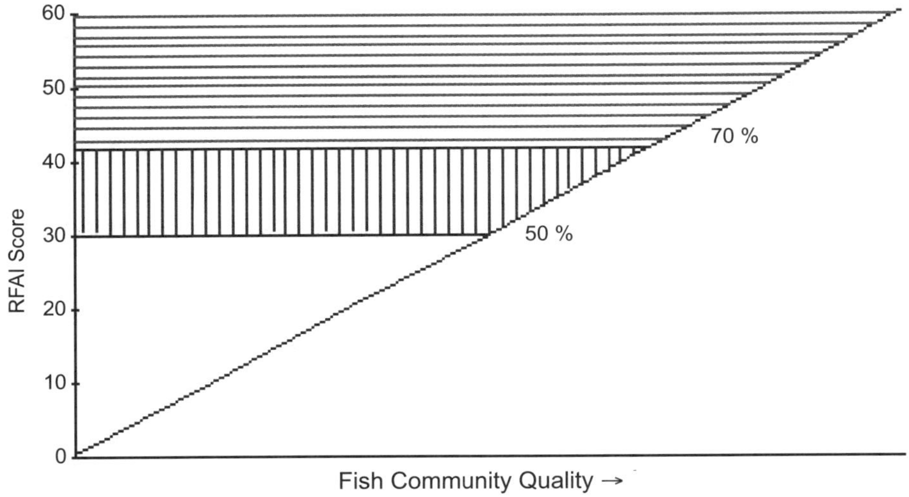

FIGURE 2. Proposed RFAI endpoints for determination of adverse environmental impact. If RFAI score is > 70% of attainable score, then the fish community is considered to "screen out" for AEI and have BIP; between 50–70% is considered potentially adversely impacted; and < 50% is considered impacted.

70th percentile was 6; the 90th percentile was 12. The mean difference between these 54 paired scores was 4.6 points with 95% confidence limits of 3.4 and 5.8. Based on these results, a difference of 6 points or less ($\pm$3) was the value selected for defining "similar" scores.

The third conservative level maintains that if more than half of the individual metrics related to impingement/entrainment impacts receive low to moderate scores, then the site fails to screen out. The same requirement is made for thermal impacts and determination of existence of BIP.

To screen out further demonstration of BIP or absence of AEI, it is proposed that the composite RFAI score must exceed 70% (based on conservative measures mentioned above) of the maximum obtainable score of 60 (i.e., RFAI = 42) for that biological zone of the water body, adjusted for defined variability. Fig. 2 graphically shows proposed endpoints. For example, if a site receives an RFAI score of 44 and the mean variability for that reservoir type and zone is $\pm$3, then that site would fail to meet the screening level criteria using the conservative aspect of the variability (+3). It would require a score above 45 to effectively screen out.

RFAI scores below this screening level do not mean that there is AEI or that BIP do not exist. The endpoint serves as a conservative screening level, i.e., any fish community that receives a score above this level is considered not to have been adversely impacted. RFAI scores below this level would require a more in-depth assessment to determine the likelihood of occurrence of AEI or lack of BIP, and potentially suggest sources of impairment. An inspection of individual RFAI metric results would be an initial step to help identify if plant operation is

TABLE 1

RFAI Metrics Potentially Affected by Impingement/Entrainment and Thermal Impacts

Species	Impingement/ Entrainment	Thermal Effects
Total Species	X	X
Average Number of Individuals	X	X
Total Centrarchid Species		
Total Benthic Invertivores	X	X
Total Intolerant Species	X	X
Percent Tolerant	X	X
Percent Top Carnivores	X	X
Percent Omnivores	X	X
Percent Dominance by One Species	X	X
Percent Nonnative	X	X
Percent Anomalies		X
Largemouth Relative Weight	X	X

contributing to lower RFAI scores. Metric scores that will help guide determination of plant operational impacts include looking at what species or groups are missing or underrepresented. When and where do these impacted groups spawn? What are the characteristics of the egg and larval stages? If overall fish densities are low, or if particular groups appear overrepresented, is there attraction to flow or temperature or unique habitat created by operational characteristics? Metrics potentially affected by impingement/entrainment and thermal releases are listed in Table 1. If the RFAI score indicates that the resident fish community has been potentially impacted, impingement and/or entrainment sampling may be required to determine if these potential impact sources are playing major roles in the status of the resident fish community.

A final possible descriptive determination regards whether a resident fish community that receives an RFAI score below a particular trigger level should be labeled as adversely impacted or failing to maintain BIP. This should largely be a site-specific determination with considerable input from the state regulatory agency. An example of an adverse impact trigger level would be a fish community score below 50% of the attainable score of 60 (i.e., RFAI = 30) with adjustment for defined variability (i.e., if variability is ± 3, then RFAI = 27). Additional sampling may be necessary to determine responsible agents.

A similar or higher RFAI score at a site downstream of a plant intake/outfall compared to an upstream site has often been used as a basis for determining the presence or absence of impact by fossil plant operation on the resident fish community. Definition of "similar" is integral to accepting the validity of these interpretations between upstream and downstream fish communities. That is, differences between the upstream and downstream fish communities must be more than the natural variation in RFAI scores. If the downstream RFAI score is within 6 points (±3) of the upstream score, the communities are considered similar, and it can be concluded that the plant has had no effect. When an impacted community is suggested by a lower RFAI score, a metric-by-metric examination can be conducted to help determine causes.

A couple of examples from Tennessee Valley reservoirs are used to help visualize how these endpoints would operate. Table 2 shows average RFAI scores from TVA's standardized reservoir monitoring program from both upstream and downstream of some TVA fossil plants from 1993–2000. These are not ideal locations to determine plant operational impacts. Future compliance sampling will be done in immediate upstream/downstream areas beginning in 2001. RFAI values at Bull Run Fossil Plant (BRF) averaged 28 upstream and 37 downstream of the plant. Both values failed the conservative screening criteria, indicating that BIP may not be present and that AEI could be occurring.

During 2000, the site upstream of BRF scored 32 and the downstream site scored 47 (Table 2). An inspection of individual metric scores revealed no metric received a higher score at the upstream control station than at the downstream station. Two metrics received low scores at both sites including: relative abundance and percent omnivores in both electrofishing and gill netting samples. Four other metrics at the downstream site received moderate scores. These included: total sucker species, total intolerant species, percent tolerant, and percent insectivores. Only four of the nine RFAI metrics potentially related to impingement/entrainment losses received either a low or moderate score, and only five of the 11 metrics potentially related to heated discharge effects received a low or moderate score.

Hickman and Hevel[17] documented a significant inverse relationship between water volume discharged during spring and early summer from upstream Norris Dam and reproductive success of warm water species in Melton Hill Reservoir, and growth of the major piscivore (largemouth bass) in the lake. The periodic releases of hypolimnetic water through Norris Dam can cause considerable fluctuation in daily water temperatures. When this occurs during spawning periods, impacts to the composition of the entire fish community are possible. Metric results tend to support this conclusion as overall Wr of largemouth bass and numbers of fish were depressed. Additionally, percentage of the community comprised of tolerant individuals and omnivores and the number of benthic invertivores were adversely influenced by the daily fluctuations in water temperatures. It is likely that the BRF heated effluent minimally enhances the community downstream of the discharge, as the fish community in this area scored higher than the upstream site during all sample years (1993–2000). The BRF discharge acts to temper the cold water discharged through Norris Dam.

TABLE 2

RFAI Scores (1993–2000) in the Vicinity of Various TVA Fossil Plants*

Plant	Site	Reservoir	Location	Year 1993	1994	1995	1996	1997	1998	1999	2000	Average
		Tennessee River										
		Upper Mainstream										
Watts Bar	Intake	Chickamauga	TRM 529	56	52	44	38	52		46	48	48
Watts Bar	Discharge	Watts Bar	TRM 531	39	43		41		45	40	45	42
Bull Run	Upstream	Melton Hill	CRM 66	22	28		36		36		32	31
Bull Run	Downstream	Melton Hill	CRM 45	43	43		38		41		47	42
Raccoon Mtn	Upstream	Nickajack	TRM 470.2	58	50	54		46		46		51
Raccoon Mtn	Downstream	Nickajack	TRM 425.5	49	45	44		35		34		41
Kingston	Upstream	Watts Bar	CRM 22	44	40		48		46		42	44
Kingston	Downstream	Watts Bar	TRM 560.8	53	46		42		48		48	47
		Lower Mainstream										
Widows Creek	Upstream	Guntersville	TRM 424	38	42		46		32		32	38
Widows Creek	Downstream	Guntersville	TRM 375.2	38	35		36		30		34	35
New Johnsonville	Upstream	Kentucky	TRM 206	38	34	36		38		44		38
New Johnsonville	Downstream	Kentucky	TRM 85	44	43	42		44		38		42
Colbert	Upstream	Pickwick	TRM 259	50	46		48		42		50	47
Colbert	Downstream	Pickwick	TRM 230	47	47		53		37		47	46

* Upstream control and downstream impact area sites.

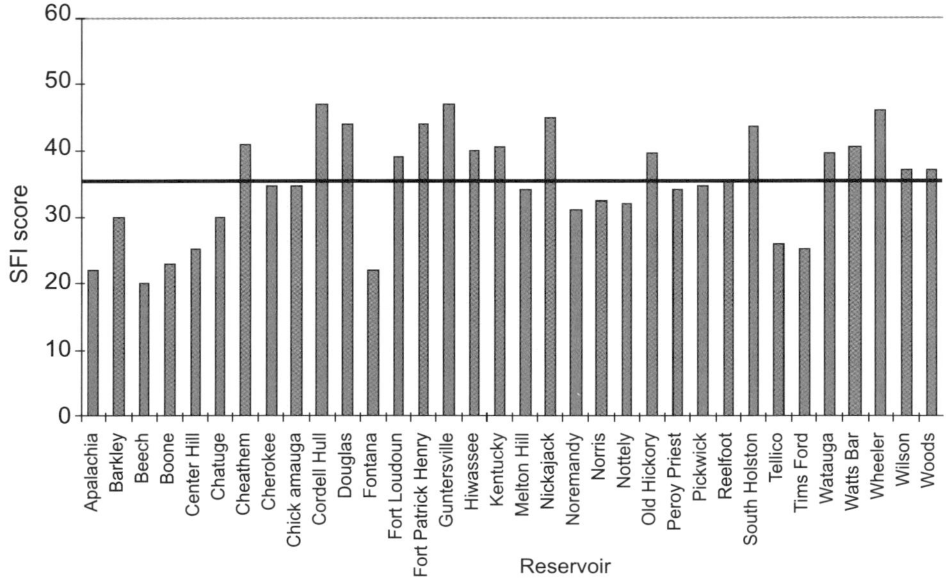

FIGURE 3. Black bass SFI scores for 2000. Line indicates overall average.

Colbert Fossil Plant (COF) provides an example of a site meeting or approaching the screening level criteria. RFAI scores averaged 48 upstream and 46 downstream of the plant, within the six-point acceptable sample variation, during 1993–2000 (Table 2). The upstream score averaged above the screening criteria and did so four out of the five years this site was sampled. The downstream site average was slightly above the screening level and scored above screening out in four of the five years. This was the only plant out of the seven Tennessee River plant sites sampled where both upstream and downstream sites exceeded the conservative screening level, indicating that resident fish communities at these locations are not adversely impacted.

The SFI provides a mechanism of screening for an individual species population within a reservoir. Fig. 3 provides an example of SFI results for black bass in Tennessee and Cumberland River reservoirs during 2000. It is proposed that any individual species population successfully screens out of additional BIP or AEI determinations if the SFI score is 10% above average for all reservoirs with SFI data for that particular year. Use of this endpoint requires a complete range of population quality (from excellent to poor).

A species population score more than 10% below average is the trigger point indicating that AEI may be occurring with regard to that species population. If the SFI score does suggest adverse impacts, inspection of individual metric scores may give insight on the potential of plant-induced impacts.

Melton Hill Reservoir SFI results, based on population data only, as no angler catch or pressure information were available for 2000, indicate a striped bass/

TABLE 3

Sport Fishing Index (SFI) Scores for Representative Important Species During 2002

Reservoir	Black Bass	Bleu-gill	Channel Catfish	Crap-pie	Large-mouth Bass	Sau-ger	Small-mouth Bass	Spot-ted Bas	Striped Bass or Hybrids	Wal-leye	White Bass
Apalachia	22*	20*	20*		20*			20*		20	
Barkley	30	42	20	43	30		20*	20*			20
Beech	20	20*	34*		32*						
Boone	23				28*		46*				
Center Hill	25	30	20	20	20		20			40	
Chatuge	30*	26*	26*		30*		20*	30*	30*	36	48
Cheathem	41	32*			44*			32*8			
Cherokee	35	26	23	41	32	24*	27	22	52	25	30
Chick amauga	35	33	29	31	32	39	22	40*	30	20	30
Cordell Hull	47				34*		26*				
Douglas	44	30	20	28	44	20	20			20	20
Fontana	22*	22*	20*		32*		26*	20*	48*	46	
Fort Loudoun	39	20*	20*		36*	42*	32*	26*	46*		
Fort Patrick Henry	44*				24*		42*				
Guntersville	47	22*	24*		32*	26*	20*	34*	20*	20	
Hiwassee	40*	24*	20*		28*		24*	32*	20*	48	
Kentucky	41	44	50	47	32	40	32	32*	20		30
Melton Hill	34	20*	24*		30*	20*	20*	20*	54	20	
Nickajack	45				50*			52*			
Noremandy	31	40	20	20	32	20*	25	32*	20*	20	
Norris	33	30	20	20	26	20	30	40*	24	24	
Nottely	32*	30*	20*		30*		20*	32*	42*		40*
Old Hickory	40	31	20	30	37	20			20		20
Peroy Priest	34	35	20	30	34		20	40*	30		20
Pickwick	35	33	20	21	27	40	37	24*	20		40
Reelfoot	36	49	20	60	33						
South Holston	44	26	20	20	36		55		20*	32	
Tellico	26*				28*		20*	32*			
Tims Ford	25	40*	20*		32*	20*	22*	24*	24*	20	
Watauga	40	27	20	30	33		52	20*		41	
Watts Bar	41	27	22	37	43	34	28*	38*	39		25
Wheeler	46	24*	20*		28*	20*	44*	20*	22*		
Wilson	37	22*			52*		56*	20*	20*		
Woods	37	42	20	20	37		20*				
Average	35.3	29.9	22.7	31.1	32.9	27.5	29.5	29.3	30.1	28.8	29.4

hybrid population well above the 10% above average screening level (Table 3). Channel catfish and largemouth bass populations were below the upper screening level, but were not low enough to indicate impacted populations. Densities of smallmouth bass and spotted bass were too low to develop accurate length frequency or relative health analyses. The bluegill population was dominated by young individuals with a PSD of only 8.9 and no fish of preferred, memorable, or trophy size. The Wr value of 75 indicates that resident bluegill are well below

TABLE 4

Sport Fishing Index Results for Melton Hill and Pickwick Reservoirs

Species	PSD (a)	PSD (b)	RSDP (a)	RSDP (b)	RSDM (a)	RSDM (b)	RSDT (a)	RSDT (b)	Wr (a)	Wr (b)	Catch Rate (a)	Catch Rate (b)	Creel Catch (a)	Creel Catch (b)	Pressure (a)	Pressure (b)	Bait Bite (a)	Bait Bite (b)	SFI
Melton Hill																			
Rock bass	37.2	2*	9.0	2	1.3	2	0.0	1	89.4	2	15.6	5					2.0	20	34
Bluegill	8.9	2	0.0	2	0.0	2	0.0	2	75.5	2	61.3	10							20
Channel catfish	83,3	2	42	4	0,0	2	0,0	2	0,0	4	1,0	10							24
Hybrid Striped bass x White bass	36,4	4	15,2	6	9,1	6	0,0	2	92,9	6	1,4	30							64
Largemouth bass	35,7	4	5,7	4	1,4	4	0,0	2	90,7	6	14,0	10							30
Sauger	0,0	2	0,0	2	0,0	2	0,0	2	0,0	2	0,1	10							20
Smallmouth bass	0,0	2	0,0	2	0,0	2	0,0	2	0,0	2	0,2	10							20
Spotted bass	0,0	2	0,0	2	0,0	2	0,0	2	0,0	2	1,4	10							20
Walleye	0,0	2	0,0	2	0,0	2	0,0	2	0,0	2	0,2	10							20
Pickwick																			
Rock bass	62,5	3	17,1	3	2,3	2	0,0	1	96,0	3	58,7	5	0,8	7,5	7,0	5	2,0	5	34,5
Bluegill	67,0	3	2,0	2	0,0	1	0,0	1	0,0	1	66,7	15	1,4	5	0,5	5			33
Crappie	0,0	1	0,0	1	0,0	1	0,0	1	80,9	2	2,2	5	0,4	5	0,5	5			21
Largemouth bass	61,2	3	17,9	3	3,0	2	0,0	1	95,3	3	44,7	5	0,2	5	7,0	5			27
Sauger													1,3	30	0,3	10			40
Smallmouth bass	70,0	3	17,5	3	2,5	2	0,0	1	93,2	3	8,7	10	0,2	10	7,0	5			37
Spotted bass	0,0	2	0,0	2	0,0	2	0,0	2	92,5	6	5,3	10							24
White bass													40	30	0,1	10			40

(a) - Attained Value Density of Memorable-Sized Individuals;
(b) - Assigned Score Based on Attained Value Density of Trophy-Sized Individuals
PSD - Proportional Stock Density Index
RSDP - Relative Stock Density of Preferred-Sized Individuals
RSDM - Relative Stock
RSDT - Relative Stock
SFI - Sport Fishing
Wr - Relative Weight

* No creel information available; thereforepopulation metric scores were doubled to obtein comparable SFI-score.

anticipated weights per unit length. Catch rate received a moderate score (see Table 4).

The channel catfish population in Melton Hill received a high PSD score indicating a lack of sufficient recruitment. A moderate number of preferred-size fish were present, but no memorable or trophy-size individuals. The Wr was slightly low and the catch rate was moderate. The aforementioned conditions resulting from the influence of Norris Dam periodic releases are revealed by these metric scores for Melton Hill Reservoir[17]. Large fluctuations in water temperatures during spawning in most years lead to large differences in year class strength as shown by these species. Striped bass and hybrids are stocked into the reservoir to maintain these populations. Only largemouth bass seem capable of maintaining an average population in the reservoir. SFI determinations are reservoir-wide and cannot be used in upstream/downstream comparisons. However, as mentioned previously, the fish community below the plant thermal discharge is superior to those found upstream of the plant, suggesting a positive influence. However, the influence is not substantial enough to improve all sport fish populations on a reservoir-wide basis.

SFI results indicate that Pickwick Reservoir provides populations of bluegill, sauger, smallmouth bass, and white bass that exceed the 10% above average screening level. (Table 3). However, crappie and spotted bass did not meet the 10% above or below average screening criteria, suggesting that these populations may be failing to reach their potential. The Pickwick spotted bass and crappie populations received low or moderate scores from all aspects. Spotted bass habitat is limited in Pickwick due to limited availability of their preferred steep, rocky banks and relatively low nutrient levels, but Pickwick does maintain adequate habitat capable of supporting a reasonable crappie population. Under these circumstances, additional sampling could be necessary to demonstrate whether or not plant operation is impacting the crappie population in Pickwick Reservoir.

CONCLUSIONS

RFAI and SFI indices can be used to define various levels of fish community/population quality within a reservoir. A "no-risk" screening level for demonstration of BIP, or no AEI, when attained RFAI scores exceed 70% of the maximum score of 60 (RFAI = 42), appears suitable to protect resident fish communities. The screening level endpoint must be adjusted for defined variability in the index score for that reservoir type and zone (i.e., with variability ± 3, RFAI = 45 would screen out). Due to the conservative manner in which index reference conditions are developed, this level minimizes the potential of screening out a fish community that is adversely impacted. If a fish community fails to exceed the RFAI screening level score, it does not mean that the community is adversely impacted, just that additional information is necessary to make that determination. A possible endpoint where the resident fish community may be considered to be adversely impacted would be if the RFAI score fell below 50% of the maximum score, adjusted for average variability (i.e., with variability ± 3, RFAI = 27).

RFAI scores were successfully used to describe fish community status in reservoirs with fossil and nuclear plant intake and thermal discharges using upstream control and downstream potentially impacted areas. Some fish communities failed to attain the conservative screening level; some were below a proposed endpoint, suggesting that they were adversely impacted; and a couple of sites did meet the screening level criteria.

Two examples used to demonstrate how the screening process works included one incident where the fish community and individual sport fish populations failed to screen out and one where most or all screen-out criteria were met. The Melton Hill Reservoir fish community in the vicinity of BRF, the upstream (RFAI = 32) site failed the screening level criteria (RFAI = 45) (Table 5). However, inspection of individual metric results and knowledge of other potential influencing factors led to the determination that plant operation was actually having a positive impact on the downstream population, although this impact was not sufficient to override the negative impacts of upstream hypolimnetic reservoir releases.

RFAI scores for Pickwick Reservoir in the vicinity of COF exceeded the screening level criteria. The upstream control site RFAI score averaged 47 and exceeded the screening level trigger of 45 in four out of the five years sampled.

TABLE 5

Individual RFAI Metric Results From Melton Hill Reservoir Samples in the Vicinity of Bull Run Steam Plant, Fall 2000

RFAI Metrics	Upstream	Downstream
Total Species	3	5
Total Centrarchid Species	3	5
Total Sucker Species	3	3
Total Intolerant Species	3	3
Percent Tolerant (EF)	1	1.5
Percent Tolerant (XGN)		2.5
Percent Dominance by One Species (EF)	3	2.5
Percent Dominance by One Species (XGN)		2.5
Number of Piscivore Species	3	5
Percent Omnivores (EF)	1	1.5
Percent Omnivores (XGN)		0.5
Percent Insectivores (EF)	3	1.5
Percent Insectivores (XGN)		2.5
Number of Lithophilic Spawning Species	3	5
Average Number of Individuals (EF)	1	0.5
Average Number of Individuals (XGN)		0.5
Percent Anomalies	5	5
Score	32	47

EF = electrofishing, XGN = gill netting.

The downstream site averaged 46, just above the necessary screening score of 45, and attained the screening level in four of the five sample years.

Upstream/downstream scores were within the six-point acceptable sample variation, indicating no appreciable difference in fish communities residing in these areas.

The SFI screening criteria of maintaining average or above-average individual sport fish populations also appears useful. Inspection of metric scores proved insight could be gained into possible factors or conditions that might be limiting a particular population. Again using Melton Hill and Pickwick Reservoirs as examples, some individual species populations screened out in both reservoirs, and some required an in-depth look at metric scores to determine possible sources of stress on these populations.

In summary, screening level endpoints would be very helpful for both regulators and utilities alike. A series of endpoints for RFAI and SFI multimetric indices can be used to determine if existing fish communities/populations are healthy and whether or not they remain that way after plant operation begins. Establishment of such endpoints, based upon sound indices, could reduce the amount of extensive sampling necessary without jeopardizing the well-being of the resident fish community or individual sport fish populations. In cases that meet the conservative screening level, periodic low-intensity fish community/population monitoring would be sufficient to determine if problem situations develop.

REFERENCES

1. Jennings, M.J., Fore, L.S., and Karr, J.R. (1995) Biological monitoring of fish assemblages in Tennessee Valley reservoirs. *Reg. Rivers: Res. Man.* **11,** 263–274.
2. Hickman, G.D. and McDonough, T.A. (1996) Assessing the reservoir fish assemblage index: a potential measure of reservoir quality. In *Reservoir Symposium – Multi-dimensional Approaches to Reservoir Fisheries Management. Reservoir Committee.* DeVries, D., Ed. Southern Division, American Fisheries Society, Bethesda, MD. pp. 85-97.
3. McDonough, T.A. and Hickman, G.D. (1999) Reservoir Fish Assemblage Index development – a tool for assessing ecological health in Tennessee Valley Authority impoundments. In *Assessing the Sustainability and Biological Integrity of Water Resources Using Fish Communities.* Simon, T., Ed. CRC Press, Boca Raton. pp. 523–540.
4. Hickman, G.D. (2000) Sport fishing index (SFI): a method to quantify sport fishing quality. *Environ. Sci. Pol.* **3(1),** 117–125.
5. Karr, J.R., Fausch, K.D., Angermeier, P.L., Yant, P.R., and Schlosser, I.J. (1986) Assessing biological integrity in running waters: a method and its rationale. *Illinois National Historic Survey Special Publication* 5, 28 pp.
6. Miller, D.L., Leonard, P.M., Hughes, R.M., Karr, J.R., Moyle, P.B., Schrader, L.H., Thompson, B.A., Daniel, R.A., Fausch, K.D., Fitzhugh, G.A., Gammon, J. R., Halliwell, D.B., Angermier, P.L., and Orth, D.J. (1988) Regional applications of an index of biotic integrity for use in water resource management. *Fisheries* **13(5),** 12–20.
7. Oberdorff, T. and Hughes, R.M. (1992) Modification of an index of biotic integrity based on fish assemblages to characterize rivers of the Seine-Normandie basin, France. *Hydrobiologia* **228,** 117–130.
8. Davis, W.S. and Simon, T.P., Eds. (1995) *Biological Assessment and Criteria: Tools for Water Resource Planning and Decision Making.* Lewis Publishers, Boca Raton, FL.
9. Karr, J.R. and Chu, E.W. (1998) *Restoring Life in Running Waters: Better Biological Monitoring.* Island Press, Washington, D.C..

10. Norris, R.H. and Thom, M.C., Eds. (1999) River health. *Freshwater Biology* **41,** 197–479.
11. Simon, T.P., Ed. (1999) *Assessing the Sustainability and Biological Integrity of Water Resources Using Fish Communities.* CRC Press, Boca Raton, FL.
12.. Jungwirth, M., Muhar, S., and Schmutz, S., Eds. (2000) Assessing the ecological integrity of running waters. *Hydrobiologia* **422/423,** 1–487.
13. Hickman, G.D. and Olmsted, L.L. (2001) Performance of the reservoir fish assemblage index (RFAI) in Catawba and Cumberland River reservoirs. Final report to Electric Power Research Institute, 43 pp.
14. Colvin, M.A. and Vasey, F.W. (1986) A method of qualitatively assessing white crappie populations in Missouri reservoirs. In *Reservoir Fisheries Management: Strategies for the '80s.* Hall, G.E. and Van Den Avyle, M.J., Eds. Reservoir Committee, Southern Division, American Fisheries Society, Bethesda, MD. pp 79–85.
15. Smogor, R.A. and Angermeier, P.L. (2001) Determining a regional framework for assessing biotic integrity of Virginia streams. *Trans. Am. Fish. Soc.***130,** 18–35.
16. Gablehouse, D.W., Jr. (1984) A length-categorization system to assess fish stocks. *North Am. J. Fish. Man.* **4(3),** 273–285.
17. Hickman, G.D. and Hevel, K.W. (1986) Effect of hypolimnetic discharge on reproductive success and growth of warmwater fish in a downstream impoundment. In *Reservoir Fisheries Management Strategies for the '80s.* Hall, G.E. and Van Den Avyle, M.J., Eds. Reservoir Committee, Southern Division, American Fisheries Society, Bethesda, MD. pp. 286–293.

BIOSKETCHES

Gary D. Hickman is a Principal Environmental Scientist with the Tennessee Valley Authority in Norris, Tennessee. He received B.S. and M.S. degrees from the University of Arkansas and is a Certified Fisheries Scientist by the American Fisheries Society.

Mary L. Brown is an Environmental Scientist with Westaff Technical (TVA contract), Norris, Tennessee. She received a B.S. degree from the University of Tennessee.

Minimizing Adverse Environmental Impact: How Murky the Waters

Reed W. Super* and David K. Gordon
Riverkeeper, Inc., 25 Wing & Wing, Garrison, NY 10524

Received November 16, 2001; Revised February 22, 2002; Accepted February 25, 2002; Published February, 2003

The withdrawal of water from the nation's waterways to cool industrial facilities kills billions of adult, juvenile, and larval fish each year. U.S. Environmental Protection Agency (EPA) promulgation of categorical rules defining the best technology available to minimize adverse environmental impact (AEI) could standardize and improve the control of such mortality. However, in an attempt to avoid compliance costs, industry has seized on the statutory phrase "adverse environmental impact" to propose significant procedural and substantive hurdles and layers of uncertainty in the permitting of cooling-water intakes under the Clean Water Act. These include, among other things, a requirement to prove that a particular facility threatens the sustainability of an aquatic population as a prerequisite to regulation. Such claims have no foundation in science, law, or the English language. Any non-trivial aquatic mortality constitutes AEI, as the EPA and several state and federal regulatory agencies have properly acknowledged. The focus of scientists, lawyers, regulators, permit applicants, and other interested parties should not be on defining AEI, but rather on minimizing AEI, which requires minimization of impingement and entrainment.

KEY WORDS: adverse environmental impact, cooling-water intake structure, entrainment; impingement, power plant, 316(b), aquatic ecology, fisheries, density-dependence, surplus production, compensation theory

DOMAINS: ecosystems and communities, environmental management and policy, environmental modeling, environmental monitoring, environmental technology, freshwater systems, marine systems, water science and technology

---

INTRODUCTION

Steam-electric–generating facilities use water for cooling and, in particular, to condense the steam used to drive the turbines. Some power plants withdraw hun-

* Corresponding author. Email: rsuper@riverkeeper.org

dreds of millions or billions of gallons of river, lake, or ocean water per day[1]. These plants and all other significant users of cooling water harm and kill large numbers of fish and other aquatic biota[2] through impingement[3] and entrainment[4]. In the early 1970s, a number of well-publicized massive fish kills occurred at U.S. power plants, such as the Brayton Point Power Station in Mt. Hope Bay, Massachusetts, which killed an astonishing 164.5 million menhaden and river herring in just one day, July 2, 1971[5]. In 1972, the U.S. Congress mandated in the Federal Water Pollution Control Act (Clean Water Act or CWA) that cooling-water intake structures (CWISs) use the best technology available (BTA) for minimizing such adverse environmental impact (AEI)[6].

Unlike other sources of degradation to aquatic ecosystems controlled under the 1972 CWA amendments, however, CWISs have uniquely avoided nationally uniform limitations. Instead, regulation of CWISs has long been relegated to *ad hoc* determination by individual permit writers exercising best professional judgment. This lack of categorical standards has resulted in uneven and conflicting regulation as well as enormous, unnecessary aquatic mortality, which runs contrary to the goals of the CWA and the direct mandate of section 316(b). The individualized assessments have typically relied on narrow and inaccurately applied population models and have ignored further impact on ecosystem health.

DISCUSSION

Congressional Intent in Enacting Section 316(b) to Minimize AEI

Congress enacted section 316(b) as part of the CWA amendments of 1972 in response to a number of well-profiled fish kills at power plants in the early 1970s.

1 The nation's largest user of cooling water, the Salem Nuclear Generating Station in New Jersey, withdraws 3.024 billion gallons of water each day from Delaware Bay.

2 For brevity, this paper uses the word "fish" to denote "fish at all life stages and other aquatic biota," unless the context clearly indicates otherwise.

3 Impingement is the trapping of adult or larger juvenile fish against an intake's screening devices.

4 Entrainment is the drawing of small fish, eggs, larvae, and other organisms through a CWIS into the plant's cooling system and heat exchanger.

5 U.S. Environmental Protection Agency, *Development Document for Best Technology Available for the Location, Design, Construction and Capacity of Cooling Water Intake Structures for Minimizing Adverse Environmental Impact*, 1976 at p. 9, table I-3. EPA reported that the fish were "mangled." *Id.* Mt. Hope Bay forms the northeast arm of the Narragansett Bay estuary.

6 Clean Water Act section 316(b); 33 U.S.C. § 1326(b), which provides:
Any standard established pursuant to [Section 301 or Section 306 of the Act] and applicable to a point source must require that the location, design, construction, and capacity of cooling water intake structures reflect the best technology available for minimizing adverse environmental impact.

For example, in addition to the Brayton Point incident, the P.H. Robinson plant in Galveston Bay, Texas, impinged more than 7 million fish in 12 months in 1969 and 1970, and the Indian Point No. 1 nuclear facility on New York's Hudson River killed 1.3 million fish over a 10-week period[7]. In the late summer of 1971, more than 2 million dead menhaden clogged the screens at the Millstone plant in Niantic Bay, Connecticut[8]. In fact, during debate over the CWA, Senator Buckley cited with approval two newspaper articles reporting a decision of the Atomic Energy Commission (AEC) to require Consolidated Edison (Con Ed) to install closed cycle cooling at Indian Point[9]. The articles noted that the plants withdrew massive amounts of water from the Hudson River, entraining thousands of organisms per minute, and that the AEC had ordered Con Ed to stop removing such large volumes of water from the River and to install cooling towers in order to abate these massive fish kills[10]. Public concern over these and other incidents prompted Congress to add section 316(b) to the CWA[11].

The structure of the Act indicates how Congress intended the section 316(b) "best technology available" standard for minimizing AEI to be implemented. The Act prohibits all discharges of pollutants to waters of the U.S. except as permitted in a National Pollutant Discharge Elimination System (NPDES) permit[12]. EPA established industry-wide, nationally uniform, technology-based control standards, without regard to site-specific water parameters (such as receiving water quality) to govern the setting of individual NPDES permit limitations[13]. Once

7 Clark and Brownell, *Electric Power Plants in the Coastal Zone: Environmental Issues* (1973), p. V-8, table V-B. See also *New York Times Abstracts*, May 24, 1972, p. 94, col. 1 ("alleged 'massive' killing of fish at [Con Ed's] No. 2 nuclear-power plant at Indian Point on the Hudson River"); *New York Times Abstracts*, March 1, 1972, p. 77, col. 3 ("more than 100,000 fish have been killed in last wk [at Indian Point]").

8 Clark and Brownell, *Electric Power Plants in the Coastal Zone: Environmental Issues* (1973), p. V-8, table V-B. See also *New York Times Abstracts*, August 16, 1972, p. 41, col. 1 ("massive fish kill in Apr at Millstone Point nuclear power complex").

9 Senate Committee on Public Works, *A Legislative History of the Water Pollution Control Act Amendments of 1972*, 93d Congress, 1st Session, at 196-197, 1973. See also *In the Matter of: Carolina Power & Light Company (Brunswick Steam Electric Plant)*, U.S. Environmental Protection Agency, Decision of the General Counsel, EPA GCO 41 (June 1, 1976) at fn. 10.

10 *Id.*

11 See e.g. Yost and Thomas, "Science in the Courtroom", in Barnthouse et al., ed., *Law, Science and the Hudson River Power Plants,* American Fisheries Society Monograph 4, 1988:

Originally [section 316] only included subsection (a), which provided the utilities with a way to avoid cooling towers if they could convince EPA that they were not needed. That section only applied to the *thermal discharge* from a power plant. Concerned citizens, environmentalists, marine and aquatic biologists, and other scientists alerted Congress to the dangers presented to the aquatic community by the mortality induced by entrainment and impingement. Thus, in a last minute amendment, Congress added subsection (b), which was directed at the *intake* portion of the power plant's generating processes.

Id. at 299 (emphasis in original). Judge Yost presided over the Hudson River power plant hearings in the late 1970s.

12 33 U.S.C. § 301(a); see also 33 U.S.C. § 1342 (NPDES program).

13 See 40 C.F.R Parts 402–699. In waters that violate ambient quality standards, a more restrictive set of limitations may apply. See 33 U.S.C. §§ 1312, 1313, 40 C.F.R. Parts 130–131.

established by EPA, these national, technology-based standards must be incorporated into every individual NPDES permit issued nationwide. The goals of technology-based standards are to bring all facilities up to state-of-the-art pollution control as quickly as possible (sometimes referred to as "technology forcing") and to ensure national consistency in NPDES permit limitations[14].

Congress chose the NPDES permitting program as the vehicle for minimizing AEI by making the provisions of § 316(b) applicable to any facility containing a point source[15]. Section 316(b)'s explicit cross-reference to sections 301 and 306 further clarifies that cooling-water intake standards are an integral component of the NPDES technology-based regulations[16]. As a result, EPA must promulgate national technology-based regulations specifying BTA for minimizing AEI, as it does for effluent limitations under sections 301 and 306[17]. This integration, along with the spare and direct "best technology available" mandate, clearly indicates Congressional intent that EPA set nationwide technology-based standards for CWISs in the same fashion as for chemical pollutants. Such standards apply to permittees across the board, despite potential claims by regulated parties that their individual discharges — or cooling-water intakes — do not cause substantial ecological impact.

On December 18, 2001, as required by Congress in section 316(b) and the District Court in the *Cronin v. Browner* consent decree, EPA promulgated BTA regulations for CWISs at new facilities. [See 66 Fed. Reg. 65256 (December 18, 2001), hereinafter referred to as "Phase I Rule."]

---

14 A primary objective of Congress in implementing nationally applicable standards was to avoid the "race to the bottom," which commonly occurred in the absence of uniform national effluent limitations prior to the adoption of the Act, where states would compete to attract and maintain industries by relaxing control requirements. See Hines, "Controlling industrial water pollution: Color the problem green," 1968, *9 B.C. Indus. and Comm. L. Rev.* **553**, p. 573; Grad, *Treatise on Environmental Law*, v.2, § 303[a-1].

15 33 U.S.C. § 1326(b).

16 Section 301 mandates the "best available technology" for existing sources while the section 306 new source performance standard must reflect the "best available demonstrated control technology." 33 U.S.C. §§ 1311(b)(2)(A), 1316(a)(1). Congress' use of substantially similar statutory language in Section 316(b) underscores its intent to incorporate that section's limitations into the categorical standards of sections 301 and 306:

[T]he regulations issued under § 316(b) are...closely related to the effluent limitations and new source performance standards of §§ 301 and 306... It bears emphasis that § 316(b)...requires § 301 and § 306 standards to deal with cooling water intake structures....[The] regulations [are] issued at least in part under the same statutory sections, some of which limit intake structures, others, effluent discharges.

Virginia Electric and Power Company v. Costle (VEPCO), 566 F.2d 446, 450 (4th Cir. 1977); see also *Cronin v. Browner*, 898 F.Supp. 1052, 1059 (S.D.N.Y. 1995).

17 See Consent Decree, October 10, 1995, *Cronin v. Browner*, No. 93 Civ. 0314 (AGS), as amended March 27, 2000; *VEPCO*, 566 F.2d at 450.

PLAIN MEANING OF THE PHRASE "ADVERSE ENVIRONMENTAL IMPACT"

In its plain and ordinary meaning, the phrase "adverse environmental impact" refers to any negative effect on any aspect of the environment, and clearly encompasses nontrivial aquatic mortality. The dictionary definitions of each word in the phrase bear that out. "Adverse" means "unfavorable or antagonistic"[18]. "Environment" means "the air, water, minerals, organisms and all other external factors surrounding and affecting a given organism at any time"[19]. "Impact" means "influence; effect"[20]. Thus, anything that affects the environment in a negative way has an adverse environmental impact.

In a gambit to create an additional technical and procedural hurdle to effective regulation, industry insists that permitting agencies must pointedly define AEI at some threshold level of ecological damage for each individual application[21]. The contention is inconsistent with the structure of the CWA and the statutory language. There is no need or sound reason to precisely define a level of acceptable impact, or ascertain the level of unnecessary killing before destabilization of population, to implement the statutory requirement to "minimize adverse environmental impact"[22]. This unremarkable and (in most cases) relatively simple injunction is directly analogous to the modifier "control" for discharge standards under section 306 of the Act. Both terms define the respective purpose of the technology (in each case required to be the best available): to control pollutant discharges under section 306 and to minimize ecological damage due to cooling-water withdrawals under section 316(b). In other words, they supply the answer to the question, "Best technology for what?", which is basic to the meaning of the respective sections. There is nothing in the statutory language that requires a separate determination of some supposed level of AEI, or a self-

---

[18] Random House Webster's College Dictionary, 1999. In the environmental law context, "adverse" is often used as the opposite of "beneficial."

[19] Random House Webster's College Dictionary, 1999.

[20] *Id.*

[21] July 11, 2000, letter from Utility Water Action Group Cooling Systems Committee Chair David Bailey to OMB Office of Information and Regulatory Affairs Deputy Administrator Don Arbuckle, at 2, attached to July 11, 2000 letter from Kristy A.N. Bulleit to EPA Office of Science and Technology Director Geoffrey Grubbs. See also Comments of the Utility Water Action Group on EPA's Proposed Section § 316(b) Rule for New Facilities and ICR No. 1973.01, November 9, 2000 ("UWAG Phase I Comment"), at 53-72.

[22] Industry and other parties have at times framed the debate as whether the CWA addresses "population level" impacts or "individual" fish. See e.g UWAG Phase I Comment at 58-68. The dichotomy is false and unhelpful. Neither the Act nor fisheries advocates maintain a serious interest in the survival of an individual fish, or even more absurdly, an individual larva. *Cf.* UWAG Phase I Comment at 60. This obvious fact does not, however, justify requiring demonstration of measurable population decline or, even worse, "unacceptable risk to the population's ability to sustain itself," *id.* at 65, as a threshold for regulating the cooling-water intake at all. Notably, as a further threshold, UWAG would require a showing that the population decline "is attributable to the operation of the cooling water intake structure." *Id.*

defeating assessment of how much unnecessary killing it will take to push a population into long-term decline, before a permitting agency can minimize the damage.

Notably, section 316(b) contains no threshold adjective such as "*significant* adverse environmental impact." That stands in marked contrast to the similar language Congress used 3 years earlier in the National Environmental Policy Act of 1969 (NEPA). NEPA requires preparation of an environmental impact statement (EIS) for every major federal action "significantly affecting the human environment"[23]. Thus, although both statutes address environmental impact, in NEPA the significance of the impact is a threshold requirement for the EIS requirement. Conversely in section 316(b), the threshold requirements are that there must be a CWIS and a point source. Additionally, in its Phase I Rule, EPA adopted a 2 million gal/day *de minimis* withdrawal threshold, below which the national new facility regulation would not apply, to avoid burden to "smaller operations that may face issues of economic affordability and are therefore more appropriately addressed on a case-by-case basis using [best professional judgment]." [See 66 Fed. Reg. at 65289; 40 C.F.R. § 125.^{81}(a)(3)][24]. Once those threshold requirements are met, the facility must use BTA for minimizing AEI. Any other construction improperly reads requirements into the statute that are not in the text and substantially complicates administration.

AEI OF CWISS

The withdrawal of large quantities of water by CWISs causes AEI in a variety of familiar ways. First, such withdrawal impinges and kills adult fish, eliminating their availability for a number of important functions. Human fishers can no longer catch these fish commercially or recreationally or for sustenance; all opportunities for observation, appreciation, photography, nature study, or research are lost as well as the potential for pecuniary profit associated with any of these uses. Ecologically, the lost fish become unavailable as prey for wildlife higher on the food chain, such as birds, mammals, and larger fish, or to serve as predators for pest insects such as mosquitoes.

Entrainment kills young fish, eggs, larvae, and smaller aquatic life such as plankton by the tens or hundreds of millions. The young fish are thus prevented from maturing to adulthood (where they would provide the benefits discussed immediately above) or even maturing to the next life stage. The vast majority that would have perished before maturity would have contributed to the aquatic

[23] 42 U.S.C. § 4332.

[24] For existing facilities, pending the Phase II rule, EPA policy allows a permittee relief from technology requirements in exceptional cases where the costs are "wholly disproportionate to the environmental benefit to be gained." In the Matter of Public Service Company of New Hampshire, et al. (Seabrook Station, Units 1 and 2), 1 EAD 332, 1977 EPA App. Lexis 16, *24-25, June 10, 1977.

ecosystem by consuming prey and ultimately providing fodder for a predator. Because of their death by entrainment, however, they are prevented from doing so. Their contribution to the ecosystem is converted from a gradual, multifaceted process to one in which all of the entrained biota immediately become detritus for decomposers. One result is that energy is transferred down the food chain from higher predators to lower decomposers.

Entrainment and impingement thus disrupts the natural function of the ecosystem, which includes fish, smaller aquatic organisms, insects, birds, mammals, and plant communities. This may broadly alter the nature, structure, and function of the ecosystem, or may damage the sustainability of one or more of its species[25]. Stresses for a particular species may include population decline, reduced prey, more difficulty locating adequate sustenance, or suffering a greater number of pests.

Accurately defining and assessing all of these ever-changing interactions in an ecological community or ecosystem and determining the impact of extraneous manmade and environmental factors changing over time is not feasible[26], especially within the constraints of a permitting proceeding. Apparent stability or robustness in a population may not reflect the impact of significant environmental stresses. A fish population in decline may be bolstered by migrants from other communities, so that the stressors on the community go unnoticed. Also, populations may stabilize at a lower equilibrium level than would be the case absent cooling-water intake mortality. Further, if some percentage of a fish population or community is killed by CWISs, the loss of fish to other causes, both natural and manmade, will aggregate to cause a greater cumulative impact. The EPA based its refusal to define an AEI threshold for the Phase I Rule, in part, on concern that:

---

[25] In the Phase I Rule, EPA determined that there are:
multiple types of undesirable and unacceptable adverse environmental impacts, including entrainment and impingement; reductions of threatened, endangered, or other protected species; damage to critical aquatic organisms, including important elements of the food chain; diminishment of a population's compensatory reserve; losses to populations, including reductions of indigenous species populations, commercial fishery stocks, and recreational fisheries; and stresses to overall communities or ecosystems as evidenced by reductions in diversity or other changes in system structure or function.
66 Fed. Reg. at 65292.

[26] EPA concurred with a National Marine Fisheries Service panel conclusion on the difficulty of assessing impact on complex fishery ecosystems:
As a recent NMFS advisory panel expressed it, "Uncertainty and indeterminancy are fundamental characteristics of the dynamics of complex adaptive systems. Predicting the behaviors of these systems cannot be done with absolute certainty, regardless of the amount of scientific effort invested." Consistent with its own Guidelines for Ecological Risk Assessment, EPA agrees with the conclusions of the NMFS panel that "Given the high variability associated with ecosystems, managers should be cognizant of the high likelihood for unanticipated outcomes."
66 Fed. Reg. at 65293, citing National Marine Fisheries Service Ecosystem Principles Advisory Panel, *Ecosystem-based fishery management: A report to Congress,* 1998.

historical overfishing increased the sensitivity of coastal ecosystems to subsequent disturbance, making them more vulnerable to human impact and potential collapse. Based on the long-term record of anthropogenic impacts to coastal ecosystems, their documented degradation, and their potential sensitivity to additional anthropogenic disturbance, as well as the admitted uncertainty associated with managing coastal fishery populations, EPA firmly believes that protective, risk-averse measures are warranted to prevent further declines or collapses of coastal and other aquatic ecosystems. EPA views impingement and entrainment losses to be one of many potential forms of disturbance that should be minimized to avoid further degradation[27].

Thus, an impact that would have been trivial can become significant because power plants consumed the reserve margin that would have allowed the population to withstand the other impact.

The Conflation of Minimization of AEI with Fisheries Management

Environmental law distinguishes between exploitation of resources for sustenance, such as killing wild organisms for food (hunting and fishing), and collateral or wasteful killing, for example as a by-product of industrial processes. Harvesting of plants and animals for food is fundamental to survival, so the law tolerates the resulting ecological impact (such as reduction in size of wild populations) to a substantial extent. Recreational taking of organisms is also accepted and encouraged in many instances as a social good. Typically, regulation of such activities as hunting and fishing attempts to maximize the availability of prey by balancing the permitted take with the long-term stability of the population[28]. The science of fisheries management focuses on the sustainability of populations, so as to allow fishers and consumers to benefit from a continuous harvest.

A critical basis for industry's approach to section 316(b), and its interpretation of AEI in particular, is the inclusion of BTA regulation within the framework of fisheries management — essentially characterizing entrainment and impingement as fishing. It asserts, "[b]ecause entrainment and impingement are forms of harvesting which (upon conversion to equivalent adults) are analogous to fishing, the methods used by fisheries scientists to evaluate the impacts of proposed harvesting regimes also can be used to evaluate the potential impacts of CWISs"[29]. Despite industry's claim, we do not know of any federal or state governmental body, legislative, executive or administrative, which considers the unnecessary killing of

27 66 Fed. Reg. at 65293.

28 See e.g. Magnuson Stevens Fishery Conservation and Management Act § 301(a)(1); 16 U.S.C. § 1851(a)(1) ("Conservation and management measures shall prevent overfishing while achieving, on a continuing basis, the optimum yield from each fishery for the United States fishing industry.")

29 UWAG Phase I Comment at p. 66.

aquatic biota as a socially beneficial harvest. Allocation of scarce fishery resources to cooling-water intake mortality and away from fishers is almost unthinkable in any area of the U.S. with declining or restricted populations and depressed fishing communities.

Nothing in the CWA reflects intent to manage collateral industrial impacts on aquatic ecosystems as a sustainable harvest. Congress's requirement to use the BTA to minimize AEI sets forth a dramatically different protective standard from, for example, the Magnuson Stevens Act provision enacted just 4 years later to allow "on a continuing basis, the optimum yield from each fishery"[30]. Other than the explicit variance for thermal discharges in section 316(a)[31], the CWA and the NPDES program in particular represent a marked Congressional determination to regulate industrial discharges and cooling-water withdrawals through technology-based standards rather than by ecological assessment.

INDUSTRY'S USE OF POPULATION-BASED AEI MODELS TO AVOID MINIMIZING FISH KILLS

Section 316(b) determinations have typically involved individualized ecological assessment and determination of BTA for each proposed or renewed CWIS. The multiplicity of these individual determinations and the combination of ecological and mathematical/statistical expertise necessary to determine the complex population dynamics for individual species has granted industry a critical strategic advantage because of superior resources in these proceedings. Industry's insistence on some arbitrary threshold of AEI would further tip the scales in its favor by requiring a detailed ecological assessment simply to subject the CWIS to the fundamental minimization requirement in section 316(b).

---

30 16 U.S.C. § 1851(a)(1).

31 Despite industry's longstanding insistence, see e.g UWAG Phase I Comment at 16-20, neither the inclusion of cooling-water intake regulation in section 316 nor the legislative history of the CWA indicates Congressional intent to regulate intakes according to the explicit ecologically based variance for thermal discharges in section 316(a). First, the plain language of section 316(b), "best technology available to minimize adverse environmental impact" bears no similarity to the 316(a) variance where "the effluent limitations [are] more than necessary to assure the [protection] and propagation of a balanced, indigenous population of shellfish, fish and wildlife in and on the body of water into which the discharge is to be made..." and does not contemplate such ecological calculation. Instead, the BTA standard much more closely resembles the BTA standards of sections 301 and 306. Moreover, the factors that led Congress to enact the section 316(a) variance for thermal discharges do not apply to fish mortality caused by entrainment and impingement. In the hearings on the CWA Congress heard "extensive and detailed testimony" about the "unique characteristics of heat as a pollutant," including its ability to "dissipate quickly, result in only local and temporary effects and...benefit the environment under circumstances." W. Anderson, II and E. Gotting, *Taken In Over Intake Structures? Section 316(b) of the Clean Water Act*, 26 *Colum. J. Envtl. L.* **1**, 83–86. None of these unique characteristics of thermal pollution are shared by industrial fish kills, particularly on a large scale.

Industry's most common analytical tools in these individualized technical determinations are density-dependent models of fisheries populations. Cooling-water users have for decades used arguments based on density-dependence to justify the destruction of large numbers of fish and crustaceans via impingement and entrainment at their CWISs. Compensatory density-dependence, in general, is a common natural adjustment in an animal population's birth, death, or survivability rates to maintain equilibrium despite potential increases or decreases in population levels. A population will continue either increasing or decreasing unless at least one of these rates is density-dependent at some life stage. Mathematical models of density-dependent compensation can predict relatively stable populations despite severe anthropogenic mortality.

In many critical cases these models have been misapplied. Typically, their use is based on a generalized assumption of density dependence; regulators rarely require an explanation or demonstration of the actual existence of any pertinent compensation mechanism. Models reflecting compensation due to environmental conditions and periods of relative population abundance may not be appropriate in less abundant circumstances, even for the same species. Nevertheless, industry consultants often utilize models derived from fisheries[32] influenced by biological factors and compensation mechanisms that are absent from the populations and life stages damaged by their cooling-water withdrawals. For example, commonly used Ricker curves, originally developed for salmonid fisheries with intense competition for spawning space, reflect compensation based on the surviving adults' greater opportunity to successfully spawn due to increased availability of suitable sites. This mechanism could not effectively compensate for entrainment mortality because it would operate prior to the development of entrainable life stages[33], and in many cases is not even found in the species subject to entrainment[34].

---

32 In deciding not to rely on population estimates to determine AEI, EPA cited the inability of sophisticated fishery models to reliably forecast even the fishing impacts for which they were derived, especially for populations in decline:

Despite the availability of state-of-the-art fish population models and considerable experience managing fisheries, NMFS recently classified 34% of their managed fishery stocks as over-utilized. EPA agrees with fisheries experts and resource managers that there is unavoidable uncertainty associated with managing fish populations...EPA and other fishery scientist [*sic*] support the concept of a precautionary approach, particularly when dealing with complex systems, as described below.

EPA recognizes that the limitations of existing population models, including models used to manage fisheries, may be related to our overall limited understanding of the complexity of aquatic ecosystems and the long-term effects of anthropogenic activities.

66 Fed. Reg. at 65293 (citations omitted).

33 EPA noted the inapposite application of standard fishery models to entrainment:

EPA considered the premises underlying MSY [maximum sustainable yield as developed under the Magnuson-Stevens Act] and the models used by National Marine Fisheries Service (NMFS) to derive MSY. Because the concept of MSY is based on harvesting adult fish, EPA generally questions whether this approach is directly relevant to egg, larvae, and juvenile losses associated with intakes.

66 Fed. Reg. at 65292.

More fundamentally, typical compensation analysis relies on an ecologically baseless concept of "surplus production." It dismisses the ecological value of the tens of millions of fish that are a critical base of the food chain whether or not they grow to adulthood — even though their predators may be populated at far below their historic values. It narrowly focuses on the larval fish only as potential adults for individual species, rather than as forage for small predators in varied parts of the larger food web. By killing off forage, power plants create the very mechanism — food limitation — that is among the most likely to lead to the density dependence they may assert for any particular species[35]. Nevertheless, "the appealing, yet narrow, perspective of the surplus production concept ... has allowed us to justify, perhaps blindly, prosecution of fisheries"[36]. Prof. Boreman notes:

> If a "surplus" is being removed by power plant operations, then something else in the ecosystem is being out-competed. Use of surplus production is essentially an allocation issue among competitors for that resource. Do we use it for supporting fisheries, for allowing the population to hedge against bad times, for providing extra sustenance for natural predators, or for supporting other uses of the resource?[37]

Moreover, industry's compensation theories have no explanation for why fish would have evolved to create long-term surplus production. High fecundity serves to maintain population levels despite multiple, variable stressors, such as climatic, environmental, and other natural causes of mortality. Production of individuals in excess of what is needed to replenish population in any particular year also cushions against intermittent excessive mortality. Destruction of seemingly excess offspring through entrainment could render the surviving class vulnerable to other stressors that it might otherwise have been able to endure; as a result, what constitutes surplus production in one year may be needed the next to counter-

---

34 See e.g. *Draft Environmental Impact Statement for State Pollutant Discharge Elimination System Permit Renewal for Bowline Point 1 & 2, Indian Point 2 & 3 and Roseton 1 & 2 Steam Generating Stations,* December 1999, Appendix VI-4-B, relying on Ricker models to estimate the impact of entrainment on Atlantic tomcod young.

35 In the preamble to the Phase I Rule, EPA repeatedly cited concern about the potential impacts of cooling-water withdrawals on the food chain, i.e., the potential loss of forage for predators. See e.g. 66 Fed. Reg at 65263, fn. 1 and 2, 65264, 65291, 65292. In the context of impingement, EPA concluded:
EPA does not believe that loss of such forage species should be viewed as having limited importance simply because they have minimal or no commercial or recreational value. From a more holistic, ecological perspective, forage species can have great importance in their role as prey for higher trophic levels, including many commercially and recreationally important fish species. In today's rule, EPA seeks to minimize impingement losses for all affected species.
66 Fed. Reg. at 65295.

36 J. Boreman, 2000, "Surplus production, compensation, and impact assessments of power plants," *Envtl. Sci. and Policy* 31, 445–446.

37 *Id.* at S447.

act changes in environmental conditions that affect cohort survival[38]. Natural selection would have reduced such high fecundity, which places great demands on fish, if it truly exceeded long-term ecological needs. It did not evolve to allow the fish to be killed in large numbers by a power station. In implementing its Phase I Rule, EPA specifically cited the diminution of a population's potential compensatory reserve due to such mortality as an AEI within the meaning of section 316(b)[39].

Ironically, the presumption of overwhelming compensation for the species of fish entrained by CWISs is the only critical ecological feature that the power industry and their consultants assume is universal, not subject to the requirement of individualized determination. An applicant's burden of proof, especially for the counterintuitive notion that the destruction of large numbers of fish is "ecologically irrelevant" should typically require greater a greater showing of density dependence than has been typically required to date.[40]

Some of the models utilized by industry make density dependence seem inevitable because they mathematically derive compensatory mechanisms based on almost any input. For example, in the joint permit renewal of three of the power stations on the Hudson River, the applicants relied on a striped bass model based on Beverton-Holt functions, which assume *a priori* strong compensation during early life stages subject to entrainment[41]. The model effectively derives density-dependent results from environmental variation and other nondensity dependent factors, which may well actually control the population. The independent consultant hired by New York State to review the Draft Environmental Impact Statement prepared by the applicant for the permit renewals concluded:

> We believe the striped bass modeling results and conclusions...are unreliable due to limitations of the data, modeling of data and model assumptions. These limitations may cause the DEIS striped bass model to estimate extremely high and counter-intuitive levels of density dependent mortality....We believe the qualitative effect of the types of errors we discuss in our full report...support an alternative hypothesis of much lower density-dependence (and higher sensitiv-

---

38 *Id.* at S446. EPA similarly found that:
a population's potential compensatory ability is affected by all stressors encountered within the population's natural range, including takes attributed to individual or multiple cooling water intake structures. Thus, even if there is little evidence that cooling water intakes alone reduce a population's compensatory reserve, EPA is concerned that the multitude of stressors experienced by a species can potentially adversely affect its ability to recover.
66 Fed. Reg. at 65294.

39 66 Fed. Reg. at 65291, 65292

40 See e.g. New York State regulations for burden of proof in environmental adjudicatory hearings at 6 N.Y.C.R.R. § 624.9(b)(1).

41 See *Draft Environmental Impact Statement for State Pollutant Discharge Elimination System Permit Renewal for Bowline Point 1 & 2, Indian Point 2 & 3 and Roseton 1 & 2 Steam Generating Stations* Appendix VI-4.

> ity to entrainment and impingement) than the results presented in the DEIS for the striped bass population[42].

The indiscriminate use of density-dependent models of fishery population has substantially allowed power generators to eviscerate statutory requirements to minimize AEI. Even where applicants fail to support or inappropriately apply these models, simply understanding their implication requires an extraordinary level of biostatistical sophistication beyond the range of most permitting agencies and members of the public, which typically do not have the resources necessary for effective critical review. Industry's use of such models will typically undergo less scrutiny than in the historic conflict over the Hudson River power plants, and thereby avoid critical questioning even though they may be no more appropriate. As a result, ubiquitous assertions of density dependence, whether overt or implicit in stock-recruitment equations, have effectively allowed applicants to utilize individualized cooling-water intake review to avoid the statute's minimization mandate.

AGENCY INTERPRETATION OF AEI

In 1977, EPA issued section 316(b) guidance for regional offices and state regulators[43], defining AEI as follows:

> Adverse aquatic environmental impacts occur whenever there will be entrainment or impingement damage as a result of the operation of a specific cooling water intake structure. The critical question is the magnitude of any adverse impact[44].

This definition properly acknowledges that harm to aquatic life through impingement and entrainment including "damage" to fish, short of death, is AEI. The guidance also recognized the difficulty in assessing the impacts of impingement and entrainment on all species in an ecosystem:

> Adverse environmental impact may be felt by many species in all trophic levels. A species need not be directly affected but nevertheless harmed due to loss of food organisms or other associated organisms in some way necessary for

---

42 ESSA Technologies Ltd., *Review of the Draft Environmental Impact Statement for SPDES Permits for Bowline Point 1 & 2, Indian Point 2 & 3, and Roseton 1 & 2 Steam Electric Generating Stations*, October 20, 2000, at v.

43 U.S. Environmental Protection Agency, Office of Water Enforcement, Permits Division, Industrial Permits Branch, *Guidance for Evaluating the Adverse Impact of Cooling Water Intake Structures on the Aquatic Environment: Section 316(b), P.L. 92-500*, Washington, D.C., May 1, 1977.

44 *Id.* at pp. 11, 15 (the quoted text appears on both cited pages).

> the well-being and continued survival of the population. It is not practicable to study all species that may be directly or indirectly harmed by intake structure operations[45].

The EPA Phase I Rule decisively rejects proposals to restrictively define AEI in terms of identifiable population level impacts. The rule contains no explicit definition, and in the preamble EPA acknowledges that:

> there are multiple types of adverse environmental impact including impingement and entrainment; reductions of threatened, endangered, or other protected species; damage to ecologically critical aquatic organisms, including important elements of the food chain; diminishment of a population's potential compensatory reserve; losses to populations, including reductions of indigenous species populations, commercial fishery stocks, and recreational fisheries; and stresses to overall communities or ecosystems as evidenced by reductions in diversity or other changes in system structure or function[46].

The EPA further clarifies:

> The Agency has long maintained that adverse environmental impact from cooling water intake structures must be minimized to the fullest extent practicable, even in cases where it can be demonstrated that the requirement applicable under section 316(a) is being met [*i.e.* that the projection and propagation of a balanced, indigenous population of shellfish, fish, and wildlife is assured in and on the receiving waterbody]. Thus, the objective of section 316(b) includes population effects but is not limited to those effects[47].

Thus, despite pointed advocacy from industry commenters[48], the rule for new facilities contains no threshold, other than the 2 million gal/day *de minimus* withdrawal, before the section 316(b) minimization requirement applies.

In most states, the federal CWA NPDES program, including section 316(b), is implemented by state regulatory agencies pursuant to a delegation agreement with the EPA. State agencies, therefore, typically determine the BTA for minimizing AEI and issue NPDES permits that incorporate those requirements. For new facilities, they will administer the Phase I Rule; for existing plants they will continue to determine requirements on a site-specific basis until EPA promulgates the Phase II Rule. It is highly significant, then, that a majority of states commenting on the EPA's proposed definitions of AEI recognize that any impingement and entrainment constitutes AEI. During the public comment period on the Phase I Rule, New

---

45 *Id.* at p. 15.
46 66 Fed. Reg. at 65291.
47 66 Fed. Reg. at 65292.
48 See e.g. 66 Fed. Reg. at 65291-65295.

York, New Jersey, Pennsylvania, Michigan, and Alaska, as well as the Atlantic State Marine Fisheries Commission, each recommended that if EPA were to define AEI, it should adopt the proposed third definition because in their experience it is accurate and workable[49].

The New York State Department of Environmental Conservation (NYSDEC), which has had extensive experience with large-scale impingement and entrainment on the Hudson River, properly recognizes that the killing of any fish or other aquatic biota negatively impacts the ecosystem, and that no more fish than necessary should be allocated to death by impingement and entrainment:

> Our program considers the death of any fish at or through a cooling water intake to be an "adverse impact". In the past the Department has stated that it is not appropriate to allocate a component of the public fish and wildlife resource to electric energy generators or other cooling water intakes. This pragmatic approach to defining adverse impact avoids the alternative of trying to determine the incremental impact of each intake upon the populations of dozens of affected species. That is, it avoids potentially endless, expensive studies that usually yield ambiguous or debatable results. The results are often debatable because it is impossible to identify, measure, and attribute the impact of each the many variables affecting populations on each of the impacted species[50].

Likewise, the State of New Jersey Department of Environmental Projection (NJDEP) asserted that:

> this third alternative [definition of AEI as impingement and entrainment] is the only practical alternative and should be adopted... This is the same definition that the Department currently uses in applying Section 316(b) policies for existing facilities. *The Department considers the death of any fish at or through a cooling water intake structure to be an "adverse impact" which must be minimized under Section 316(b). This position makes sense and simplifies an already complex analysis.* State agencies and permitting authorities could engage in a debate for years as to the population measure of a given fish species, let alone many fish species. The results of biological population studies and modeling can be very subjective because it is difficult to identify, measure, and attribute the impact of each of the many variables...affecting populations

---

[49] EPA's proposed third definition was "AEI would be deemed to occur whenever aquatic organisms are impinged or entrained as a result of the operation of a cooling water intake." 65 Fed. Reg. at 49074-49075.

[50] New York State Department of Environmental Conservation, Division of Fish, Wildlife, and Marine Resources, Clean Water Act section 316(b), *Statement provided to U.S. Environmental Protection Agency, June 29, 1998: Public Meeting to Discuss Adverse Environmental Impacts resulting from Cooling Water Intake Structures*, at p. 1. In addition, NYSDEC states that the goal of CWIS regulation should be zero mortality. This is consistent with minimizing AEI.

> of the impacted species. Rather than engage in this kind of biological debate, time and resources would be better spent focusing on the magnitude of the impingement and entrainment losses in relation to the costs and benefits of implementing various technologies to avoid or minimize the impact. This focus is appropriate for section 316(b) which the Department feels is a technology-driven provision[51].

And the Michigan Department of Natural Resources (MDNR)

> support[s] the third alternative, which is a similar approach taken by the State of New York that defines adverse environmental impact as any impingement or entrainment of aquatic organisms....The State of Michigan has experienced considerable inaction in the adoption of technology because of disagreement among power producers and agency biologists if operation of the facility is causing adverse impact. The adoption of the new language would make the definition of adverse impact very clear and ultimately better protect the aquatic resource[52].

CONCLUSION

Industry's insistence on a separate definition of AEI is a stratagem to create a significant procedural and technical roadblock to effective regulation of CWISs. Such a definition would set an inappropriate threshold before regulation could even take place, and would ironically complicate the task of determining BTA to minimize AEI. It would also confer substantial advantages on industrial applicants due to their superior resources. AEI includes any aquatic mortality due to impingement and entrainment, and EPA should by regulation require the BTA to minimize such impact. EPA properly refused to burden its Phase I new facility rule with this unnecessary threshold determination and should similarly eschew it in promulgating the Phase II Rule for existing facilities.

---

51 *State of New Jersey, Department of Environmental Protection. Letter from Dennis Hart, Assistant Commissioner, to USEPA dated November 9, 2000* at p. 4 [emphasis added].

52 *MNDR letter to USEPA dated Nov 7, 2000* at p. 2. See also *Commonwealth of Pennsylvania Department of Environmental Protection Comments on U.S. EPA's Proposed Regulations Addressing Cooling Water Intake Structures for New Facilities; August 10, 2000 (65 FR 49060) dated Nov 7, 2000* at p. 3–4 ("We would recommend that, at this time, EPA only reference in the regulation that adverse environmental impact is considered to be a level of impingement or entrainment of aquatic organisms that is recurring and/or non-trivial"); *Alaska Department of Fish and Game November 1, 2000 letter to USEPA* ("We support the proposed third alternative that defines [AEI] as 'any impingement or entrainment of aquatic organisms'"); and *Atlantic States Marine Fisheries Commission letter of November 8, 2000 to USEPA*: "defining [AEI] as impingement and entrainment [ ] is the most appropriate [definition]."

BIBLIOGRAPHY

Alaska Department of Fish and Game. (2000). Letter to U.S. Environmental Protection Agency. November 1.
Anderson, W. A. and Gotting, E. P. (2001). Taken in over intake structures? Section 316(b) of the Clean Water Act. *Columbia J. Environ. Law* **26**, No. 1, p. 1–79.
Atlantic States Marine Fisheries Commission. (2000). Letter to U.S. Environmental Protection Agency. November 8.
Bailey, D. (2000). Letter from Utility Water Action Group Cooling Systems Committee Chair David Bailey to OMB Office of Information and Regulatory Affairs Deputy Administrator Don Arbuckle, attached to July 11, 2000, letter from Kristy A.N. Bulleit to EPA Office of Science and Technology Director Geoffrey Grubbs.
Boreman, J. (2000). Surplus production, compensation, and impact assessments of power plants. *Environ. Sci. Policy* **31**, 445–449.
Clark, J. and Brownell, W. (1973). *Electric Power Plants in the Coastal Zone: Environmental Issues*. American Littoral Society, Special Publication No. 7.
Clean Water Act section 316(b).
Comments of the Utility Water Action Group on EPA's Proposed Section § 316(b) Rule for New Facilities and ICR No. 1973.01. November 9, 2000.
Commonwealth of Pennsylvania Department of Environmental Protection. (2000). Comments on U.S. EPA's Proposed Regulations Addressing Cooling Water Intake Structures for New Facilities; August 10, (65 FR 49060), November 7.
Consent Decree, *Cronin v. Browner*, No. 93 Civ. 0314 (AGS), October 10, 1995, as amended March 27, 2000.
Cronin v. Browner, 898 F. Supp. 1052 (S.D.N.Y. 1995).
CWA § 301(a).
CWA § 402.
Draft Environmental Impact Statement for State Pollutant Discharge Elimination System Permit Renewal for Bowline Point 1 & 2, Indian Point 2 & 3 and Roseton 1 & 2 Steam Generating Stations.
ESSA Technologies Ltd. (2000). Review of the Draft Environmental Impact Statement for SPDES Permits for Bowline Point 1 & 2, Indian Point 2 & 3, and Roseton 1 & 2 Steam Electric Generating Stations. October 20.
Grad, F. (1973). *Treatise on Environmental Law*. M. Bender, New York.
Hines, N. W. (1968). Controlling industrial water pollution: Color the problem green. *9 B.C. Indus. and Comm. L. Rev.* **553**.
Hill, T. (2000). Letter from NEFMC Chairman Thomas Hill to U.S. Environmental Protection Agency, November 17.
Magnuson Stevens Fishery Conservation and Management Act § 301(a)(1).
Michigan Department of Natural Resources. (2000). Letter to USEPA dated November 7, 2000.
Nagle, D. G. and Morgan, J. T. (2000). *Environ. Sci. Policy* **3**.
New York State Department of Environmental Conservation, Division of Fish, Wildlife, and Marine Resources, Clean Water Act section 316(b). (1998). Statement provided to U.S. Environmental Protection Agency, June 29: Public Meeting to Discuss Adverse Environmental Impacts resulting from Cooling Water Intake Structures.
New York Times, p. 41, August 16, 1972 (Abstr.).
New York Times, p. 77, March 1, 1972 (Abstr.).
New York Times, p. 94, May 24, 1972 (Abstr.).
NOAA. (2000). Comments on the Proposed Rule for Cooling Water Intake Structures for New Facilities, provided to U.S. EPA on December 18.
Random House Webster's College Dictionary. (1999).
Senate Committee on Public Works. (1973). A Legislative History of the Water Pollution Control Act Amendments of 1972, 93d Congress, 1st Session, at 196-197.
State of New Jersey, Department of Environmental Protection. (2000). Letter from Dennis Hart, Assistant Commissioner, to U.S. Environmental Protection Agency, November 9.
U.S. Environmental Protection Agency, Decision of the General Counsel, EPA GCO 41. (1976). *In the Matter of: Carolina Power & Light Company (Brunswick Steam Electric Plant)*. June 1.

U.S. Environmental Protection Agency. (1977). *In the Matter of Public Service Company of New Hampshire, et al. (Seabrook Station, Units 1 and 2)*. 1 EAD 332, App. Lexis 16, *24-25.
U.S. Environmental Protection Agency. (1977) Development Document for Best Technology Available for the Location, Design, Construction and Capacity of Cooling Water Intake Structures for Minimizing Adverse Environmental Impact. *Virginia Electric and Power Company v. Costle* (VEPCO). 566 F.2d 446 (4th Cir.).
U.S. Environmental Protection Agency, Office of Water Enforcement, Permits Division, Industrial Permits Branch. (1977). Guidance for Evaluating the Adverse Impact of Cooling Water Intake Structures on the Aquatic Environment: Section 316(b), P.L. 92-500. Washington, D.C.
16 U.S.C. § 1851(a)(1).
33 U.S.C. § 1326(b).
33 U.S.C. §§ 1312, 1313.
40 C.F.R § 125.81(a)(3).
40 C.F.R Parts 402–699.
40 C.F.R. Parts 130–131.
42 U.S.C. § 4332.
6 N.Y.C.R.R. § 624.9(b)(1).
65 Fed. Reg. 49059 (August 10, 2000).
66 Fed. Reg. 28853 (May 25, 2001).
66 Fed. Reg. 65256 (December 18, 2001).

BIOSKETCHES

Reed W. Super is a Senior Attorney at Riverkeeper, Inc., a not-for-profit environmental organization based in Garrison, New York. Riverkeeper is dedicated to preserving the ecological integrity of the Hudson River, and Riverkeeper's National Fisheries and Power Plant Project focuses on the aquatic impacts of cooling water withdrawals. Mr. Super obtained his J.D. and M.B.A. degrees from the University of Virginia in January 1992, and has a BA (1985) from Duke University. Mr. Super practices, teaches, and writes about environmental law.

David K. Gordon is also a Senior Attorney at Riverkeeper, and has served there since 1990. Mr. Gordon has a J.D. from the University of Wisconsin Law School (1986), an LL.M. in Environmental Law from Pace University Law School, and a B.S. in economics from Binghamton University. He currently concentrates on reducing impacts to the Hudson River from power plants and other industrial facilities. Prior to this, he served as Reservoirkeeper and helped negotiate the landmark $1.4 billion 1997 agreement to protect the New York City watershed with federal, state, city, and upstate municipal officials. He is also a member of the Town of Lloyd Planning Board and the Vice President of the Hudson Valley Rail Trail Association.

Measurement Error Affects Risk Estimates for Recruitment to the Hudson River Stock of Striped Bass

Dennis J. Dunning[1,*], Quentin E. Ross[1], Stephan B. Munch[2], and Lev R. Ginzburg[2]

[1]*New York Power Authority, 123 Main Street, White Plains, NY 10601;*
[2]*Applied Biomathematics, 100 North Country Road, Setauket, NY 11733*

Received November 13, 2001; Revised March 25, 2002; Accepted March 26, 2002; Published February, 2003

We examined the consequences of ignoring the distinction between measurement error and natural variability in an assessment of risk to the Hudson River stock of striped bass posed by entrainment at the Bowline Point, Indian Point, and Roseton power plants. Risk was defined as the probability that recruitment of age-1+ striped bass would decline by 80% or more, relative to the equilibrium value, at least once during the time periods examined (1, 5, 10, and 15 years). Measurement error, estimated using two abundance indices from independent beach seine surveys conducted on the Hudson River, accounted for 50% of the variability in one index and 56% of the variability in the other. If a measurement error of 50% was ignored and all of the variability in abundance was attributed to natural causes, the risk that recruitment of age-1+ striped bass would decline by 80% or more after 15 years was 0.308 at the current level of entrainment mortality (11%). However, the risk decreased almost tenfold (0.032) if a measurement error of 50% was considered. The change in risk attributable to decreasing the entrainment mortality rate from 11 to 0% was very small (0.009) and similar in magnitude to the change in risk associated with an action proposed in Amendment #5 to the Interstate Fishery Management Plan for Atlantic striped bass (0.006) – an increase in the instantaneous fishing mortality rate from 0.33 to 0.4. The proposed increase in fishing mortality was not considered an adverse environmental impact, which suggests that potentially costly efforts to reduce entrainment mortality on the Hudson River stock of striped bass are not warranted.

KEY WORDS: measurement error, ecological risk assessment, recruitment, striped bass, Hudson River, adverse environmental impact, entrainment, Section 316(b), Clean Water Act, mitigation, sustainability

DOMAINS: environmental management and policy

---

* Corresponding author.

INTRODUCTION

The U.S. Environmental Protection Agency (EPA) is required to issue regulations for implementing Section 316(b) of the Clean Water Act (CWA), 33 U.S.C. Section 1326(b). Section 316(b) provides that any standard established pursuant to Sections 301 or 306 of the CWA and applicable to a point source shall require that the location, design, construction, and capacity of the cooling-water intake structures reflect the best technology available (BTA) for minimizing adverse environmental impact (AEI). Early guidance provided by the EPA indicated that AEI occurs whenever there is entrainment or impingement of aquatic organisms resulting from the operation of a cooling-water intake structure[1]. However, this policy could require costly mitigation like a cooling tower that produces little benefit if an alternative definition of AEI is adopted, e.g., one based on populations. In such high-stakes cases, the degree of environmental protection and the associated cost should be reconciled with scientific data and methods[2].

Recently, the EPA began a process to update and formalize its early guidance for defining and assessing AEI. Factors that the EPA is considering include new approaches and tools developed since the early guidance was issued[3]. One of the tools being considered is ecological risk assessment. It is used to evaluate the likelihood that adverse ecological effects may occur or are occurring as a result of exposure to one or more stressors and includes an evaluation of uncertainty[4].

Two types of uncertainty affect risk assessments for populations[5]. One is intrinsic to the populations, reflecting natural variability in abundance. The other reflects variability in abundance due to sampling (i.e., measurement error). This distinction is not usually recognized in risk analyses of population extinction[6,7,8,9], but the distinction may be very important. Using very simplified and idealized numerical examples, Ferson and Ginzburg[5] demonstrated that failure to partition natural variability and measurement error could produce biased estimates of risk.

Large measurement errors, which are present in most fisheries data, result in substantial uncertainty in abundance estimates[10], often overwhelming effects of density dependence in stock-recruitment relationships[11]. To improve the credibility of scientific advice and to provide better information, measurement error must be considered explicitly[12]. Analysis of the effects of measurement error usually involves bootstrapping simulation studies because multiple, independent estimates of specific parameters, needed to estimate measurement error, are rarely available. For the Hudson River stock of striped bass, multiple, independent indices of abundance are available and measurement error can be estimated.

The effects of entraining fish, especially striped bass, at power plants operating on the Hudson River have been of considerable interest to regulators, electric utilities, and the public[13,14,15]. Currently, the New York State Department of Environmental Conservation (the Department) is reviewing applications to renew the State Pollutant Discharge Elimination System (SPDES) permits for power plants operating on the Hudson River at the Bowline Point, Indian Point, and Roseton sites. In the review process, the Department will consider the level of uncertainty that can be accommodated in making a decision on the SPDES permit renewals (i.e., what level of risk to the fishery resource is acceptable)[16]. The Department will also consider issuing consecutive SPDES permits covering a time horizon of up to 15 years as an alternative to issuing a single permit for a 5-year period[17].

The choice of time horizon can strongly affect both the outcome and reliability of risk assessments[18]. For shorter time horizons, the risks of alternative actions may not differ appreciably. In such cases, having estimates of measurement error would be less critical than for longer time horizons where measurement error may hide real differences in risk. Our objectives were to identify a measure of risk for the Hudson River stock of striped bass that could be used to evaluate the effects of entrainment mortality at the Bowline, Indian Point, and Roseton power plants, assess the effects of measurement error and time on the risk estimates, and compare the risks due to entrainment mortality with those due to increased fishing mortality recommended under Amendment #5 to the Interstate Fishery Management Plan for Atlantic striped bass.

METHODS

We evaluated risk by examining changes in recruitment to the Hudson River stock of striped bass due to entrainment mortality. Indices of recruitment serve as input to the spawning stock model used by the Atlantic States Marine Fisheries Commission (ASMFC) to estimate future population levels and as an early warning signal to fishery managers[19]. The measure of recruitment accepted by the ASMFC is juvenile (age 0) abundance. However, we used abundance of age-1+ (i.e., yearling) striped bass in the Hudson River to represent recruitment because abundance estimates for juvenile striped bass in the Hudson River appear to be affected by emigration and because the age-1+ index has more values than other postjuvenile indices[20].

The Model

We projected the number of age-1+ Hudson River striped bass using an age-structured Leslie matrix model with random temporal variation in survival and fecundity. Age specific rates of natural mortality (M_x) were 1.12 for fish of ages 1+ and 2+[21]. For ages 3+ and older, natural mortality was assumed constant at an average value of $M_x = 0.15$. Age-specific values of fecundity are shown in Table 1[21]. Recruitment (R) to age-1+ was assumed to follow Beverton-Holt type density dependence

$$R = \frac{1}{r + \frac{k}{\sum_a f_a N_a}}$$

where f_a is the average fecundity of fish at age a and N_a is the number of fish at age a. The values for parameters r (6.94E-04) and k (4.82E+04) were derived from the relationship between the abundance of age-1+ fish and post yolk-sac larvae (Fig. 1)[17].

In our simulations, annual values of survival and fecundity were independent log-normal random variables. Values for each were obtained by multiplying the mean by

$$e^{v \ln (CV^2 + 1)^{1/2} - \frac{1}{2} \ln (CV^2 + 1)}$$

where v is a standard normal random variate and CV is the coefficient of variation[21]. The CVs for fecundity and survival for all age classes were 0.34 and 0.321, respectively[17]. The estimation of variance in survival to age 1 is described in the section on measurement error.

TABLE 1

Estimates of Gear Selectivity, Fraction of Fish of Legal Size, and Fecundity (Number of Eggs per Mature Female) by Age*

Age Class	Gear Selectivity	Fraction Legal	Mean Fecundity at Age	Initial Age Distribution
1	0.00	0.08	0	1,207,500
2	0.02	0.08	0	393,800
3	0.14	0.08	0	128,100
4	0.43	0.08	11,190	109,300
5	0.65	0.08	111,150	92,900
6	0.84	0.14	411,680	78,600
7	0.93	0.46	747,000	65,400
8	0.97	0.69	1,083,950	50,400
9	0.98	0.93	1,451,120	36,600
10	0.99	0.97	1,676,160	25,100
11	1.00	0.99	2,022,000	16,900
12	1.00	1.00	2,301,000	11,400
13	1.00	1.00	2,285,000	7,600
14	1.00	1.00	2,342,000	5,100
15+	1.00	1.00	2,591,000	10,400

* Also shown is the initial age distribution used in model simulations.

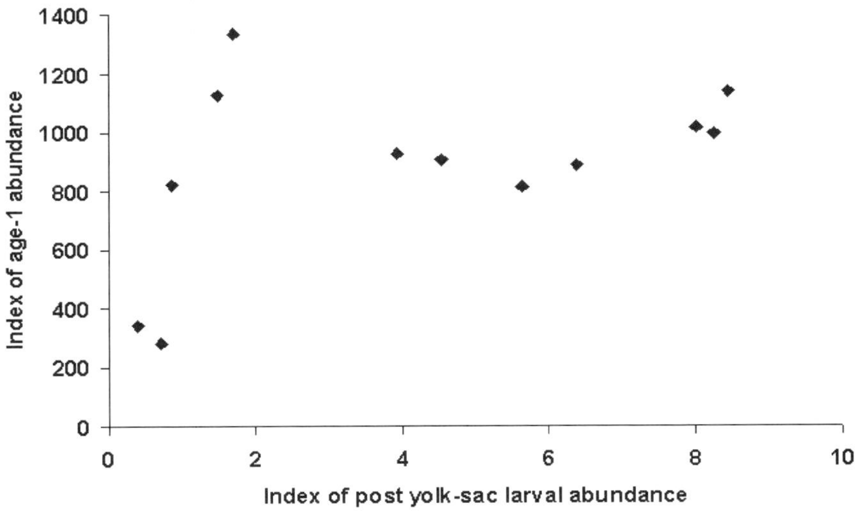

FIGURE 1. Recruitment, as measured by an index of age-1+ abundance, vs. stock size, as measured by an index of post-yolk-sac larval abundance, for the Hudson River stock of striped bass from 1984 through 1995, i.e., the years when values are available for both indices.

Entrainment Mortality

Annual rates of entrainment and impingement were included in survival to age 1+. The average survival of entrainment was calculated as

$$S = \sum_{i}^{n} (1\text{-}M_i) / n$$

where M_i is the mortality rate in year i due to entrainment, and n is the the number of years for which data were available. The expected number of recruits to age 1+ when the effects of entrainment mortality were included was given by $R*S_E$. This method for calculating S will produce estimates that are biased low if the mortality between post-yolk-sac larvae and age-1+ fish is nonlinear due to density dependence.

The annual conditional entrainment mortality rate (CEMR) due to the operation of the Bowline Point, Indian Point, and Roseton power plants was about 11% on average from 1974 through 1997[17]. To assess the relative effect of a CEMR of 11% on recruitment, we also ran the model using a CEMR of 0%.

Fishing Mortality

To compare the effects of fishing and entrainment, simulations were also conducted, which included age-specific estimates of fishing mortality. These were obtained by multiplying the assumed population rate of fishing mortality by an estimate of gear selectivity for each age and the fraction of fish of legal size in each age class. Specifically, the survival of each age class (S_a) was calculated as

$$S_a = e^{-[M_a + Fg(a)l(a)]}$$

where M and F are the natural and fishing mortality rates, $g(a)$ is the age specific gear selectivity, and l(a) is the fraction of fish of legal size in each age class (Table 1).

Under Amendment #5 to the Interstate Fishery Management Plan for Atlantic Striped Bass, the interim target for fishing mortality is an F of 0.33 for the recovering stock and an F of 0.4 for the recovered stock. We used an F of 0.33 in combination with a CEMR of 11% to represent current conditions. For comparison with a management action recommended by Amendment #5, we used an F of 0.4 and a CEMR of 11%.

Measurement Error

Measurement error represents the uncertainty in the estimates of abundance for age-1+ striped bass. Natural variability, also referred to as process error[22], reflects year-to-year changes in the conditions for survival of striped bass from juvenile (i.e., age 0) to age 1+. Partitioning natural variability from measurement error requires at least two independent, empirical estimates of abundance for age-1+ striped bass generated during the same time period; something unavailable for age-1+ striped bass. As a surrogate, we used an estimate of measurement error calculated from estimates of abundance for juvenile striped bass in the Hudson River. The location, time, and gear used to generate the abundance estimates for juvenile striped bass are different from those used to generate the estimates for age-1+ striped bass. Although the estimate of measurement error for juvenile abundance was not directly comparable, we used it to select an initial upper limit for analysis of recruitment to age 1+.

Estimates of abundance for juvenile striped bass have been calculated from two beach seine survey programs conducted on the Hudson River: the Juvenile Striped Bass Survey (JSBS) conducted by the New York State Department of Environmental Conservation and the Beach Seine Survey (BSS) conducted on behalf of four electric utilities. The JSBS sampled on alternate weeks from August through November between river-miles 25 and 40 using a 200-ft beach seine. The BSS sampled on alternate weeks from June through October along the entire length of the Hudson River using a 100-ft beach seine[17].

We represented the fraction of the population sampled by each beach seine survey as

$$\mathrm{BSS}_y = q_y R_y$$
and
$$\mathrm{JSBS}_y = p_y R_y$$

where R_y is the real abundance of age-0 striped bass in year y, while q and p are the proportions of the population caught by the BSS and JSBS, respectively. While the surveys are expected to sample a constant fraction of the population on average, the proportions vary each year due to measurement error. We used a log transform to obtain additive errors:

$$\ln(\mathrm{BSS}_y) = \ln(q_y) + \ln(R_y)$$
and
$$\ln(\mathrm{JSBS}_y) = \ln(p_y) + \ln(R_y)$$

If there is no covariance between annual fluctuations in $\ln(R)$ and measurement errors in p and q, the variance in each index is

$$\mathrm{Var}(\ln(\mathrm{BSS})) = \mathrm{Var}(\ln(q)) + \mathrm{Var}(\ln(R))$$
and
$$\mathrm{Var}(\ln(\mathrm{JSBS})) = \mathrm{Var}(\ln(p)) + \mathrm{Var}(\ln(R))$$

If there is no covariance between the measurement errors across indices, the covariance in the log transformed indices may be used as an estimate of the interannual variability in $\ln(R)$, i.e.,

$$\mathrm{Cov}(\ln(\mathrm{BSS}), \ln(\mathrm{JSBS})) = \mathrm{Var}(\ln(R))$$

The variance due to measurement error may then be estimated by subtracting this covariance from the variance in each index. This general approach has been used for abundance estimates[23,24] and estimates of survival[25].

A portion of the total variance in juvenile abundance is due to changes in reproductive effort among years. Reproductive effort for the Hudson River stock of striped bass, as measured by an index of post-yolk-sac larval abundance[20], was considerably different during the years from 1989 through 1997, compared with the years from 1976 through 1988. To avoid the potential confounding effect of changing reproductive effort, we only considered the years from 1989 through 1997 in our analysis (Fig. 2).

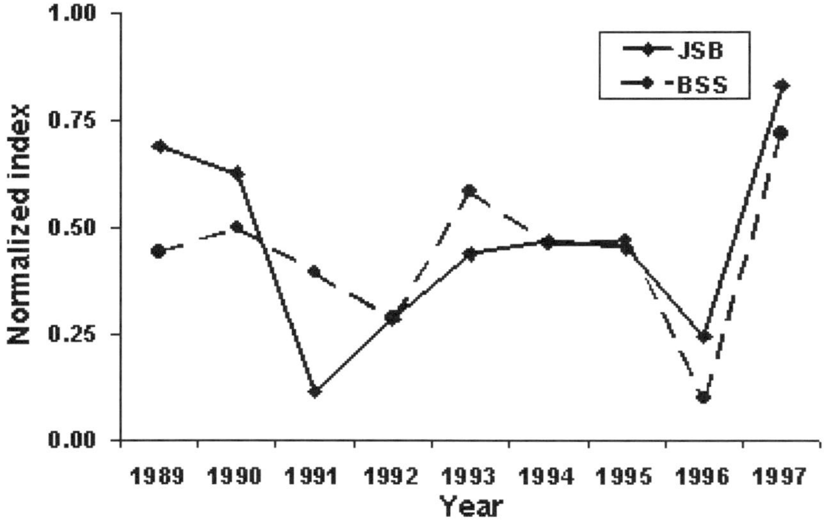

FIGURE 2. Indices of abundance for juvenile striped bass in the Hudson River, normalized to the highest value, caught in the Beach Seine Survey (BSS) and the Juvenile Striped Bass Survey (JSBS) from 1989 through 1997.

We used the covariance in log-transformed indices as the estimate of annual variability in abundance of age-1+ fish. Estimates of measurement error for each each index were then calculated by subtracting this estimate from the total observed variance. This method of partitioning measurement error and natural variability assumes that measurement errors are neither correlated with the actual abundance nor across indices. Measurement error could be correlated with abundance if the proportion of the population sampled varies with population size. For example, an aggregation of fish in a particular location tends to cause the proportion of fish sampled to increase as the population decreases. This is not likely the case for the indices we used because sampling is done at multiple locations during a time of the year when juvenile striped bass are migrating downriver. Furthermore, the index values we used came from years when the population was increasing in size.

Measurement error could be correlated across indices if the capture rate of juveniles by the BSS and the JSBS positively covaried because of common environmental conditions or the fraction of juveniles sampled by each index was influenced by migration among the sampling sites. This is not likely because the sampling dates and locations for the BSS and JSBS are different. The BSS covers the entire Hudson River, while the JSBS covers the lower part of the river.

TABLE 2

Total Variance, Covariance, Variance Due to Measurement Error, and Standard Error for the Log-Transformed Beach Seine Survey (BSS) and Juvenile Striped Bass (JSBS) Indices of Striped Bass Abundance in the Hudson River during the Period 1989 through 1997

	Value	
Statistic	**ln (BSS)**	**ln (JSB)**
Total variance	0.3360	0.3783
Covariance	0.1664	0.1664
Variance due to measurement error	0.1696	0.2119
Standard error	0.1373	0.1534

Measurement error accounted for 50% of the log transformed variance in the BSS index and 56% of the log transformed variance in the JSBS index, for an average of 53%(Table 2). These estimates are comparable to analogous estimates for other species[25]. Therefore, we used 50% of the observed variability as the upper bound for measurement error. We used 0% as the lower bound and considered values of 10, 20, 30, and 40% to assess the sensitivity of risk to measurement error.

The assumed level of measurement error was subtracted from the total variance in age-1+ survival to arrive at an estimate of interannual, natural variability. To reflect the uncertainty associated with the estimate of the mean survival rate, upper and lower bounds were generated for each level of measurement error by performing two additional simulations. The upper bound was based on the the estimate of natural variability plus one standard error. The lower bound was based on the natural variability minus one standard error. The standard error for mean survival (Table 2) was calculated as

$$S.E. = \sqrt{\frac{Total\ Variance\ *\ \%\ Measurement\ Error}{n}}$$

where n is the number of years of data used to estimate the mean survival rate ($n = 9$).

Risk of a Decline in Recruitment

For each combination of measurement error, CEMR, and F, we ran the model 1,000 times for a period of 1, 5, 10, and 15 years starting from the equilibrium age distribution. The 1-year period was the smallest whole-year increment we could examine. The 5-year period corresponded to the length of a SPDES permit. The 10-year period represented issuance of two consecutive SPDES permits, a condition of the Hudson River Cooling Tower Settlement Agreement[26]. The 15-year period corresponded to issuance of three consecutive SPDES permits and the number of years remaining before the license issued by the Nuclear Regulatory Commission for the Indian Point 3 Nuclear Power Plant expired.

Ecological risk assessments are conducted to evaluate the likelihood of adverse ecological effects and not simply to calculate the probability of a common ecological occurrence. Therefore, we wanted to use a rare event as our criterion for calculating risk. The criterion we selected was unusually low recruitment expressed as a probability equal to the proportion of 1,000 model runs in which age-1+ abundance falls below 20% of the initial value at least once during the time period selected. The 20% threshold corresponds to the approximate difference between the lowest and highest estimates in the index of age-1+ abundance for Hudson River striped bass during the period from 1984 through 1997. Although other threshold values could have been selected, there are no established criteria. So rather than selecting a threshold arbitrarily, we used empirical data. Regardless of the value selected, the relative effect of measurement error would be the same.

Analysis

We tested the effect of measurement error and time on recruitment using a non-parametric Friedman rank sums test[27].

RESULTS

Effect of Time Horizon

Time horizon had a significant effect on the probability that recruitment of age-1+ striped bass in the Hudson River stock would fall below 20% of the initial value ($p = 0.0007$ with a CEMR of 0% and $p = 0.0004$ with a CEMR of 11%). This effect was most pronounced when all of the variation was assumed to be natural (Figs. 3 and 4). Under this condition, the probability that recruitment would fall below the 20% threshold was less than 0.016 after 1 year. After 5 years it increased to 0.075 with a CEMR of 0% and to 0.090 with a CEMR of 11%. After 10 years, it more than doubled. Over a 15-year period, the probability that recruitment would fall below the 20% threshold was more than three times higher than for a 5-year period with a CEMR of 11% (0.308) and about 4 times higher than for a 5-year period with a CEMR of 0% (0.287).

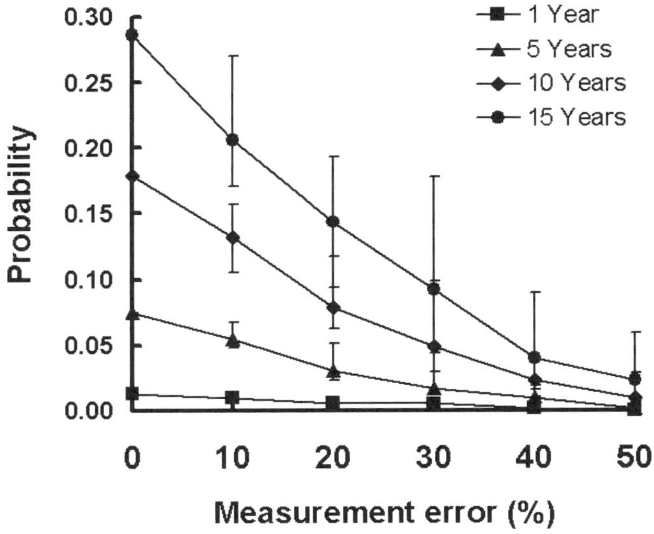

FIGURE 3. Probability that recruitment of age-2 striped bass in the Hudson River stock would fall below 20% of the initial value at least once for levels of measurement error ranging from 0 to 50% of the observed variability after 1, 5, 10, and 15 years with a conditional entrainment mortality rate of 0% and an F of 0.33. The bounds for each estimate are based upon simulations involving the specified level of measurement error plus and minus one standard error for mean survival. When measurement error was 0%, the bounds were so small that they were not displayed.

Effect of Measurement Error

The range of measurement errors evaluated in this study had a significant effect on the probability that recruitment would fall below the 20% threshold ($p = 0.0015$ with a CEMR of 0% and a CEMR of 11%). A measurement error of 50%, the empirically derived estimate from the abundance measures for juvenile striped bass, caused a large reduction in the probability that recruitment would fall below the 20% threshold. Over a 15-year period, the probability that recruitment would fall below the 20% threshold was 0.023 with a measurement error of 50% and CEMR of 0% and was 0.032 with a measurement error of 50% and a CEMR of 11%. These are one-tenth the values with a measurement error of 0% (Table 3). For periods less than 15 years, the probability that recruitment would fall below the 20% threshold with a measurement error of 50% was substantially less than that observed with a measurement error of 0%.

The error bounds on the risk of an 80% decline in recruitment overlapped, after 10 and 15 years, with a measurement error of 20% or higher. The error bounds for all time periods overlapped with a measurement error of 40% or higher.

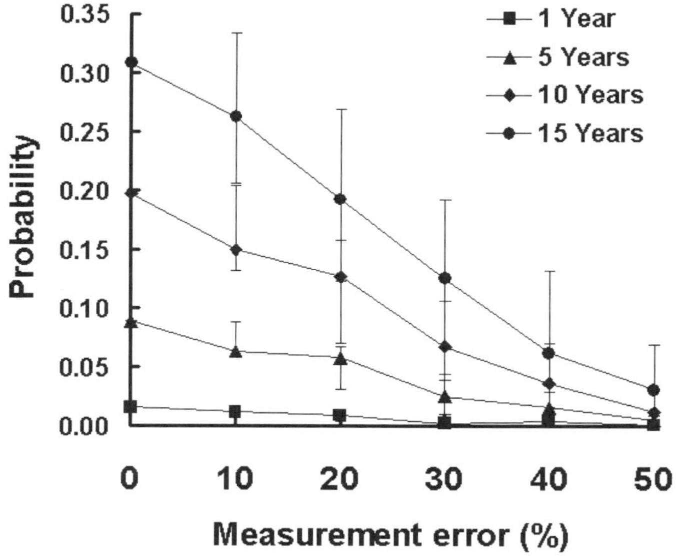

FIGURE 4. Probability that recruitment of age-2 striped bass in the Hudson River stock would fall below 20% of the initial value at least once for levels of measurement error ranging from 0 to 50% of the observed variability after 1, 5, 10, and 15 years with a conditional entrainment mortality rate of 11% and an F of 0.33. The bounds for each estimate are based upon simulations involving the specified level of measurement error plus and minus one standard error for mean survival. When measurement error was 0%, the bounds were so small that they were not displayed.

Comparison of CEMR and Fishing

With an F of 0.33, CEMR had a small effect on the probability that recruitment would fall below the 20% threshold. Over a 15-year period, an increase in the CEMR from 0 to 11% increased the probability that recruitment would fall below the 20% threshold from 0.287 to 0.308 with a measurement error of 0% and from 0.023 to 0.032 with a measurement error of 50% (Table 3). With a CEMR of 11%, an increase in F from 0.33 to 0.40 had a small effect on the probability that recruitment would fall below the 20% threshold. The increase in the probability that recruitment would fall below the 20% threshold, averaged across all levels of measurement error, was only 0.023 over a 15-year period (Table 3).

TABLE 3

The Probability that Recruitment of Age-1+ Striped Bass in the Hudson River Stock would Fall below 20% of the Initial Value at Least Once after 1, 5, 10, and 15 Years by Level of Measurement Error, Fishing, and Conditional Entrainment Mortality Rate (CEMR)

		Probability		
	Measurement	F = 0.33		F = 0.4
Years	Error	CEMR = 0%	CEMR = 11%	CEMR = 11%
1	0	0.013	0.016	0.016
1	10	0.010	0.012	0.015
1	20	0.006	0.010	0.014
1	30	0.001	0.003	0.005
1	40	0.002	0.005	0.001
1	50	0.000	0.002	0.001
5	0	0.075	0.090	0.110
5	10	0.055	0.064	0.076
5	20	0.031	0.059	0.055
5	30	0.017	0.026	0.027
5	40	0.010	0.016	0.008
5	50	0.002	0.005	0.008
10	0	0.179	0.198	0.231
10	10	0.132	0.150	0.176
10	20	0.079	0.127	0.122
10	30	0.049	0.068	0.073
10	40	0.023	0.036	0.033
10	50	0.010	0.013	0.018
15	0	0.287	0.308	0.384
15	10	0.206	0.263	0.280
15	20	0.144	0.192	0.210
15	30	0.093	0.125	0.133
15	40	0.040	0.062	0.072
15	50	0.023	0.032	0.038

DISCUSSION

The results from this study are consistent with the conclusions of Ginzburg and Ferson[5]. Measurement error at the level generated from indices of abundance for juvenile striped bass in the Hudson River (50%) had a significant effect on risk. When all of the variation was assumed to be natural and, thus, there was no measurement error, the probability that recruitment would fall below the 20% threshold overestimated risk about tenfold after 15 years. Overestimates of this magnitude could produce conservative impact assessments and require costly efforts to reduce entrainment mortality that may not measurably reduce risk.

Accurate estimates of risk are necessary but not sufficient for defining AEI. A change in risk should be related to a previously established benchmark, such as the one provided by Amendment #5 to the Interstate Fishery Management Plan for Atlantic striped bass. Amendment #5 recommended an increase in F from 0.33 to 0.40, which had about the same effect on the risk of a decline in recruitment to the Hudson River stock of striped bass as an increase in CEMR from 0 to 11%. Thus, if sustainability of the Hudson River stock of striped bass is not reduced by the change in risk associated with the increased fishing mortality, it is not reduced by the change in risk associated with entrainment.

It is important to know if sustainability of the Hudson River stock of striped bass would be reduced if consecutive SPDES permits were issued to the Bowline, Indian Point, and Roseton power plants for a period of up to 15 years. Although risk increases with time, the differences in risk among time horizons of 5, 10, or 15 years were smaller than the uncertainty associated with the estimates of risk when measurement error was equal to 50%. If the estimate of measurement error based on juvenile striped bass in the Hudson River (50%) corresponds to the level of measurement error for age-1+ fish, consecutive discharge permits should not reduce sustainability of the Hudson River stock.

ACKNOWLEDGMENTS

The New York Power Authority and EPRI (Electric Power Research Institute) of Palo Alto, CA, supported this research under contract CF6871-001-12191.

REFERENCES

1. U.S. Environmental Protection Agency (1977) Interagency 316(a) Technical Guidance Manual and Guide for Thermal Effects Sections of Nuclear Facilities Environmental Impact Statements. Office of Water Enforcement, Permits Division, Industrial Permits Branch, Washington, D.C.
2. Foster, K.R., Vecchia, P., and Repacholi, M.H. (2000) Science and the precautionary principle. *Science* **288,** 279–371.
3. Nagle, D.G. and Morgan, J.T., Jr. (2000) A draft regulatory framework for analyzing potential adverse environmental impact from cooling water intake structures. *Environ. Sci. Policy* **4**, 1–8, ix–xiv.
4. U.S. Environmental Protection Agency (1998) Guidelines for ecological risk assessment. *Federal Register* **63**, 26846–26923.
5. Ferson, S. and Ginzburg, L.R. (1996) Different methods are needed to propagate ignorance and variability. *Reliab. Eng. Sys. Safe.* **54**, 133–144.
6. Stephan, T. and Wissel, C. (1999) The extinction risk of a population exploiting a resource. *Ecol. Model.* **115**, 217–225.
7. Foley, P. (1994) Predicting extinction times from environmental stochasticity and carrying capacity. *Conserv. Biol.* **8**, 124–137.
8. Lande, R. (1993) Risks of population extinction from demographic and environmental stochasticity and random catastrophes. *Am. Nat.* **142**, 911–927.
9. Pimm, S.L., Jones, H.L., and Diamond, J. (1988) On the risk of extinction. *Am. Nat.* **132**, 757–785.
10. Ludwig, D. and Walters, C.J. (1981) Measurement errors and uncertainty in parameter estimates for stock and recruitment. *Can. J. Fish. Aquat Sci.* **38**, 711–720.
11. Walters, C.J. and Ludwig, D. (1981) Effects of measurement errors on the assessment of stock-recruitment relationships. *Can. J. Fish. Aquat. Sci.* **38**, 704–710.
12. Rosenberg, A.A. and Restrepo, V.R. (1994) Uncertainty and risk in stock assessment advice for U.S. Marine Fisheries. *Can. J. Fish. Aquat. Sci.* **51**, 2715–2720.
13. Barnthouse, L.W., Boreman, J., Christensen, S.W., Goodyear, C.P., Van Winkle, W., and Vaughn, D.S. (1984) Population biology in the courtroom: the Hudson River controversy. *BioScience* **34**, 14–19.
14. Barnthouse, L.W., Klauda, R.J., and Vaughn, D.S. (1988) Introduction to the monograph. *Am. Fish. Soc. Monogr.* **4**, 1–8.
15. Dunning, D.J., Ross, Q.E., and Merkhofer, M.W. (2000) Multiattribute utility analysis for addressing section 316(b) of the Clean Water Act. *Environ. Sci. Policy* **4**, S7–S14.
16. Colquhoun, J. (2000) Hudson River Estuary Management Advisory Committee 31 January 2000 briefing by NYSDEC on: Status of the Hudson River Settlement Agreement Power Plant regulation (SPDES and EIS). New York State Department of Environmental Conservation, Albany.
17. Central Hudson Gas & Electric Corp., Consolidated Edison Company of New York, Inc., New York Power Authority, and Southern Energy New York (1999) Draft Environmental Impact Statement for State Pollutant Discharge Elimination System Permits for Bowline Point 1 & 2, Indian Point 2 & 3, and Roseton 1& 2 Steam Electric Generating Stations. New York State Department of Environmental Conservation, Albany.
18. Goldwasser, L., Ferson, S., and Ginzburg, L. (2000) Variability and measurement error in extinction risk analysis: the northern spotted owl on the Olympic Peninsula. In *Quantitative Methods for Conservation Biology*. Ferson, S. and Burgman, M., Eds. Springer-Verlag, New York. pp. 169–187.
19. Atlantic States Marine Fisheries Commission (1975) Amendment #5 to the Interstate Fishery Management Plan for Atlantic Striped Bass. Atlantic States Marine Fisheries Commission, Washington, D.C.
20. Ross, Q.E., Dunning, D.J., Young, J., and Heimbuch, D.G. Impact of entrainment on recruitment of Hudson River striped bass: an empirical approach, in preparation.
21. Saila, S., Martin, B., Ferson, S., Ginzburg, L., and Millstein, J. (1991) Demographic modeling of selected fish species with RAMAS. EPRI (Electric Power Research Institute) Palo Alto, CA, EPRI Research Project 2553, EN-7178.
22. Hilborn, R. and Mangel, M. (1997) *The Ecological Detective: Confronting Models with Data.* Princeton University Press, Princeton.

23. Skalski, J.R. and Robson, D.S. (1992) *Techniques for Wildlife Investigations: Design and Analysis of Capture Data*. Academic Press, San Diego.
24. Link, W.A. and Nichols, J.D. (1994) On the importance of sampling variance to investigations of temporal variation in animal population size. *Oikos* **69**, 539–544.
25. Gould, W.R. and Nichols, J.D. (1998). Estimation of temporal variability of survival in animal populations. *Ecology* 7**9**, 2531–2538.
26. Barnthouse, L., Boreman, J., Englert, T.L., Kirk, W.L., and Horn, E.G. (1988) *Am. Fish. Soc. Monogr.* **4**, 267–273.
27. Sokal, R.R. and Rohlf, F.K. (1969) *Biometry*. W.H. Freeman, San Francisco.

Use of Equivalent Loss Models Under Section 316(b) of the Clean Water Act

William P. Dey
ASA Analysis & Communication, Inc., 291 County Route 62, New Hampton, NY 10958

Received November 3, 2001; Accepted March 15, 2002; Revised March 15, 2002; Published February, 2003

Equivalent loss models encompass a variety of life table–based approaches that can be used to convert age- and life stage–specific estimates of entrainment and impingement loss to a common, easily understood currency. This common currency can be expressed in terms of numbers of individuals, yield to the fishery, or biomass to the ecosystem. These models have at least two key uses in the Section 316(b) assessment process: screening for adverse environmental impact (AEI) and determination of environmental benefits associated with intake alternatives. This paper reviews the various forms of equivalent loss models, their data input requirements, and their assumptions and limitations. In addition, it describes how these models can be used as a second-level screening tool as part of the assessment of the potential for AEI. Given their relative simplicity and ease of use, equivalent loss models should prove to be an important tool in the arsenal of impact assessment methods for Section 316(b).

KEY WORDS: impact assessment, population modeling, cooling water intakes, 316(b), fish

DOMAINS: freshwater systems, marine systems, ecosystems and communities, water science and technology, environmental management and policy, environmental modeling

INTRODUCTION

Section 316(b) of the Clean Water Act requires that a cooling-water intake reflect the best technology available (BTA) to minimize adverse environmental impact (AEI). This section of the Act has traditionally been addressed in two steps. First, there is the issue of whether or not the intake as proposed or constructed will result or has resulted in an AEI. Although there is currently no clear regulatory

Email: wdey@asaac.com

guidance as to what constitutes an AEI, assessments have most commonly focused on effects to populations of aquatic organisms inhabiting the source water body for the cooling water[1]. Such population level effects can result from the loss of organisms through one of two processes: *entrainment,* the passage of smaller, typically planktonic organisms through the cooling system along with the cooling-water flow, and *impingement,* the entrapment of larger aquatic organisms against the intake screens. Both of these processes can result in the mortality of organisms.

The second step in the 316(b) determination process is selection of the BTA to minimize any AEIs expected to occur. As with the concept of AEI, little regulatory guidance exists for the selection of the BTA. However, based on case law and practice, "best technology" has been typically interpreted to mean a proven intake technology that could be installed at a cost not wholly disproportionate to the environmental benefits.

Both steps in the 316(b) determination process require biological information about the aquatic populations in the source water body. Over the years, a variety of modeling approaches have been used in each step of the determination process. One approach has been to use a class of models to estimate the equivalent losses resulting from entrainment and impingement. While specific variations of these models have been used for 316(b) determinations for many years, these techniques have been recently expanded to make them even more relevant for both impact assessment and estimation of the environmental benefits of installing cooling-water intake structures. The purpose of this paper is to provide a brief overview of this class of models, to discuss their strengths and weaknesses, to provide some guidance for selection of input parameters, and to provide recommendations as to their most appropriate incorporation into the determination process under 316(b).

BACKGROUND

Use of equivalent loss models for the assessment of power plant impacts was first suggested by Horst in his review of methods for assessing impacts of entrainment of ichthyoplankton[2]. Horst's proposed method was described as a "simplistic approach … to translate the number of ichthyoplankters lost to entrainment into the number of equivalent adults that would have resulted assuming no compensatory mechanisms in the population." If we assume a population in equilibrium, then total fecundity produced by a breeding pair over their lifetime would result in the average survival of two breeding adults to the next generation. In other words, the lifetime fecundity of a single female is expected to result in the replacement of that female and a mate if the population is to neither increase nor decrease. Under such a scenario, Horst reasoned that overall average survival across a generation could be estimated as follows:

$$S_{e\to a} = \frac{2}{FEC_L} \tag{1}$$

where $S_{e\to a}$ is the overall survival from egg to adult, 2 is the average number of surviving adults, and FEC_L is the lifetime fecundity of a breeding pair.

Consequently, if the entrained organisms are all eggs, then the number of equivalent adults (NA) expected to result from the entrained eggs can be defined as:

$$NA = NE_{eggs} \times S_{e\to a} \tag{2}$$

where NE_{eggs} is the number of eggs lost to entrainment.
Horst further reasoned that if $S_{e\to a}$ is the survival from egg to larval stage and $S_{l\to a}$ is survival from larval to adult, then

$$S_{e\to a} = S_{e\to l} \times S_{l\to a} \tag{3}$$

and

$$S_{l\to a} = \frac{S_{e\to a}}{S_{e\to l}} = \frac{2}{F \times S_{e\to l}} \tag{4}$$

Thus, if the entrained organisms are larvae instead of eggs, the number of equivalent adults becomes:

$$NA = NE_{larvae} \times S_{l\to a} \tag{5}$$

where NE_{larvae} is the number of larvae lost to entrainment.

Horst concluded that the resulting number of equivalent adults could be compared to some reference, such as catch statistics for commercial or sport species, as part of a population-level impact assessment.

Subsequently, Goodyear expanded on Horst's model to include multiple life ages or stages entrained as follows[3]:

$$NA = \sum_{i=1}^{n_e} \left(NE_i \times S_{i\to a} \right) \tag{6}$$

where NA is the total number of equivalent adults, NE_i is the number of life stage or age (i) entrained, $S_{i\to a}$ is the survival from life stage or age (i) to adult, and n_e is the total number of life stages or ages entrained.

Further, Goodyear identified that ages or life stages for this analysis could be defined on either an age or length basis. He also established that the lifetime fecundity used to estimate total eggs-to-adult survival (FEC_L) should be based on the expected lifetime fecundity of a female entering the adult population, as follows:

$$FEC_L = \sum_{J=a}^{m} (FM_j \times S_{r\to j} \times FEC_j) \tag{7}$$

where FM_j is the fraction of females that are mature in age class (j), $S_{r\to j}$ is the survival from recruitment to adult age class (j), FEC_j is the average fecundity of mature female of age class (j), a is the age at recruitment to adult, and m stands for the oldest age classes in the population.

Finally, Goodyear stated that the equivalent number of fish lost to the fishery (NF) could be estimated from the number of equivalent adults (NA) as follows:

$$NF = \frac{(NA \times F_a)}{Z_a} \tag{8}$$

where Z_a is the instantaneous total mortality rate for adults and F_a is the instantaneous fishing mortality rate for adults.

Horst's and Goodyear's model, commonly referred to as the Equivalent Adult Model (EAM), has been widely adopted as part of the suite of techniques used to assess the potential for AEI of cooling-water withdrawals[4] .

Subsequent to these two seminal publications, impact assessors realized that the EAM approach was equally useful for assessment of potential effects of impingement as well as entrainment. In addition, it was determined that the EAM framework could be used to estimate the equivalent loss of individuals at any selected life stage, not just adults. For example, the EAM framework could be used to estimate the equivalent loss in reproductive effort (e.g., eggs) resulting from entrainment or impingement of older life stages[5]. Further, this approach could be used to estimate the number of individuals at a specific life stage (e.g., juveniles or fingerlings) that could be replaced through stocking or habitat improvements[6]. However, it is important to recognize that the number of equivalent individuals is dependent on the age endpoint selected for the calculation. For example, the loss of 1 million larvae might be equivalent to the loss of 100 individuals at age 1 but only 1 individual at age 5. Thus, it is important that the age of equivalency selected be most relevant to the impact assessment goals.

Further, assessors recognized that this same framework could be extended to address two additional assessment endpoints beyond the equivalent number of adults: equivalent yield to the fishery and equivalent amount of forage lost. Equivalent yield to the fishery allows estimation of total yield (in weight) that could have accrued to a commercial or recreational fishery from those individuals lost to entrainment or impingement in the absence of compensatory changes in total mortality. Calculation of equivalent yield integrates Baranov's catch equation[7], similar to the concept of the equivalent number of fish lost to the fishery as defined by Goodyear, with estimates of the mean weight by age. This equivalent yield is estimated as follows:

$$EY = \sum_{i=1}^{n_L} \left[\frac{NA_i \times V_i \times F_i \times A_i}{Z_i} \times W_i \right] \tag{9}$$

where *EY* is the equivalent yield to the fishery, NA_i is the equivalent number at the beginning of each age estimated using the EAM sequentially for each Age (i), V_i is the vulnerability of Age (i) to fishing, F_i is the instantaneous fishing mortality rate for Age (i), Z_i is the instantaneous total mortality rate for Age (i), A_i is the total mortality rate for Age (i) (equal to $1\text{-}e^{-Zi}$), W_i is the average weight for individual of Age (i), and n_f is the maximum number of Ages (i) vulnerable to fishery.

This method, the Equivalent Yield Model (EYM), results in an estimate of yield defined in the same units used to describe the average weight of the individuals (e.g., lb or kg) and integrates yield across the entire lifetime of surviving individuals. This method is clearly most relevant for species with active commercial or recreational fisheries. As with the EAM, the results assume no compensatory changes in natural mortality rates. This model has been used to address the effects of entrainment and impingement at several power plants[8,9].

For aquatic organisms whose principal ecological role is to serve as food for larger predators (e.g., minnows, anchovies) or otherwise provide energy for other trophic levels, the number of individuals lost expressed as the number of adults is a measure of little direct relevance to man. Further, without any commercial or recreational harvest, the potential yield to a fishery is also not relevant. For these species, then, what is important is the amount of biomass that could be used as energy for other trophic levels, including many predators that are directly harvested by man.

For these species, it is the cumulative mortality of the population across all life stages and ages that provides the biomass for other trophic levels, assuming this mortality is largely a result of predation. Thus, for such species, a useful and relevant measurement endpoint is the total cumulative biomass, which otherwise would have been consumed by other trophic levels, that was lost to the system as a result of entrainment and impingement at cooling-water intakes. Using a framework similar to both EAM and EYM, it is then possible to estimate the mortality occurring in each life stage and multiply the result by the average weight of each life stage to determine the total amount of biomass that would have resulted from the subsequent consumption of the individuals had entrainment or impingement not occurred. This equivalent biomass lost is calculated as follows:

$$BL = \sum_{i=1}^{n_L} \left\{ \sum_{j=1}^{n_e} \left[N_j \times \left(S_{j,i} - S_{j,i+1} \right) \times W_i \right] \right\} \tag{10}$$

where BL is the equivalent biomass lost, N_j is the number of life stage or age (j) lost to entrainment or impingement, $S_{j \rightarrow i}$ is the cumulative survival from life stage or age (j) to beginning of age (i), $S_{j \rightarrow i+1}$ is the cumulative survival from life stage or age (j) to beginning of age ($i+1$), W_i is the average weight of life stage (i), n_e is the total number of life stages or ages (j) entrained or impinged, and n_L is the total number of life stages or ages (i) up to maximum life span.

This method, the Biomass Lost Model (BLM), results in an estimate of biomass lost defined in the same units used to describe the average weight of the individuals and integrates this loss across all ages. While this method is specifically designed to address the loss of forage species, the BLM can also be applied to the earlier life stage of commercial and recreational species when natural mortality rates (presumably as a result of predation) are high. As with both the EAM and the EYM, the results assume no compensatory changes in natural mortality rates. The BLM is conceptually similar to the Production Foregone Model proposed by Rago[10] and Jensen[11]. The BLM has been used to estimate the effects of entrainment and impingement at several power plants[9,12,13,14,15].

As a result of these advances, there now exist three variations of equivalent loss models — the EAM, the EYM, and the BLM — all of which are based on the approach originally proposed by Horst[2] and Goodyear[3]. These three models address different measurement endpoints that result from the three possible fates that can befall an individual passing through a life stage: (1) surviving to next stage, (2) being caught by a fisherman, or (3) being consumed by other trophic levels. Each of these endpoints can have relevance to the assessment of AEI and to the determination of ecological benefits of potential alternative intake technologies. Each model can be implemented in spreadsheet software with minimal programming expertise.

SELECTION OF MODEL INPUTS

All three versions of equivalent loss models require three common life stage/age–specific input parameters: estimates of entrainment and impingement loss, estimates of rate of mortality for each life stage/age in the population, and estimates of the duration of each life stage/age. In addition, the EYM and BLM both require estimates of life stage/age–specific average weights and the EYM requires estimates of age-specific fishing vulnerability and mortality rates. Each of these input parameters is described below.

Entrainment and Impingement Loss Estimates

Estimates of entrainment and impingement loss are most commonly made on an annual basis and are generated for each vulnerable life stage of each species that is the target of the assessment. Typically, these estimates of loss are based on site-specific sampling that is scaled up to the total flow of the intake and adjusted for collection efficiency, potential recirculation, and entrainment/impingement mortality. The general form of this calculation is as follows:

$$NL_i = \sum_{s=1}^{q} \left[\frac{D_{si} \times CW_s}{CE_{si}} \times PM_{si} \right] \tag{11}$$

where NL_i is the estimated number life stage/age (i) lost to entrainment or impingement, D_{si} is the density life stage/age class (i) entrained or impinged during sampling period (s), CE_{si} is the collection efficiency of life stage/age class (i) collected during sampling period (s), PM_{si} is the entrainment or impingement mortality at the plant for life stage/age class (i) during sampling period (s), CW_s is the total cooling water flow for the plant during sampling period (s), and q is the total number of sampling periods (s) in the estimation interval (typically 1 year).

Details on collecting site-specific entrainment and impingement data and the subsequent estimation of losses are not discussed further as they are highly site specific.

Population Mortality Rates

Mortality rates refer to the probability of death of an individual. Mortality rates are often expressed as instantaneous rates[7] and the total instantaneous mortality rate combines the effects of mortality from fishing and from all other sources (lumped under *natural mortality*) such that:

$$Z_i = F_i + M_i \tag{12}$$

where Z_i is the instantaneous total mortality rate for life stage/age(i), F_i is the instantaneous fishing mortality rate for life stage/age(i), and M_i is the instantaneous natural mortality rate for life stage/age(i).
Obviously, for species and/or life stages that are not fished, then $F_i = 0$, and the total mortality rate equals the natural mortality rate (i.e., $Z_i = M_i$).
The complement of mortality is survival, such that:

$$S_i = e^{-(Z_i \times t_i)} \tag{13}$$

where S_i is survival during life stage/age (i) and t_i is the duration of life stage/age (i).

Estimates of life stage– and age-specific mortality rates, particularly for the older ages, are often available from the scientific literature. This is especially true for species of commercial and/or recreational importance, where there is an increasing desire to manage these species through the use of quantitative models that require much of the same information as required by the models described in this paper.

However, it is often the case that reliable population mortality rates are not available for all life stages and ages. Thus, it is up to the assessor to select the most appropriate mortality rates for equivalent loss estimation. One commonly used tool for this selection process is a *life table*. A life table is a technique used to track life stage- and age-specific population parameters, such as mortality, maturity, sex ratios, and fecundity[16]. Also displayed in a life table is the integration of all parameters in their effects on subsequent population behavior.

One common simplifying assumption for selection of life stage – and age-specific mortality rates is that the population is at equilibrium – that is, that the population is neither increasing nor decreasing. This is the assumption used by Horst[2] and Goodyear[3] in their development of the EAM. Yet it is clear that populations are rarely, if ever, at equilibrium, particularly when considered on a short-term basis. Instead, they fluctuate to higher and lower levels of abundance as a result of a variety of abiotic and biotic factors. However, assuming the population is neither going extinct nor increasing to significantly higher levels, most populations tend to fluctuate around some long-term average[16]. It is this long-term average that represents equilibrium conditions. Thus, use of an equilibrium assumption appears appropriate for determining the long-term effects of entrainment and impingement over the life of a power plant (typically 20 to 30 years or more).

As noted above, under equilibrium conditions the total survival *(S)* across a generation is fixed at:

$$S_{e\to a} = \frac{2}{FEC_L} \tag{14}$$

This occurs when the expected survival of a female egg is 1 (i.e., when a female just replaces herself each generation):

$$1 = \sum_{i=1}^{n_L} \left(S_{e\to i} \times PF_i \times FM_i \times FEC_i \right) \tag{15}$$

where $S_{e\to i}$ is the cumulative survival from egg to life stage/age (i), FM_i is the fraction of life stage/age (i) females that are mature, PF_i is the proportion of life stage/age (i) that are female, and F_i is the mean fecundity of life stage/age (i).

Using this relation, it is possible to adjust the mortality rates within the life table so that the cumulative survival is S and the population comes into equilibrium. There are a variety of techniques that could be used for this adjustment process. For example, it is likely that the assessor will have greater confidence in some of the estimates of mortality than others. In fact, it is common that estimates for some life stages and ages might be missing altogether. One approach, then, would be to fix the estimates with the highest degree of certainty and iteratively vary the others until arriving at internally consistent and biologically meaningful estimates of mortality. Another approach would be to assume some underlying functional relationship between natural mortality and a known biological measure such as size[17,18]. This functional relationship could then be used as guidance to adjust the available estimates of life stage–specific mortality to generate the inputs needed for equivalent losses estimation.

STAGE/AGE DURATIONS

Typically, the older ages are defined on an annual basis (e.g., age 1, age 2, etc.). For these ages, durations are fixed at 1 year (i.e., 365 days). Younger individuals are commonly categorized by developmental stage (e.g., egg, yolk-sac larvae, post yolk-sac larvae, etc.). The durations of these stages are dependent on the development rate of the individual, and hence are typically a function of water temperature. For the purposes of equivalent loss modeling, average stage durations are commonly used, although it is possible to have variable life stage durations as well. Finally, it is possible to assign the early life stages of fish to specific ages (e.g., days) through the use of microstructure analysis of otoliths[19]. While this approach could reduce the uncertainty resulting from variable state durations, such a practice is not common owing to the labor-intensive requirements of the otolith analysis.

It is important to determine the age of the individuals lost to entrainment or impingement in addition to the total duration of each life stage/age. This is especially important for the larval stages with high natural mortality rates. For example, substantially different equivalent loss estimates could result depending on whether entrained post yolk-sac larvae came from the beginning, middle, or end of the total duration of this life stage. Typically, three approaches have been used to estimate the specific age of individuals within a life stage/age category. First, use of otolith analysis can provide actual ages of fish. However, as previously noted, this practice is not common because of high labor requirements. Second, analysis of length-frequency distributions within individual life stages/ages can provide insight as to whether the individuals came from early or late within the stage/age category. Finally, one could assume that all individuals within a stage/age category are equally vulnerable. In that case, the age could be assigned to the median age of surviving individuals within the category. This median age is a function of the mortality rate within the category and is calculated as follows[8]:

$$ma_i = \frac{\ln 2 - \ln(1+e^{-Z_i t_i})}{Z_i} \tag{16}$$

where ma_i is the median age of life stage/age (i), t_i is the duration of life stage/age (i), and Z_i is the instantaneous mortality rate for life stage/age (i).

Regardless of the method used to estimate the age of individuals with a life stage/age, for the purposes of estimating equivalent loss, individuals entrained or impinged are assumed to be exposed to this mortality only from the estimated age of the individuals through the end of that life stage/age (i.e., $t_i - ma_i$).

Average Weights

Both the EYM and the BLM require estimates of life stage/age–specific average weights. However, the weight requirements of each model are slightly different conceptually. The EYM requires average weights of individuals harvested by the fishery, whereas the BLM requires average weights of those consumed as prey. Depending on the nature of the fishery and of predation, these weights could be slightly different for the same life stage/age. For example, principal harvests for many anadromous fish species occur during spawning runs. In that case, the weights for the EYM would be heavily weighted towards individuals in the early part of the annual growth cycle. On the other hand, these same individuals are likely to be equally vulnerable to predation throughout the year with an average weight equal to the median weight of individuals passing through that life stage/age.

Information on average weights for life stages or ages is readily available for many species from the scientific literature. Alternatively, weights can be derived by combining known life stage/age–specific lengths and length-weight relationships to calculate the geometric mean of the average weight at the beginning and end of the interval. For time-specific fisheries, average life stage/age–specific weights can be estimated from fishery monitoring studies.

Fishing Vulnerability and Mortality Rates

Estimates of life stage/age–specific fishing and vulnerability rates are needed for the EYM. Since this model only applies to species that are actively harvested, estimates of these two rates can often be obtained from fishery management plans or from local resource management agencies. For species with specific size limits, vulnerability can often be estimated from age-specific growth rates or length frequency distributions.

EXAMPLES OF USE

This section presents three examples of how equivalent loss models might be used as part of the overall 316(b) determination process. All examples are hypothetical and do not reflect data from any specific power plants. Population input parameters (e.g., mortality rates) were selected to reflect possible values for each species. However, the author makes no warranty as to their accuracy. Each is designed to illustrate one of the three measurement endpoints of equivalent loss: equivalent adults, equivalent yield to fishery, and biomass lost. Each will also show how this information might be used in the 316(b) process.

Equivalent Adults

A power plant withdraws its cooling water from nearshore marine waters along the Southern California Bight. This area is inhabited by a typical complement of nearshore marine fish species, including queenfish, a small member of the drum family. Queenfish are a popular target of pier fishermen along the coast of Southern California. Entrainment and impingement sampling at the power plant results in annual estimates of queenfish lost as follows:

Life stage	Number lost
Eggs	300,000,000
Yolk-sac larvae	100,000,000
Post yolk-sac larvae	50,000,000
Age 0	24,000
Age 1	10,000

Queenfish typically mature at the end of their second year of life and this age was used to define the adult stage for purposes of equivalent adult estimation.

Based on the EAM, the estimates of entrainment and impingement loss at this hypothetical power plant are equivalent to slightly more than 68,000 adults (Table 1). The regulatory agency was concerned that this level of could not be easily dismissed and required additional assessment at a population level before they could make a determination as to the potential for AEI for queenfish at this power plant.

Equivalent Yield

A power plant withdraws its cooling water from a large (1,000-acre) lake inhabited by a typical complement of warm-water fish species, including bluegill sunfish. Entrainment and impingement sampling at the power plant results in annual estimates of bluegills lost as follows:

Life stage	Number lost
Eggs	0
Yolk-sac larvae	0
Post yolk-sac larvae	1,500,000
Young of year	2,500
Age 1	400

In this lake, bluegills are a popular target of recreational fishermen and there was concern that the cooling-water withdrawals would reduce yield to the fishermen. Bluegills are presumed to enter the fishery when they are 6 in. long and 4 years old; the current annual instantaneous fishing mortality rate (F) is assumed to be 0.2.

TABLE 1

Example of the EAM for Queenfish Based on Entrainment and Impingement Losses at a Hypothetical Power Plant

Life Stage/ Age	Duration in days[1]	Instantaneous Mortality Rate[1]	Life Stage Total Survival Rate[2]	Cumulative Survival to Adult[3]	Number Lost to Cooling-Water Withdrawals[1]	Equivalent Adults[4]
Eggs	2	0.250000	0.606530660	0.000014326	300,000,000	4,298
YSL	16	0.217432	0.030840963	0.000036811	100,000,000	3,681
PYSL	28	0.119659	0.035068962	0.001188686	50,000,000	59,434
Age 0	335	0.010000	0.035084354	0.033895164	24,000	813
Age 1	365	0.001899	0.500000000	0.666666667	10,000	6,667
					Totals	**68,227**

[1] Model inputs.
[2] Total survival across life stage = Exp($-Z_i$ * t_i).
[3] Calculated from median age (d_i) to adult.
[4] Number lost times cumulative survival to adult.

Estimates of equivalent yield, using a combination of site-specific data and information from similar water bodies, produced estimates of equivalent yield of 53 kg/year (Table 2). The permitting authority then compared this estimate of lost yield, equal to less than 0.05 kg/acre, to the current annual recreational harvest (0.5 kg/acre). Based on this comparison, the authority concluded that entrainment and impingement losses were not likely to result in an AEI as such losses were a tiny fraction (approximately 10%) of the sustained annual harvest by recreational fishermen.

Biomass Lost

A power plant withdraws its cooling water from the mesohaline section of an estuary located along the mid-Atlantic coast. Within this estuary, bay anchovy is an important prey for a variety of predatory fish, many of which support valuable commercial or recreational fisheries. Entrainment and impingement sampling at the power plant results in annual estimates of bay anchovies lost as follows:

Life stage	Number lost
Eggs	600,000,000
Yolk-sac larvae	300,000,000
Post yolk-sac larvae	750,000,000
Age 0	8,000,000
Age 1	250,000
Age 2	50,000

TABLE 2

Example of the EYM for Bluegill Based on Entrainment and Impingement Losses at a Hypothetical Power Plant

Life Stage/ Age	Duration in days[1]	Instantaneous Mortality[1]		Fishing Vulnera-bility Rate[1]	Numbers Lost to Cooling-Water Withdrawal[1]	Equivalent Number Entering Stage/Age			Equiva-lent Catch[4]	Weight per Fish (g)[1]	Equiva-lent Yield (kg)[5]
		Natural	*Fishing*			Non-vulner -able[2]	Vulner-able[3]	Total			
Eggs	2.5	0.250000	0.000000	0.0000	—	—	—	—	—	0.1	—
YSL	5.5	0.250000	0.000000	0.0000	—	—	—	—	—	0.1	—
PYSL	22	0.250000	0.000000	0.0000	1,500,000	—	—	—	—	0.5	—
Age 0	335	0.006560	0.000000	0.0000	2,500	12,210	—	12,210	—	3.0	—
Age 1	365	0.001096	0.000000	0.0000	400	1,856	—	1,856	—	12.0	—
Age 2	365	0.001096	0.000000	0.0000	—	1,565	—	1,565	—	37.0	—
Age 3	365	0.001096	0.000000	0.0000	—	1,049	—	1,049	—	73.0	—
Age 4	365	0.001096	0.000500	1.0000	—	—	703	703	97	116.0	11.3
Age 5	365	0.001096	0.000500	1.0000	—	—	471	471	65	159.0	10.4
Age 6	365	0.001096	0.000500	1.0000	—	—	316	316	44	200.0	8.7
Age 7	365	0.001096	0.000500	1.0000	—	—	212	212	29	238.0	7.0
Age 8	365	0.001096	0.000500	1.0000	—	—	142	142	20	272.0	5.3
Age 9	365	0.001096	0.000500	1.0000	—	—	95	95	13	301.0	4.0
Age 10	365	0.001096	0.000500	1.0000	—	—	64	64	9	325.0	2.9
Age 11	365	0.001096	0.000500	1.0000	—	—	43	43	6	345.0	2.0
Age 12	365	0.001096	0.000500	1.0000	—	—	29	29	4	365.0	1.4
Total											53.0

1 Model inputs.
2 Equivalent number surviving not vulnerable to fishing assuming that actual loss is median age for life stage (d_i).
3 Equivalent number surviving vulnerable to fishing assuming actual loss is median age for life stage (d_i).
4 Expected number harvested based on number vulnerable and Baranov's catch equation.
5 Expected number harvested multiplied by average weight.

TABLE 3

Example of the BLM for Bay Anchovy Based n Entrainment and Impingement Losses at a Hypothetical Power Plant

				Eggs	YSL	PYSL	Age 0	Age 1	Age 2			
Number Lost to Entrainment/Impingement[1]				60,000,000	30,000,000	75,000,000	800,000	25,000	5,000			
Life Stage/ Age	Duration in days[1]	Instantaneous Total Mortality Rate[1]	Life Stage Total Survival Rate[2]	Number That Would Have Survived to Each Age[3]						Number Consumed[4] (gm)[1]	Average Weight (kg)[5]	Biomass Lost
Eggs	1	1.044000	0.3520437							28,754,529	0.0010	28.8
YSL	2	0.78300	0.2088790	31,245,471						44,351,705	0.0010	44.4
PYSL	32	0.19140	0.0021879	6,526,524	10,367,242					91,529,331	0.0040	366.1
Age 0	330	0.00445	0.2302706	14,280	22,683	327,473				781,043	0.1130	88.3
Age 1	365	0.00445	0.1970594	3,288	5,223	75,407	299,473			324,610	1.5730	510.6
Age 2	365	0.00445	0.1970594	648	1,029	14,860	59,014	8,231		88,782	3.9270	348.6
											Total	1,386.7

1 Model inputs.
2 Total survival across life stage = $Exp(-Z_i * t_i)$.
3 Equivalent number surviving to each age assuming actual loss is median age for life stage (d_i).
4 Assuming all dead are consumed.
5 Number consumed times average weight.

The permitting authority wanted to know if an alternative to the existing intake, which could reduce these losses but would cost an average of $1.5 million/year, could be justified from an economic perspective.

Using a combination of site-specific data and information from similar water bodies, the BLM produced estimates of biomass lost from entrainment and impingement of approximately 1,387 kg/year (Table 3). Since the permitting authority wanted to make an economic comparison and direct economic values of bay anchovy are uncertain, this biomass lost was converted to an equivalent weight of recreational and commercial species for which economic values are available. Assuming a 10% conversion efficiency, this biomass lost equates to a biomass of 1,387 kg of commercial or recreational fish, assuming that all bay anchovies would have been consumed by these species. Using an average value of $20/kg of commercial or recreational fish, the biomass of bay anchovy lost would have an economic value of $27,800/year expressed as commercial or recreational fish.

Presuming the intake alternative has the potential to reduce these losses by 80%, then the economic benefit of this alternative to bay anchovy would be $22,240/year. However, by comparing this expected annual benefit to the estimated annual cost for the intake alternative ($1.5 million), the permitting authority concluded that the intake alternative could not be justified on a cost-benefit basis.

CONCLUSIONS AND RECOMMENDATIONS

The class of models discussed in this paper (equivalent loss models) can be a useful and relatively simple tool for making determinations under 316(b) for several reasons.

First, this approach provides loss measures in common currency (numbers, fishery yield, or biomass) that can be used to address the question of AEI. For a relatively few high-profile cases in which population-level losses are of concern, these models are unlikely to be sufficient for final determination. In these cases, more complex population-based assessment techniques will be required (as we saw in the Equivalent Adult hypothetical example). However, for many cooling-water intakes, losses are relatively small and the use of equivalent loss models can be all that is required to demonstrate that these losses are relatively small compared to acceptable levels of harvest (as we saw in the Equivalent Yield hypothetical example).

Second, equivalent loss models provide measures of loss that are easily recognizable by the lay public (e.g., number or pounds of fish). This allows the public to better understand management decisions that are being made regarding public resources.

Third, the results of equivalent loss modeling are defined in units that can be directly translated into a consideration of cost and benefits of management decisions (as we saw in the Biomass Lost hypothetical example). While the role of explicit cost-benefit analysis in the 316(b) determination process remains to be determined, some consideration of relative costs vs. resulting environmental benefits is likely to be involved in these decision-making processes, as it is with many other environmental management questions.

Fourth, all of the models described in this paper can be easily implemented using readily available spreadsheet software. Not special knowledge of ecological modeling or computer programming is required.

It is important to recognize, however, that despite the advantages of equivalent loss modeling listed above, these approaches require biological information not readily available for many aquatic populations (e.g., life stage specific mortality rates). Uncertainty in these model inputs can be addressed through sensitivity analysis, Monte Carlo simulations, or fuzzy arithmetic[20]. In addition, I think it would be valuable for the electric utility industry and the regulatory agencies to work together to develop a mutually acceptable range for each input parameter for various categories of species. These ranges could then be combined with site-specific estimates of entrainment and impingement loss to estimate equivalent loss.

References

1. Dey, W.P., Jinks, S.M., Lauer, G.L. (2000) The 316(b) assessment process: evolution towards a risk-based approach. *Environ. Sci. Policy* **3,** S15–S23.
2. Horst, T.J. (1975) The assessment of impact due to entrainment of ichthyoplankton. In *Fisheries and Energy Production: A Symposium.* Saila, S.B., Ed. D.C. Heath, Lexington, MA. pp. 107–118.
3. Goodyear, C.P. (1978) Entrainment Impact Estimates Using the Equivalent Adult Approach. Report No. FWS/OBS-78/65. U.S. Fish and Wildlife Service, Washington, D.C.
4. EPRI. (1999) Catalog of Assessment Methods for Evaluating the Effects of Power Plant Operations on Aquatic Communities. Electric Power Research Institute. TR-112013. Palo Alto, CA.
5. Tenera, Inc. (2000) Diablo Canyon Power Plant 316(b) Demonstration Report. Prepared for Pacific Gas and Electric Co., San Francisco, California. Doc. No. E9-055.0.
6. PSEG. (1993) Supplemental Permit Renewal Application: Salem Generating Station, NJPDES Permit No. NJ0005622. Vol. I and II. Prepared for New Jersey Department of Environmental Protection.
7. Ricker, W.E., Ed. (1975) Computation and Interpretation of Biological Statistics of Fish Populations. Bulletin 191. Department of the Environment, Fisheries, and Marine Service, Ottawa. 382 p.
8. PSEG. (1999) Permit Renewal Application. NJPDES Permit No. NJ0005622. Public Service Electric and Gas Company Salem Generating Station, March 4, 1999. PSE&G, Newark, NJ.
9. PSEG. (2001) Mercer Generating Station Supplemental 316(b) Demonstration. Prepared for New Jersey Department of Environmental Protection.
10. Rago, P.J. (1984) Production foregone: an alternative method for assessing the consequences of fish entrainment and impingement at power plants and water intakes. *Ecol. Model.* **24,** 79–111.
11. Jensen, A.L., Reider, R.H., and Kovalak, W.P. (1988) Estimation of production forgone. *N. Am. J. Fish. Manage.* **8,** 191–198.
12. PSEG. (1988a) Bergen Generating Station Units 1 and 2 Supplemental 316(b) Demonstration. Prepared for New Jersey Department of Environmental Protection.
13. PSEG. (1988b) Hudson Generating Station Units 1 and 2 Supplemental 316(b) Demonstration. Prepared for New Jersey Department of Environmental Protection.
14. PSEG. (1989a) Linden Generating Station Units 1 and 2 Supplemental 316(b) Demonstration. Prepared for New Jersey Department of Environmental Protection.

15. PSEG. (1989b) Sewaren Generating Station Supplemental 316(b) Demonstration. Prepared for New Jersey Department of Environmental Protection.
16. Krebs, C.J. (1985) *Ecology: The Experimental Analysis of Distribution and Abundance*. 3rd ed. HarperCollinsPublishers, New York. 800 p.
17. Ware, D.M. (1975) Relation between egg size, growth, and natural mortality of larval fish. *J. Fish. Res. Bd. Can.* **32,** 2503–2512.
18. Boudreau, P.R. and Dickie, L.M. (1989) Biological model of fisheries production based on physiological and ecological scaling of body size. *Can. J. Fish. Aquat. Sci.* **46,** 614–623.
19. Secor, D.H., Dean, J.M., and Laban, E.H. (1991) Manual for Otolith Removal and Preparation for Microstructural Examination. Electric Power Research Institute and the Belle W. Baruch Institute for Marine Biology and Coastal Research, Palo Alto, CA.
20. Saila, S.B., Lorda, E., Miller, J.D., Sher, R.A., and Howell, W.H. (1997) Equivalent adult estimates for losses of fish eggs, larvae, and juveniles at Seabrook Station with use of fuzzy logic to represent parametric uncertainty. *N. Am. J. Fish. Manage.* **17,** 811–825.

A Blueprint for the Problem Formulation Phase of EPA-Type Ecological Risk Assessments for 316(b) Determinations

Webster Van Winkle[1,*], William P. Dey[2], Steve M. Jinks[2], Mark S. Bevelhimer[3], and Charles C. Coutant[3]

[1]Van Winkle Environmental Consulting Co., 5163 N. Backwater Ave., Boise, ID 83703; [2]ASA Analysis & Communication, Inc., 291 County Road 62, New Hampton, NY 10958; [3]Environmental Sciences Division, Oak Ridge National Laboratory, Oak Ridge, TN 37831-6036

Received November 26, 2001; Revised May 15, 2002; Accepted May 20, 2002; Published February, 2003

The difference between management objectives focused on sustainability of fish populations and the indigenous aquatic community, and a management objective focused on minimizing entrainment and impingement losses accounts for much of the ongoing controversy surrounding §316(b). We describe the EPA's ecological risk assessment framework and recommend that this framework be used to more effectively address differences in management objectives and structure §316(b) determinations. We provide a blueprint for the problem formulation phase of EPAtype ecological risk assessments for cooling-water intake structures (CWIS) at existing power plant facilities. Our management objectives, assessment endpoints, conceptual model, and generic analysis plan apply to all existing facilities. However, adapting the problem formulation process for a specific facility requires consideration of the permitting agency's guidelines and level of regulatory concern, as well as site-specific ecological and technical differences. The facility-specific problem formulation phase is designed around the hierarchy of biological levels of organization in the generic conceptual model and the sequence of cause-effect events and risk hypotheses represented by this model. Problem formulation is designed to be flexible in that it can be tailored for facilities where §316(b) regulatory concern is low or high. For some facilities, we anticipate that the assessment can be completed based on consideration of susceptibility alone. At the other extreme, a high level of regulatory concern combined with the availability of extensive information and consideration of costly CWIS mitigation options may result in the ecological risk assessment relying on analyses at all levels. Decisions on whether to extend the ecological risk assessment to additional levels should be based on whether regulatory or generator concerns merit additional analyses and whether available information is adequate to support such analyses. In making these decisions, the

* Corresponding author. E-mails: wvanwinkle@juno.com; wdey@asaac.com; sjinks@asaac.com; mb2@ornl.gov; ccc@ornl.gov

functional dependence between levels of analysis must be considered in making the transition to the analysis phase and risk estimation component of the ecological risk assessment. Regardless of how the generic analysis plan is modified to develop a facilityspecific analysis plan, the resulting plan should be viewed as a tool for comparing representative species and alternative CWIS options by focusing on relative changes (i.e., proportional or percent changes) in various measures. The analysis plan is specifically designed to encourage consideration of multiple lines of evidence and to characterize uncertainties in each line of evidence. Multiple lines of evidence from different levels of analysis, obtained using both prospective and retrospective techniques, provide a broader perspective on the magnitude of potential effects and associated uncertainties and risks. The implications of the EPA's recent (April 2002) proposed regulations for existing facilities on the applicability of this blueprint are briefly considered.

KEY WORDS: analysis plan, assessment endpoint, conditional mortality rate, cooling-water intake system, ecological risk assessment, entrainment, equivalent loss, exposure and effects, fish population, fractional loss, impingement, individual loss, management objective, measure, prospective analysis, power plants, problem formulation, retrospective analysis, representative species

DOMAINS: freshwater systems, marine systems, ecosystems and communities, organisms, water science and technology, environmental management and policy, environmental technology, modeling, environmental modeling, environmental monitoring, information management

---

INTRODUCTION

Impingement and entrainment at cooling-water intake systems (CWIS) are two sources of potential mortality for fish. Impingement occurs when fish are trapped or pinned by the force of the intake flow against the intake screens at the entrance of a facility's CWIS. Mortality can be high, but numerous technologies have been developed to successfully reduce at a reasonable cost both number of fish impinged and mortality of those fish that are impinged[1]. Entrainment occurs when fish eggs and larvae are taken into a facility's CWIS, pass through its heat exchanger, and are pumped back to the water body with the discharge from the facility. Mortality can approach 100% for sensitive species and life stages. However, for many species, mortality for those eggs and larvae entrained can be reduced when facilities are operated to reduce exposure of entrained organisms to potentially lethal high temperature, to large changes in temperature, and to toxic chemicals[2,3]. Substantially reducing the number of eggs and larvae entrained, however, is difficult to achieve at a reasonable cost for existing facilities with once-through cooling systems. This cost difference between mitigation technologies for entrainment as compared to impingement, in combination with the uncertain ecological impact created by entrainment, has led to a good deal of the difficulty and controversy surrounding §316(b) determinations.

The entire §316(b) text from the 1972 Clean Water Act is brief[4]: "Any standard established pursuant to Section 301 or Section 306 of this Act and applicable to a point source shall require that the location, design, construction, and capacity of cooling-water intake structures reflect the best technology available for minimizing adverse environmental impact." The terms "best technology available" (BTA), "minimizing", and "adverse environmental impact" (AEI) are not defined.

The U.S. Environmental Protection Agency (EPA) published §316(b) assessment guidelines in 1977 that were remanded in court due to procedural issues. Nonetheless, state regulators essentially followed the unofficial guidelines into the 1990s, with several hundred §316(b) determinations made during the 1970s and 1980s. In the absence of EPA regulations clearly defining AEI, BTA, or an assessment process, state and federal permitting authorities generated their own definitions on a case-by-case basis, relying on past decisions, administrative findings, scientific advances, and site-specific considerations. Several recent papers trace the history of §316(b) assessments[2,5,6,7]. Renewed interest in §316(b) assessments has been triggered by a 1995 Consent Decree that establishes a timetable for the EPA to propose and take final action with respect to addressing impacts from existing and new facilities. Final §316(b) regulations for CWIS for new facilities and proposed §316(b) regulations for large existing facilities have recently been released[4,8]. Our paper applies primarily to these large existing facilities.

In this paper, we briefly describe the EPA's ecological risk assessment framework and recommend its use to more effectively guide §316(b) determinations. We focus on developing a blueprint for the problem formulation phase of the ecological risk assessments. This blueprint includes generic assessment endpoints, a conceptual model and analysis plan, and guidance on how to modify these three generic products to develop a facility-specific problem formulation plan. In addition, we discuss the transition from problem formulation to the analysis phase and risk estimation step of a §316(b) ecological risk assessment, methods of analysis available for §316(b) ecological risk assessments, and the implications of the EPA's recent proposed regulations for existing facilities[8] on the applicability of this blueprint.

THE EPA'S ECOLOGICAL RISK ASSESSMENT FRAMEWORK

The contribution of science to the §316(b) decision making could be increased if the §316(b) determination process adhered to an accepted overall risk assessment framework. All attempts to develop regulatory tools for §316(b) need to be viewed in the context of a dichotomy of definitions of AEI. Mayhew et al.[9] effectively summarize the history of eight definitions. This dichotomy has its basis, however, in a more fundamental difference than definitions of AEI. Differences in management objectives, assessment endpoints, and measures (defined below) for assessing CWIS effects cloud every step of the §316(b) regulatory effort[10,11].

The EPA ecological risk assessment process provides an effective framework for addressing these differences (Fig. 1)[12,13]. The EPA's Guidelines call for ecological risk assessments to be conducted in three sequential phases: problem formulation, analysis, and risk characterization. Alternative frameworks are used in other countries and by other organizations within the U.S.[14,15]. We have focused on the EPA framework because the EPA has responsibility for §316(b). In addition, others have recently suggested using the EPA ecological risk assessment framework for §316(b) assessments and for environmental decision making in general[2,16,17,18,19].

The EPA framework includes a hierarchy of terms, which we have adhered to throughout the paper[12,13].

Management Goal. A management goal is a general statement of the desired condition or direction of preference for the entity to be protected. It is often developed independently of any specific risk assessment, such as part of federal or state legislation. The enabling legislation for §316(b) is the Clean Water Act (1972). The management goal for this legislation [and thus for §316(b)] is "to protect and restore the chemical, physical, and biological integrity of the nation's waters."

Management Objective. A management objective is a specific statement about something one desires to achieve that includes an ecological entity targeted for protection, a direction of preference, and a decision context of place and time. It is commonly derived from a management goal and is focused on a particular regulation in the legislation. For the purpose of this paper, we define the ecological management objective relating to CWIS under §316(b) as follows: to maintain and ensure the sustainability of populations of species in the source water body and the beneficial uses these populations support[20,21,22].

An ecological management objective of maintaining sustainability of fish populations subjected to harvesting is favored by many scientists and used by most resource management agencies[23,24]. The focus on sustainability is favored for several reasons. First, this focus is premised on a view that the population level is the proper ecological level of biological organization for managing fishery resources. The reason for this is that all individual organisms have finite life spans; only populations and higher levels persist through time. As long as fish populations of concern are relatively stable and the mix of species present remains relatively constant, sustainability can be maintained in spite of the deaths of individuals. Second, while acknowledging uncertainty, fisheries resource management agencies believe they have the ecological understanding, experience, and scientific and sociopolitical tools to monitor, forecast, and adjust regulations sufficiently well to protect fish populations.

There are ample precedents in legislation and management guidelines for this focus on populations and even higher levels. The EPA's Guidelines for Ecological Risk Assessment[12] identifies "ecological relevance" as a key criterion for selecting management objectives, assessment endpoints, and specific entities. Regardless of how management objectives are established, those that explicitly

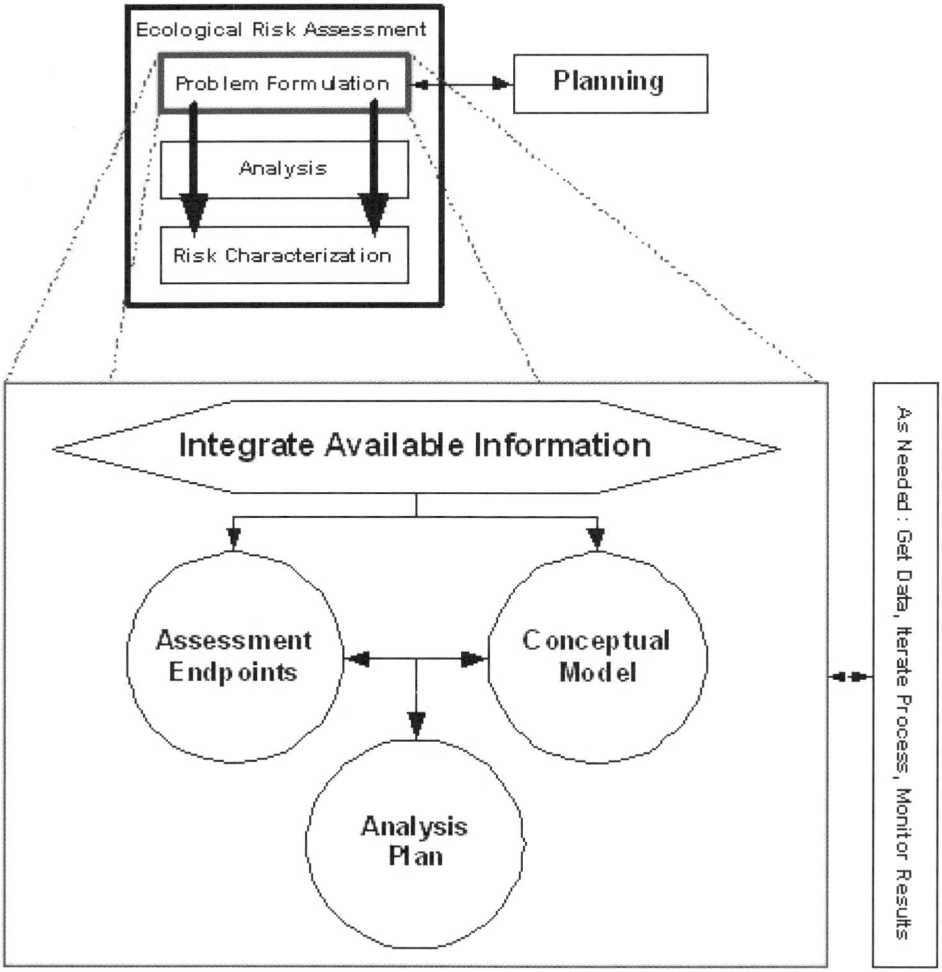

FIGURE 1. Flowchart illustrating the EPA's Ecological Risk Assessment process, including the three phases of problem formulation, analysis, and risk characterization, leading to communication of results and risk management[12].

define ecological values to be protected provide the best foundation for identifying actions to reduce risk and generating risk assessment objectives[25].

The focus on populations is also fundamental to natural resource management. The Magnuson-Stevens Fishery Conservation Act, for example, focuses on maintenance of sustainable yields from exploited populations. In fact, the concept of sustainable development implicitly focuses on populations and communities, because only populations and communities are persistent and therefore sustainable. Even for the Endangered Species Act, the management objective is preservation, conservation, and protection of endangered species, and not individual organisms.

Assessment Endpoint. An assessment endpoint is an explicit expression of what is to be protected. It is defined by an ecological entity and the entity's attributes, ideally including spatial and temporal extent. We define a hierarchy of population and community level assessment endpoints relating to CWIS under §316(b) later in this paper (Table 1).

Measures. EPA defines three classes of measures. Collectively, these measures are used to describe an assessment endpoint or factors affecting risk to that endpoint. *Measures of exposure* describe the existence and movement of a stressor in the environment and its contact or co-occurrence with the assessment endpoint or its surrogate. *Measures of effect* describe a change in an attribute of an assessment endpoint, or its surrogate, in response to a stressor to which it is exposed. *Measures of ecosystem characteristics and receptor characteristics* describe factors that influence the behavior and location of ecological entities, the distribution of a

TABLE 1

Generic §316(b) Ecological Assessment Endpoints: Entities and Their Attributes

Level Within the Hierarchy of Management Objectives	Ecological Entity	Attributes of the Entity
Level 1 – Indigenous community	Fish and macroinvertebrate communities	Species composition; species richness; species diversity
Level 2 – Populations	All individual populations in the community	Population abundance; population reproductive success
Level 3 – Populations of species selected as representative species	Populations of representative species selected on a sitespecific basis	Population abundance; population reproductive success

stressor, and life-history characteristics of the assessment endpoint that may affect exposure to, or effect of, the stressor. For the purpose of this paper, we define measures relevant for §316(b) in terms of characteristics of the facility/CWIS, characteristics of the source water body, and characteristics of the fish inhabiting the water body.

Risk Thresholds. A risk threshold (or decision criterion, target, or benchmark) is defined as the level or value for a measure beyond which is thought to result in an unacceptable level of ecological risk. Risk thresholds can be useful at low levels of regulatory concern when used as part of a tiered screening process[7,12]. Examples are risk thresholds for measures of exposure, sensitivity, number killed by entrainment and impingement, and equivalent losses. Risk thresholds for measures at higher levels of ecological organization (i.e., at the population and community levels) will always be controversial and thus not useful for screening.

THE EPA'S PROBLEM FORMULATION PHASE

Problem formulation, the first major phase of the EPA ecological risk assessment framework, is an extension of the planning process (Fig. 1). Planning and problem formulation provide the foundation for the following analysis and risk characterization phases of the ecological risk assessment. Whereas planning defines the overall responsibilities, available resources, and objectives for the ecological risk assessment, problem formulation identifies the cause-effect relationships, assessment endpoints, and measures that will be used in conducting the assessment.

Problem formulation results in three products[12]:

- Assessment endpoints that adequately reflect management goals, management objectives, and the ecosystem they represent;
- Conceptual model(s) that describe key relationships between stressors and assessment endpoints; and
- An analysis plan that documents the assessment endpoints, measures, and methods to be used in the analysis phase of the risk assessment.

The first step toward developing these products is to integrate available information. In practice, information needs are identified as part of the process of developing the above products, such that needed information is acquired and reviewed iteratively throughout the problem formulation phase. Each of the three products contains uncertainty. The explicit treatment of uncertainty during problem formulation is particularly important because it will have repercussions throughout the remainder of the ecological risk assessment.

The products of problem formulation are the scientific bases for analyzing exposure to, and effects of, a stressor on an ecological entity. Ensuring that these products are linked to the management objectives hierarchy is of utmost importance, so that the risk assessment yields indicators of risk relevant to the established values of concern.

PROBLEM FORMULATION FOR §316(B) ECOLOGICAL RISK ASSESSMENTS

We describe problem formulation for ecological risk assessments of CWIS in two steps. First, we develop generic versions of the three products listed above that are appropriate for all §316(b) ecological risk assessments/determinations. Second, we describe a facility-specific process for problem formulation based on these three generic products.

Generic Products for Problem Formulation in §316(b) Ecological Risk Assessments

On the surface, all §316(b) determinations might appear to be relatively straightforward. The source (CWIS), stressors (entrainment and impingement), receptors (typically fish and macroinvertebrates), and immediate effect (mortality) are well defined, have been studied in detail for decades, and are conceptually the same at all power plant facilities. As discussed earlier in this paper, however, the past quarter century history of §316(b) determinations amply demonstrates that environmental decision making that might appear to be relatively straightforward has commonly been controversial, time-consuming, expensive, and site specific. Value-based differences among regulators, generators, and other interested parties and site-specific differences at existing facilities explain why §316(b) determinations have not been straightforward. These differences also highlight why using the EPA's ecological risk assessment framework for §316(b) determinations merits consideration.

Final §316(b) regulations and guidelines for existing facilities need to be applicable nationwide and by water body type, as well as allowing for important facility-specific differences. Fortunately, the process of assessing entrainment and impingement impacts is fundamentally the same for all existing facilities. Consequently, generic versions of the products of problem formulation are needed that are appropriate as starting points to facilitate development of problem formulation plans for specific existing facilities.

Generic Assessment Endpoints for §316(b)

Assessment endpoints for §316(b) determinations should be consistent with available guidance for selecting such endpoints, as discussed above, and with ecological principles and practice. Assessment endpoints that directly support the management objectives in the hierarchy may be established at both the community and population levels (Table 1). Assessment endpoints at the community level are more closely linked to the management goal. However, population-level endpoints are more directly linked to potential cause-effect consequences of entrainment and impingement losses [see next section on conceptual model for §316(b)]. Selecting endpoints at both these levels of biological organization is encouraged as part of a multiple-lines-of-evidence approach to reduce overall uncertainty in the risk assessment. Both retrospective and prospective methods of analysis are readily available at the population level, whereas only retrospective methods of analysis have been effective at the community level[26,27]. Nonetheless, community-level assessments alone may, in some cases, provide sufficient information for decision making, especially where extensive water body data are available and the level of regulatory concern is low.

Numerous field and laboratory studies and assessments of power plant impacts conducted on freshwater, estuarine, and marine systems over more than 3 decades have indicated that fish, and to a lesser extent nektonic macroinvertebrates, are the biological communities primarily susceptible to entrainment and impingement. Most other community components of a water body have either low exposure to the CWIS (e.g., benthic infauna and epifauna, vascular aquatic plants), or low sensitivity to effects from exposure (e.g., phytoplankton, zooplankton).

Recommended entities and their attributes for generic §316(b) assessment endpoints are listed in Table 1. These endpoints can be used to address the upper and lower levels of the management objectives hierarchy. The recommended community-level assessment endpoint is ecologically relevant by definition, because the endpoint is the community structure itself. Susceptibility to entrainment and impingement stresses at the community level is assured if, and only if, community attribute information used in the assessment is from, or relevant to, the water body segment affected by the CWIS. In contrast, ecological relevance and susceptibility to the CWIS is established for the population-level assessment endpoint by the process of selecting representative species[28] for the specific site in question.

Attributes of an assessment endpoint determine what to measure. Where direct measures of effect can be collected on the attribute(s) of concern (e.g., direct measures of population abundance), the assessment endpoint and measure of effect are the same[12]. Otherwise, surrogate measures of effect that are readily monitored or modeled[29] must be used (e.g., organism losses from entrainment and impingement), and the effect on the endpoint (i.e., population) must be projected, introducing further uncertainty into the risk assessment. Endpoints and associated measures, if carefully selected and defined, can provide a basis for comparing the effects of a range of stressors, with effects expressed in the same units[30]. For example, using susceptible representative species populations as assessment endpoints, rather than the water body community, has the additional advantage that measures of effect can be directly compared for various CWIS hardware or operational alternatives. Surrogate measures of population-level effects,

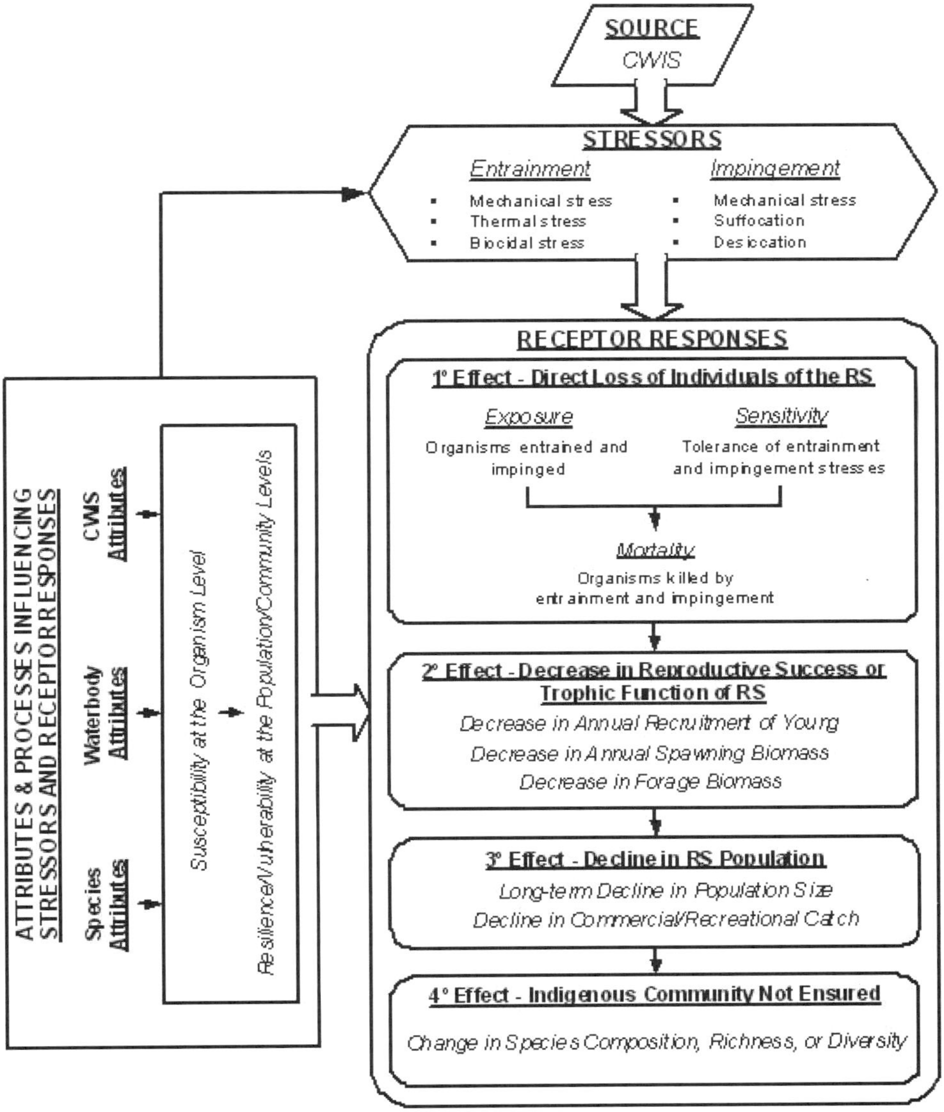

FIGURE 2. Generic conceptual model for §316(b) ecological risk assessments.

such as entrainment and impingement losses, are useful for such relative comparisons, which do not require interpretation of effect on the assessment endpoint itself.

Generic Conceptual Model for §316(b)

Factors specific to each facility and water body will influence the formulation of conceptual models appropriate for each specific §316(b) permitting action. However, commonality in the nature of the stressor and potential effects on assessment endpoints allow one to formulate a generic conceptual model for §316(b), which can be used to facilitate the development of facility-specific conceptual models and analysis plans. The generic conceptual model diagram (Fig. 2) shows the relationships among source, stressors, receptors and receptor responses, and processes influencing receptor responses. The figure also identifies risk hypotheses along the cause-effect path from stressors to potential responses by assessment endpoints.

Components of the generic conceptual model that should be described by the risk assessment are summarized below.

Source

The CWIS is the source of the stressors addressed in §316(b) assessments. CWIS characteristics that affect the nature and magnitude of entrainment and impingement exposure include cooling-water flow, intake approach velocity, intake screen system design and location, and condenser temperature elevation (ΔT). The CWIS hardware and operation, as well as the electric generating levels of the facility, may all influence entrainment and impingement exposure.

Stressors

Entrainment and impingement are the two major categories of stressors that need to be considered in §316(b) risk assessments. Stresses from entrainment can be of three types: mechanical (e.g., pressure, shear forces), thermal (heat shock from condenser ΔT), and chemical (e.g., biocides such as chlorine). The process of impingement can expose organisms to mechanical (e.g., abrasion), suffocation, and desiccation stresses. Thermal stresses are influenced not only by CWIS design and operation, but also by seasonally varying ambient temperature characteristics of the water body.

Receptors and Receptor Responses

The receptors for the §316(b) assessment are the populations of aquatic organisms that reside in the source water body, as represented by the representative species. Receptor response is primarily a function of factors that determine susceptibility[31] of the representative species by influencing either exposure or sensitivity, or both. The cause-effect linkages between stressors, receptors, and assessment endpoints is represented in Fig. 2 as a sequence of responses that constitute the hierarchy of risk hypotheses for the generic conceptual model. That is, the deaths of individuals of the representative species from entrainment and impingement (first-order effect) may cause a decline in reproductive success of their populations (second-order effect) that, in turn, could lead to long-term declines in population abundance (third-order effect) and changes in species composition at the community level (fourth-order effect). Measures of effect at any of these four levels may be used to characterize ecological risks.

Processes Influencing Receptor Responses

Characteristics of the CWIS, water body (ecosystem), and representative species ultimately determine the effect of the CWIS on assessment endpoints by their influence on the nature and magnitude of entrainment and impingement stressors, exposure and sensitivity of the organisms to entrainment and impingement stresses, and vulnerability of the representative species to the entrainment and impingement losses incurred (Fig. 2).

Generic Analysis Plan for §316(b)

The generic analysis plan for §316(b) is based on the §316(b) ecological management objectives hierarchy (Table 1), generic assessment endpoints, entities and their attributes (Table 1), and the generic conceptual model (Fig. 2). The plan is flexible and can be tailored for facilities for which the level of §316(b) regulatory concern is low or high. It is designed around the following six levels of analysis and associated scientific/management decision points (SMDP)[32]. The first five levels apply (potentially) to each representative species.

1. Describe exposure and susceptibility to entrainment and impingement by life stage,
2. Describe number killed annually by entrainment and impingement by life stage,
3. Describe annual equivalent losses,
4. Describe effects on annual reproductive success,
5. Describe multiyear effects on population abundance and beneficial uses, and
6. Describe multiyear effects on community composition.

These levels of analysis may be considered sequentially as indicated in the conceptual model (Fig. 2). However, one or more levels may be bypassed because of inadequate information or other reasons. An example is population projection modeling in the absence of retrospective estimates of effects on annual reproductive success. In either case, a scientific/management decision should be made prior to undertaking an analysis at a new level. This decision should be guided

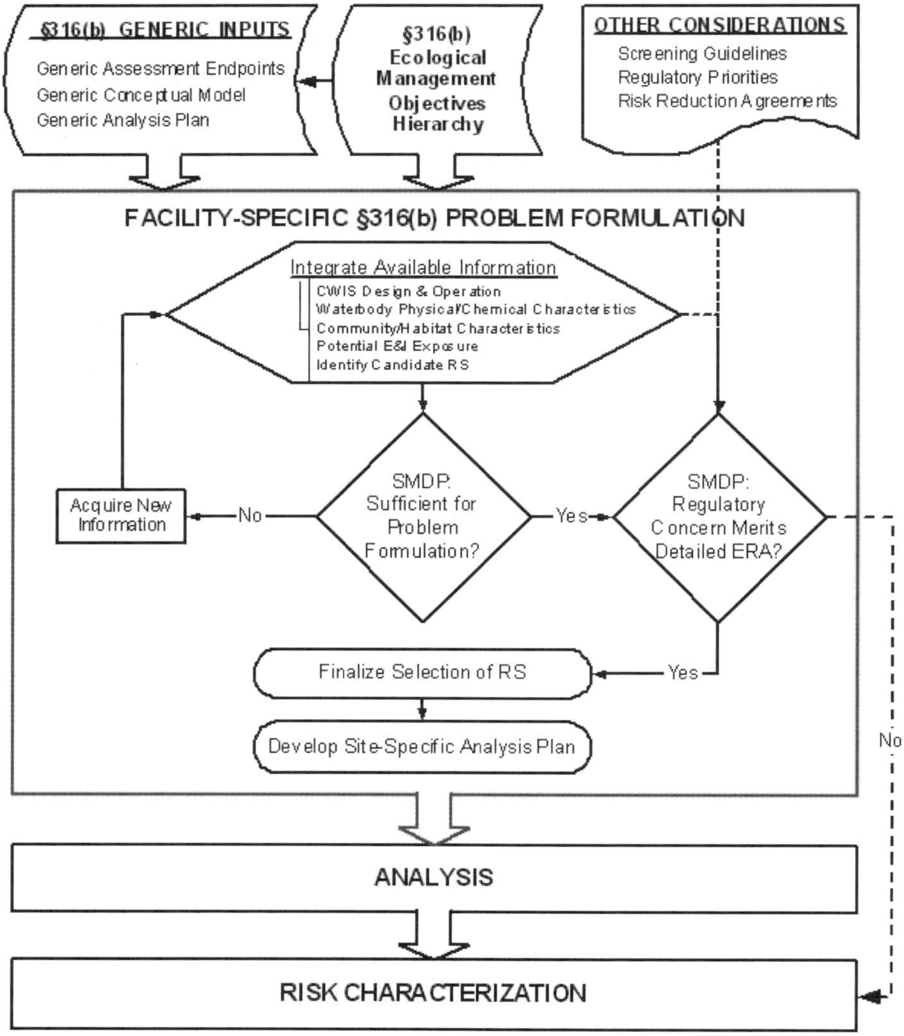

FIGURE 3. Flowchart of the facility-specific problem formulation process for §316(b) ecological risk assessments.

by consideration of regulatory guidelines, level of regulatory concern, and available information. In other words, why is analysis at this level needed, what will it contribute to the overall assessment, and can it be done in a scientifically credible manner with reasonable effort?

Regardless of how this generic plan is modified during the facility-specific §316(b) problem formulation process, the resulting analysis plan should be viewed as a blueprint for comparing alternative CWIS options and comparing representative species by focusing on relative changes (i.e., proportional or percent changes) in various measures. The analysis plan is specifically designed to encourage the regulatory agency to consider multiple lines of evidence in making a §316(b) determination.

Facility-Specific §316(b) Problem Formulation

Figure 3 is a flowchart for problem formulation for existing facilities. Adapting the generic products to develop a facility-specific analysis plan requires consideration of specific regulations and guidelines of the permitting agency, as well as site-specific ecological and technical differences including:

- Characteristics of the facility and its CWIS,
- Characteristics of the water body from which the CWIS withdraws water,
- Characteristics of the fish species in the water body,
- Magnitude of entrainment and impingement losses, and
- Quantity and quality of information available to characterize the preceding four items and to evaluate the ecological consequences of entrainment and impingement losses.

The facility-specific §316(b) problem formulation process (Fig. 3) is based on the ecological management objectives hierarchy and the generic assessment endpoints, conceptual model, and analysis plan described in the preceding sections, and includes two SMDP. The facility-specific problem formulation process is intended to be iterative, involving both planning and cycling through the following five steps as needed.

- Integrate available information;
- Determine if available information is sufficient to support problem formulation, including selection of assessment endpoints, representative species, and preparation of a site-specific analysis plan;
- Determine if the level of regulatory concern for entrainment and impingement losses merits a detailed ecological risk assessment;
- Select representative species to be the focus of the risk assessment; and
- Complete and document the facility-specific analysis plan.

Integrate Available Information

During the problem formulation process (Fig. 3), site-specific information would be compiled and evaluated to develop an analysis plan that addresses some or all of the risk hypotheses identified in the generic conceptual model. The process would typically be iterative, in which information is integrated with evolving plans for analysis, new information requirements are identified, and the process of compiling, evaluating, and integrating that information is repeated. Experienced risk assessors recognize the importance of performing this task well: "The foundation for problem formulation is based on how well available information on stressor sources and characteristics, exposure opportunities, characteristics of the ecosystem potentially at risk, and ecological effects are integrated and used"[12].

At this point in the risk assessment, the focus of compilation and review is on information needed for preliminary evaluation of the level of regulatory concern, selection of the representative species to serve as assessment endpoints, and preparation of a site-specific analysis plan. A second, more focused, compilation and evaluation of available information occurs at the beginning of the analysis phase to support tasks in analysis and risk characterization (Fig. 1). Information necessary at this point, however, is limited to a combination of CWIS design and operating data, as well as information on the source water body and its aquatic inhabitants. Examples of specific information needed for this step are:

- Design and construction of the CWIS, including dimensions, capacities, and equipment for reducing entrainment and impingement losses;
- Typical operation of the CWIS, including seasonal patterns in coolingwater flow, operation of the intake screens, and operation of the equipment for reducing entrainment and impingement losses;
- Physical characteristics of the source water body, including size, depths, and general hydrologic conditions;
- Environmental conditions within the source water body, including physical, chemical, and habitat characteristics; and
- Composition and status of the aquatic community expected to be in the vicinity of the intake, with special focus on those components typically most susceptible to entrainment and impingement (e.g., fish and macroinvertebrates).

Determine if Available Information is Sufficient to Support Site-Specific Problem Formulation

Following compilation and review of available information, the §316(b) problem formulation process (Fig. 3) includes a determination as to whether the currently available information is sufficient to continue with the problem formulation phase. Typically, available information necessary to complete this step will be obtained from the plant operator, from the natural resource management agency responsible for the source water body, and from the general literature. If available data are deemed insufficient (e.g., no information on fish species occurring in the vicinity of the intake), a work plan should be developed and implemented to gather such information.

Determine if Level of Regulatory Concern Merits a Detailed Ecological Risk Assessment

When available information is deemed sufficient to support problem formulation, the regulatory agency determines if the level of regulatory concern for entrainment and impingement losses merits a detailed ecological risk assessment. More than 500 existing electric generating stations with CWIS are subject to regulation under §316(b)[8]. However, based on the results of detailed §316(b) assessments conducted at several larger generating stations throughout the country, it is likely that many of these existing facilities have little risk of AEI to aquatic populations or the community in the source water body. Therefore, an important function of using the ecological risk assessment process for §316(b) determinations is to provide a screening mechanism that identifies those facilities that pose little risk of AEI. Such low-risk facilities, then, will not be required to conduct a detailed analysis of potential CWIS effects, but rather the ecological risk assessment process will proceed directly to a limited risk characterization (Fig. 3). A brief summary report will then document the conclusions of the assessment and include the basis for the conclusions that will be part of the regulatory record.

The bases for screening would include the preliminary indications of potential risks obtained by integrating available information, as well as other considerations brought by parties to the permitting process. Such considerations may include screening guidelines codified in regulations, priorities established in written regulatory policies, CWIS fish protection technologies existing at the facility, and agreements for risk reduction actions that have been offered by the permit applicant (Fig. 3). It seems reasonable to expect that the ecological risk assessment for those facilities that screen out at this point in the process could be completed in a few months with relatively small commitment of agency time and resources. Inclusion of such a screening process will help to ensure that the limited resources of the regulatory agencies are most effectively utilized to address §316(b). It is only when the regulatory agency deems the risk of AEI sufficiently high that a commitment of agency and utility resources is warranted to complete a full ecological risk assessment (Fig. 1).

It is up to the discretion of the regulatory agency, acting in consultation with all stakeholders, to determine whether a detailed ecological risk assessment is warranted for any specific facility. Some site-specific factors that might be considered in such a determination include:

- Capacity of the intake relative to the magnitude of the source water body,
- Frequency of operation of the CWIS (e.g., rarely used),
- Incorporation of CWIS features that minimize the use of cooling water (e.g., cooling towers),
- Incorporation of CWIS features that minimize exposure of aquatic organisms (e.g., barriers or screening),
- Incorporation of CWIS features that maximize survival of exposed aquatic organisms (e.g., fish return systems),
- Water quality in the vicinity of the intake that minimizes aquatic organism exposure (e.g., anoxic area), and
- Health of the communities of potentially exposed aquatic organisms in the vicinity of the intake provides no evidence for concern (e.g., biocriteria).

Table 2 provides an expanded list of selected characteristics of the facility/CWIS, the water body, and representative species that may be considered in the screening process, as well as in determining the level of assessment complexity. These characteristics influence available information and level of regulatory concern and would provide a basis for a tiered approach to risk assessment.

Select Representative Species

The representative species are the ecological entities that are the focus of the §316(b) ecological risk assessment. Changes in selected attributes of these species become the measures of effect that are at the heart of the assessment.

Fish species susceptible to entrainment and impingement losses may or may not be identified in the current NPDES/ SPDES permit for an existing facility. If they are not, regulators and the generator need to address this issue. Guidelines and assumptions for selection of representative species are available[27,33]. Typically, representative species include those that are susceptible to entrainment and impingement, are representative of important functional roles in the community, are representative of important beneficial uses (e.g., commercial and recreational fisheries), or are species of special concern. The intent of focusing on the representative species is to ensure that potential consequences of entrainment and impingement losses to the aquatic community of the source water body as a whole are addressed.

Complete and Document Facility-Specific Analysis Plan

Following final selection of the representative species, the facility-specific analysis plan is completed and documented (Fig. 3). In addition to the basic requirements of any analysis plan, as discussed above, the following two topics are of particular importance for the facility-specific analysis plan for §316(b) assessments.

TABLE 2

Factors to be Considered in a Tiered Approach for Determining the Level of Regulatory Concern/Priority, and thus the Extent/Complexity of §316(b) Ecological Risk Assessment that is Appropriate/Needed

Factor	Level of Regulatory Concern/Priority and Extent/Complexity of ERA that is Appropriate/Needed Low		High
Facility/CWIS Attributes			
Intake flow per day	Low		High
Proportional intake flow (%)	Low		High
Intake velocity	Low		High
Years of operation remaining	Few		Many
Attributes that minimize exposure or maximize survival	Many		Few
Water Body Attributes			
Critical habitat function	Low		High
Size	Large		Small
Other stressors[1]	Low		High
Value of beneficial uses[2]	Low		High
Apparent community health	High		Low
Fish Attributes			
Life history strategy[3]	E		O, P
Number of species susceptible	Few		Many
Obtainable Information			
Quality and quantity	Low		High
Levels of biological organization[4]	Ind	Ind and YC	Ind, YC, and Pop

[1] Index of strength of other anthropogenic and nonanthropogenic stresses on the water body, e.g., fishing, water quality, temperature regime, drought.
[2] Economic, social, and cultural value of commercial, recreational, and subsistence fisheries and other beneficial uses combined.
[3] Key to life-history strategies: E = equilibrium (e.g., largemouth bass); O = opportunistic (e.g., bay anchovy, threadfin shad); P = periodic (e.g., striped bass) [Winemiller, K.O. and Rose, K.A. (1992) *Can. J. Fish. Aquat. Sci.* **49**, 2196–2218.]
[4] Key to levels of biological organization: Ind = individual; YC = year class; Pop = population.

- For each representative species, consider the levels of analysis that available information will support and that regulatory concern merits. Because analysis at each additional level will require additional resources (money, time, technical expertise), both available information for the species and level of regulatory concern for the species should be evaluated.
- Select CWIS options of interest to regulators, generators, or interested parties. Selection can be based on considering the extent to which each option is likely to reduce entrainment and impingement losses based on experience at other facilities, as well as economic and other impacts.

The facility-specific analysis plan summarizes the problem formulation phase of the §316(b) ecological risk assessment. It is the blueprint for how the analysis for that existing facility relates to the management objectives and to the NPDES/SPDES permit decisions that must be made. The facility-specific analysis plan needs to be documented for the public record in the final ecological risk assessment report to regulators.

We deliberately have not proposed risk thresholds for measures at any level of ecological organization. Proposing such values would have shifted the focus of this paper from proposing a blueprint for the problem formulation phase of §316(b) ecological risk assessments to justifying the thresholds we selected. Selection of these thresholds is only partly a scientific process. In making a §316(b) determination for an existing facility, agency regulators are ultimately the ones who operationally define AEI for that facility (Fig. 1). Scientists must be willing and able, however, to make species- and facilityspecific estimates of uncertainty and risk of adverse effects, and then describe and communicate these risks to agency risk managers. Science alone is not and never will be in a position to provide an acceptable definition for AEI under §316(b) or for any other type of ecosystem modification[19,34,35,36]. Final determination of AEI under §316(b) is made by the regulatory agency based on joint consideration of risks of AEI and other factors, as strongly emphasized elsewhere in this volume[16,37,38,39,40,41].

TRANSITION FROM PROBLEM FORMULATION TO ANALYSIS AND RISK ESTIMATION

The analysis phase follows the problem formulation phase in the EPA's ecological risk assessment framework (Fig. 1). The two products of the analysis phase are an exposure profile and a stress-response profile[12]. The generic conceptual model for §316(b) developed in problem formulation represents a chain of cause-effect ecological connections (Fig. 2). The functional dependence between some of the levels in this model merits emphasis and should be reflected in making the transition from problem formulation to actually performing analyses. This organization also parallels the organization for prospective methods under the headings of individual loss, fractional loss, and population projections[26,27].

- Describe susceptibility to entrainment and impingement, where susceptibility includes exposure, sensitivity, and mortality due to entrainment and impingement;
- Describe annual number killed by entrainment and impingement, and the resulting effects on annual equivalent losses; and
- Describe effects on annual recruitment, and the resulting multi-year effects on population abundance and beneficial uses.

We recommend for §316(b) assessments that characterization of exposure, summarized in an exposure profile, be replaced by characterization of susceptibility, summarized in a susceptibility profile, which is a combination of the exposure and stressor-response profiles. Fig. 4 highlights the conceptual limitations in assuming that exposure and susceptibility are equivalent. We propose defining an index of susceptibility for each representative species as a weighted sum of species-specific attributes, facility/CWIS attributes, and water body attributes that are applicable for that species (Fig. 2), as part of a multiplelines-of-evidence approach. As an example, the likelihood of fish being withdrawn by hydropower intakes was determined primarily by the ecological zone from which water is withdrawn by hydropower intakes (e.g., littoral, pelagic, bathy-pelagic, etc.) and secondarily by the life-history strategy of the affected fish species[42].

Describing the number killed annually by entrainment and impingement (by species and life stage) is an important step for proceeding to any of the higher levels of analysis. If the numbers killed annually with the existing CWIS are not of regulatory concern, the regulatory decision may be to summarize the results of analysis up to that point and go to risk estimation. If the numbers killed are of regulatory concern, it is essential from the ecological perspective that the analysis progress to the next higher level to describe the resulting annual equivalent losses by species. The numbers of eggs and larvae killed by entrainment, while they may be very large numbers, do not reflect the very low natural survival rates, which can differ substantially by life stage and species. Compared to survival rates with which the public is familiar, survival rates for fish eggs and larvae of species experiencing entrainment and impingement losses are extremely low[43,44]. Extrapolating entrainment and impingement losses of these early life stages to equivalent losses (as numbers or biomass) at some older life stage (e.g., juvenile age 0, age 1, or adults) avoids the problem of mixing "apples and oranges" and provides a measure of effect of greater relevance both ecologically and socially.

Describing effects on annual recruitment and on the resulting multiyear effects on population abundance and beneficial uses typically involves both retrospective and prospective methods for those existing facilities where analysis at these higher levels is merited. These extrapolations to a large extent are independent of extrapolating annual entrainment and impingement losses to annual equivalent losses. For both retrospective and prospective methods, describing multiyear effects at the population level, as compared to describing annual effects on repro-

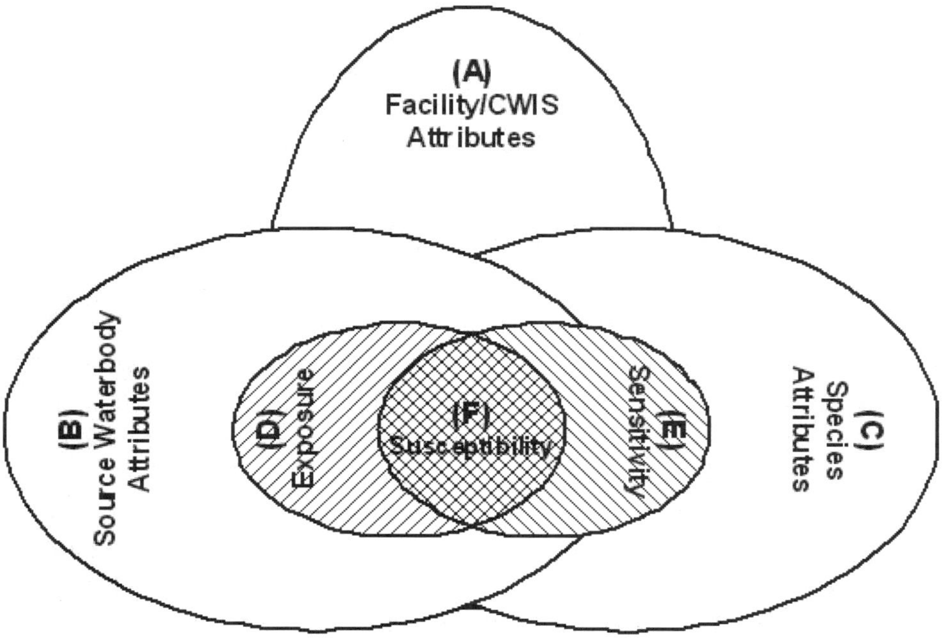

FIGURE 4. Venn diagram for conceptualizing attributes that influence susceptibility to entrainment and impingement. The three ovals represent attributes of (A) the facility/CWIS, (B) the source water body, and (C) the fish species in the water body. Oval (D) represents the reduced subset of attributes from (A), (B), and (C) that together determine exposure. Oval (E) represents the reduced subset of attributes from (A), (B), and (C) that together determine sensitivity. The intersection of ovals (D) and (E) [i.e., area (F)] represents mortality due to entrainment and impingement. Technological advances continue to modify CWIS attributes at existing facilities, resulting in reductions in the size and overlap of ovals (D) and (E), and thus reductions in entrainment and impingement mortality and annual losses.

ductive success, will be more relevant ecologically and socially because they reflect cumulative effects of entrainment and impingement losses over a period of years and are closer to the assessment endpoint. On the other hand, multi-year measures of effect at the population level are unavoidably more uncertain than annual measures of effect at the yearclass level (Table 2).

Methods of Analysis Available for §316(b) Ecological Risk Assessments

Given the above discussion of levels of analysis, a brief overview of methods available to estimate measures at these various levels of analysis is constructive. Depending on the magnitude of perceived effects and the desires and capabilities of the regulatory agencies, generators, and interested parties, a variety of measures (and associated methods to estimate these measures) have been used to describe susceptibility and ecological effects at the individual, year-class, population, and community levels. The methods have developed and evolved over the past 30 years and can be broadly grouped into two categories, prospective and retrospective.

Prospective methods attempt to forecast what effects cooling-water withdrawals will have by combining information on the magnitude of actual or predicted entrainment and impingement losses with information on the life history characteristics and population dynamics for each susceptible species. This category of methods is consistent with the EPA's category of process modeling techniques[12]. With prospective methods, magnitude of entrainment and impingement loss is typically based on site-specific information on the density of susceptible life stages in the vicinity of the intake and the mortality rates associated with the entrainment and impingement processes. Where necessary, information on the life history characteristics and population dynamics of a species can usually be obtained from the scientific literature for the same or for closely related species. The range of prospective methods extends from generally qualitative to highly sophisticated quantitative models of population dynamics. Generally, estimation of effects of cooling-water withdrawals using prospective methods is conducted in a phased manner beginning with the simplest methods. The more complex modeling exercises are typically limited to those cases where the potential for adverse population effects is expected to be high and, consequently, the concern of the regulatory agencies and the public at large is heightened.

Retrospective methods include those procedures that analyze empirical data from the source water body for evidence that entrainment and impingement losses may be having a demonstrable effect at the year-class and population levels. This category of methods is consistent with the EPA's category of empirical modeling techniques[12]. In general, retrospective methods provide direct measures of potential effects. These methods become increasingly applicable the greater the number of years an existing facility has been operating. It is important to remember that use of retrospective methods requires assessing to what extent observed effects may be caused by entrainment and impingement losses vs. unrelated natural changes in the ecosystem and other anthropogenic stresses.

THE EPA'S PROPOSED REGULATIONS FOR CWIS AT EXISTING FACILITIES

The recently proposed regulations for existing facilities[8] shift the emphasis for 316(b) determinations from an assessment process that characterizes risk of AEI to an assessment process that characteristics benefits and costs of alternative technologies to meet technology-based performance standards for CWIS. The primary proposed assessment endpoints are the CWIS itself, percent reduction in impingement mortality, and percent reduction in number entrained. Regardless of the regulatory emphasis, the great majority of the blueprint for problem formulation proposed in this paper is applicable for 316(b) determinations at existing facilities. Two implications, however, are obvious. First, the focus of assessments will shift toward comparing CWIS options and representative species and away from estimating multiyear population effects for the existing CWIS. Second, methods of analysis will shift toward describing exposure, susceptibility, entrainment, and impingement losses by life stage and equivalent losses.

In conclusion, the problem formulation process for §316(b) ecological risk assessments we propose capitalizes on the considerable body of scientific knowledge and regulatory experience that has accumulated from §316(b) determinations over the past 3 decades. Our blueprint for existing facilities provides the structure and process that can substantially improve the current repermitting procedure for regulators, generators, and other interested parties. For existing facilities, both prospective and retrospective methods should be used to provide multiple lines of evidence. Multiple lines of evidence from different levels of analysis, and using both prospective and retrospective methods, provide a broader perspective concerning magnitude of effects and associated uncertainties and risks.

ACKNOWLEDGMENTS

The authors thank D.E. Bailey, J. Kadvany, J. Loos, and W.A. Richkus for their constructive comments during the preparation and peer review of this paper.

REFERENCES AND ENDNOTES

1. Taft, E.P. (2000) Fish protection technologies: a status report. *Environ. Sci. Policy* **3(Suppl. 1)**, S349–S360.
2. Dey, W.P., Jinks, S.M., and Lauer, G.J. (2000) The §316(b) assessment process: evolution towards a risk-based approach. *Environ. Sci. Policy* **3(Suppl. 1)**, S15–S24.
3. Mayhew, D.A., Jensen, L.D., Hanson, D.F., and Muessig, P.H. (2000) A comparative review of entrainment survival studies at power plants in estuarine environments. *Environ. Sci. Policy* **3(Suppl. 1)**, S295–S302.
4. USEPA (United States Environmental Protection Agency). (2001) National Pollutant Discharge Elimination System. Final Regulations Addressing Cooling Water Intake Structures for New Facilities; Final Rule. Federal Register, Environmental Documents, December 18, 2001, pp. 65255–65345. www.epa.gov/owm/316b.htm.
5. Anderson, W. and Gotting, E. (2001) Taken in over intake structures? Section §316(b) of the Clean Water Act. *Columbia J. Environ. Law* **26**, 1–79.
6. May, J.R. and van Rossum, M.K. (1995) The quick and the dead: fish entrainment, entrapment, and the implementation and application of Section §316(b) of the Clean Water Act. *Vt. Law Rev.* **20(2)**, 375–493.
7. Nagle, D.G. and Morgan, J.T., Jr. (2000) Forward. A draft regulatory framework for analyzing potential adverse environmental impact from cooling water intake structures. *Environ. Sci. Policy* **3(Suppl. 1)**, ix–xiv.
8. USEPA (United States Environmental Protection Agency). (2002) National Pollutant Discharge Elimination System – Proposed Regulations to Establish Requirements for Cooling Water Intake Structures at Phase II Existing Facilities; Proposed Rule. Federal Register, Environmental Documents, April 9, 2002. pp. 17221–17225 and 17171–17220. www.epa.gov/owm/316b.htm.
9. Mayhew, D.A., Muessig, P.H., and Jensen, L.D. (2002) Adverse environmental impact (AEI): 30-year search for a definition. In Defining and Assessing Adverse Environmental Impact Symposium 2001. *TheScientificWorldJOURNAL* **2(S1)**, 21–29.
10. EPRI (Electric Power Research Institute). 2002. Evaluating the Effects of Power Plant Operations on Aquatic Communities. An Ecological Risk Assessment Framework for Clean Water Act §316(b) Determinations. EPRI, Palo Alto, CA. EPRI Report 1000758
11. Gordon, D.K. and Super, R.W. (2002) Minimizing adverse environmental impact: how murky the waters. *TheScientificWorldJOURNAL*, in press.
12. USEPA (United States Environmental Protection Agency). (1998) Guidelines for Ecological Risk Assessment. EPA/630/R-95/002F.
13. USEPA (United States Environmental Protection Agency). (2001) Planning for Ecological Risk Assessment: Developing Management Objectives. EPA/630/R-01/001A.
14. McCarty, L.S. and Power, M. (2000) Approaches to developing risk management objectives: an analysis of international strategies. *Environ. Sci. Policy* **3**, 299–304.
15. Power, M. and McCarty, L.S. (1998) A comparative analysis of environmental risk assessment management frameworks. *Environ. Sci. Technol.* **32**, 224A–231A.
16. Bailey, D. and Bulleit, K. (2002) Defining adverse environmental impact: a fisheries management approach. In Defining and Assessing Adverse Environmental Impact Symposium 2001. *TheScientificWorldJOURNAL* **2(S1)**, 147–168.
17. Gentile, J.H. and Harwell, M.A. (1998) The issue of significance in ecological risk assessments. *Human Ecol. Risk Assess.* **4(4)**, 815–828.
18. Harwell, M. and Gentile, J. (2002) Overcoming barriers to the use of models in decision making. In *Ecological Modeling for Resource Management*. Dale, V.H., Ed. Springer- Verlag, New York, in press.
19. Van Winkle, W., and Kadvany, J. (2002) Modeling fish entrainment and impingement impacts and the policy-science bridge. In *Ecological Modeling for Resource Management*. Dale, V.H., Ed. Springer-Verlag, New York, in press.

20. The ecological management objective is expressed here using the population level because the population level is the lowest level of biological organization that persists through time, and because it is more directly linked than higher levels to potential consequences from impingement and entrainment losses. However, the broader management objective of protecting the communities, of which the populations are a part, is implicit in this definition. See also Table 1.

21. "In a democratic society, the values represented in Federal law are often good indicators of widely held values. Except for endangered species, no case was found in which an individual nonhuman organism, or even a small number of individuals, was protected by a regulatory decision. However, effects somewhere between the individual and population levels, such as widespread mortality in fish or birds, have been used as the basis for decisions" (See endnote 24).
22. The controversy associated with some §316(b) determinations involves the difference between this ecological management objective and the alternative management objective of minimizing entrainment and impingement losses. This alternative objective reflects the societal value that it matters how we kill fish, not just that they are being killed. For some regulatory agencies and interested parties, the management objective is to minimize 'collateral losses' associated with entrainment and impingement at CWISs (see endnotes 4 and 8).
23. National Research Council. (1998) Improving Fish Stock Assessments. National Academy Press, Washington, D.C.
24. USEPA (United States Environmental Protection Agency). (1997) Priorities for Ecological Protection: An Initial List and Discussion Document for EPA. EPA/600/S-97/002.
25. McDaniels, T. (2002) Creating and using objectives for ecological risk management. *Environ. Sci. Policy* **3**, 299–304.
26. EPRI (Electric Power Research Institute). (1999) Catalog of Assessment Methods for Evaluating the Effects of Power Plant Operations on Aquatic Communities. EPRI, Palo Alto, CA, TR-112013.
27. EPRI (Electric Power Research Institute). (2002) Evaluating the Effects of Power Plant Operations on Aquatic Communities. Guidelines for Selection of Assessment Methods. EPRI, Palo Alto, CA. TR-1005176.
28. Representative species (RS) – Species selected during problem formulation on a sitespecific basis that are the focus of the ecological risk assessment. Equivalent terms used in other reports and published papers are focal species (FS), representative indicator species (RIS), representative and important species (RIS), species of concern (SOC).
29. However, while established measurement protocols are convenient and useful, they do not justify establishing assessment endpoints that are equivalent to the readily available measure. Data availability alone is not an adequate criterion for selection of assessment endpoints (see endnote 12).
30. Barnthouse, L.W., Suter, G.W., II, and Rosen, A.E. (1990) Risks of toxic contaminants to exploited fish populations: influence of life history, data uncertainty, and exploitation intensity. *Environ. Toxicol. Chem.* **9**, 297–311.
31. The reader should note that selection criteria for the representative species include susceptibility to the entrainment and/or impingement stressors, so that representative species would generally be more susceptible than the average for all the fish or macroinvertebrate populations comprising the community.
32. SMDP is USEPA'S term for Scientific/Management Decision Point, defined as: "A time during the ecological risk assessment when a risk assessor communicates results or plans at that stage to a risk manager. The risk manager decides if information is sufficient to proceed with risk management strategies or whether more information is needed to characterize risk" [USEPA (United States Environmental Protection Agency). (1999) Risk Assessment Guidance for Superfund: Volume 3 – (Part A, Process for Conducting Probabilistic Risk Assessment). Draft, Revision No. 5. U.S. Environmental Protection Agency, Washington, D.C. www.epa.gov/superfund/pubs.htm.
33. USEPA (United States Environmental Protection Agency). (1977) Guidance for Evaluating Adverse Environmental Intake Structures on the Aquatic Environment: Section 316(b). (Draft). U.S. Environmental Protection Agency, Washington, D.C.
34. Lackey, R.T. (1994) Ecological risk assessment. *Fisheries* **19(9)**, 14–18.
35. Lackey, R.T. (1998) Fisheries management: integrating societal preference, decision analysis, and ecological risk assessment. *Environ. Sci. Policy* **1**, 329–335.
36. Lackey, R.T. (1999) Salmon policy: science, society, restoration, and reality. *Environ. Sci. Policy* **2**, 369–379.
37. Kadvany, J. (2002) Decision theory and adverse environmental impacts in Section §316(b) of the Clean Water Act. In Defining and Assessing Adverse Environmental Impact Symposium 2001. *TheScientificWorldJOURNAL* **2(S1)**, 106–138.

38. McLean, R., Richkus, W.A., Schreiner, S.P., and Fluke, D. (2002) Maryland power plant cooling-water intake regulations and their application in evaluation of adverse environmental impact. In Defining and Assessing Adverse Environmental Impact Symposium 2001. *TheScientificWorldJOURNAL* **2(S1)**, 1–11.
39. Strange, E.M., Lipton, J., Beltman, D., and Snyder, B. (2002) Scientific and societal considerations in selecting assessment endpoints for environmental decision-making. In Defining and Assessing Adverse Environmental Impact Symposium 2001. *TheScientificWorldJOURNAL* **2(S1)**, 12–20.
40. Veil, J.A., Puder, M.G., Littleton, D.J., and Johnson, N. (2002) A holistic look at minimizing adverse environmental impact under Section §316(b) of the Clean Water Act. In Defining and Assessing Adverse Environmental Impact Symposium 2001. *TheScientificWorldJOURNAL* **2(S1)**, 41–57.
41. Wells, A.W. and Englert, T.L. (2002) AEI assessments: a consultant's perspective. In Defining and Assessing Adverse Environmental Impact Symposium 2001. *TheScientificWorldJOURNAL* **2(S1)**, 190–203.
42. Pavlov, D.S., Lupandin, A.I., and Kostin, V.V. (1999) Downstream migration of fish through dams of hydroelectric power plants. Moscow, Nauka Russian Academy of Sciences (in Russian; translation available from G.F. Cada or C.C. Coutant, Oak Ridge National Laboratory, Oak Ridge, TN 37831).
43. Houde, E.D. (1987. Fish early life dynamics and recruitment variability. *Am. Fish. Soc. Symp.* **2**, 17–29.
44. Miller, T.J., Crowder, L.B., Rice, J.A., and Marshall, E.A. (1988) Larval size and recruitment mechanisms in fishes: toward a conceptual framework. *Can. J. Fish. Aquat. Sci.* **45**, 1657–1670.

BIOSKETCHES

Webster Van Winkle, Ph.D., is self-employed with Van Winkle Environmental Consulting Co. He retired from Oak Ridge National Laboratory after 27 years in the Environmental Sciences Division. His research interests include data analysis and development and application of models as part of applied research projects and environment assessments involving aquatic ecosystems, fish populations in particular.

William Dey, M.S., is a Senior Scientist and Vice President of ASA Analysis & Communication, Inc. Mr. Dey has 28 years of experience conducting ecological risk assessments of man's activities on the aquatic environment. He has ecological risk assessments of power plant cooling water intake systems and toxic chemical releases to freshwater, marine, and estuarine habitats throughout much of the U.S. Mr. Dey directed the development and implementations of mathmatical models to assess the population-level consequences of large-scale cooling water withdrawals.

Steven M. Jinks, Ph. D., is a Senior Scientist and President of ASA Analysis & Communication, Inc. Dr. Jinks has over 25 years of experience conducting research on the ecological effects of power plant cooling water systems on freshwater, estuarine, and marine water bodies. He has assessed the impacts of cooling water intake systems and thermal discharges at power plants located in the Northeast, Southeast, Midwest, and West Coast of the United States. Dr. Jinks directed the design and development of entrainment abundance and survival sampling methods for the Empire State Electric Energy Research Council. Most recently he has been applying his experience to benefit and cost evaluations of cooling water intake system alternatives.

Mark S. Bevelhimer, Ph.D., is a Research Scientist at Oak Ridge National Laboratory. He received his M.S. from Ohio State University and his Ph.D. from the University of Tennessee. He has 18 years experience in aquatic ecology/fisheries biology. Dr. Bevelhimer's work has been a combination of field observation, laboratory experimentation, and computer modeling. He has used laboratory and field studies to investigate the effects of environmental changes on fish growth, contaminant accumulation, food habits, movement, and population dynamics. He has supplemented his empirical research with simulation modeling to examine fish movement, growth, food consumption, and contaminant uptake. Modeling experience includes the development and application of bioenergetics models of fish growth, individual-based population models, and hydrologic models of stream flow and water quality. Most recently he has been using these skills to investigate the impacts of power-plant operations on resident and miagratory fish.

Charles C. Coutant, Ph.D., is Distinguished Research Scientist, Oak Ridge National Laboratory. Dr. Coutant has conducted laboratory and field research on the effects of power plant cooling systems on aquatic life since 1959 in the Delaware River (PA/NJ), Columbia River (WA/OR), Tennessee Valley reservoirs (TN), and has conducted power plant assessments for facilities in Oregon, Michigan, New York, New Jersey, Georgia, Sweden, and New Zealand. He has been an advisor on power plant effects to the International Atomic Energy Agency and UNESCO and has authored USEPA water temperature criteria. Dr. Coutant was past President of the American Fisheries Society. His current research interests lie in fish behavior as related to water intakes and fish bypasses for thermal electric power plants and hydropower dams, with continuing interest in water temperature effects on fish and temperature in the aquatic landscape. Awards and honors received include the Award of Excellence, Southern Division of the American Fisheries Society and Distinguished Service Award of the American Fisheries Society.

Author index